Statistical Turbulence Modelling for Fluid Dynamics — Demystified

An Introductory Text for Graduate Engineering Students

Statistical Turbulence Modelling for Fluid Dynamics — Demystified

An Introductory Text for Graduate Engineering Students

Michael Leschziner

Imperial College London, UK

ICP

Imperial College Press

Published by

Imperial College Press
57 Shelton Street
Covent Garden
London WC2H 9HE

Distributed by

World Scientific Publishing Co. Pte. Ltd.

5 Toh Tuck Link, Singapore 596224

USA office: 27 Warren Street, Suite 401-402, Hackensack, NJ 07601

UK office: 57 Shelton Street, Covent Garden, London WC2H 9HE

Library of Congress Cataloging-in-Publication Data
Leschziner, Michael.
Statistical turbulence modelling for fluid dynamics, demystified : an introductory text for graduate engineering students / Michael Leschziner (Imperial College London, UK).
pages cm
Includes bibliographical references and index.
ISBN 978-1-78326-660-9 (hc : alk. paper) -- ISBN 978-1-78326-661-6 (pbk : alk. paper)
1. Turbulence--Mathematical models. 2. Unsteady flow (Fluid dynamics)--Mathematical models. 3. Fluid dynamics--Mathematical models. 4. Eddies. I. Title.
TA357.5.T87L467 2015
532'.0527015195--dc23
2015017459

British Library Cataloguing-in-Publication Data
A catalogue record for this book is available from the British Library.

In-house Editors: Thomas Stottor /Sree Meenakshi Sajani

Typeset by Stallion Press
Email: enquiries@stallionpress.com

Printed and bound in Great Britain by Marston Book Services Ltd, Oxfordshire

To our grandchildren

Maya, Ava, Sienna and Ethan

our legacy

Foreword: What is This Book About, and for Whom?

This book is **not** about *turbulence*, but about how the time-mean phenomena of turbulence are described by means of statistical turbulence models that are implemented in the large majority of computational fluid dynamics (CFD) codes applied to industrial fluid-flow problems. While it does introduce some essential aspects of the fundamental physical processes at play, it does so only to the extent that pertains to subsequent material, and then in basic, mostly descriptive, terms. This apart, the book focuses squarely on a route that starts with the general Navier–Stokes equations, and ends in closure approximations for the Reynolds stresses and scalar fluxes that appear as unknowns in the Reynolds-averaged Navier–Stokes (RANS) equations. With this route followed in the book, readers can safely assume that they need not have a background in turbulence or turbulence modelling. They must, however, have a general graduate-level understanding of fluid mechanics and of the basic mathematical framework that describes the behaviour of momentum and scalars in a fluid flow — i.e. the transport equations governing conserved flow properties.

The book was motivated by my observations, over many years of teaching on regular post-graduate university courses and industrial short courses, that CFD practitioners and graduate students often lack critical insight into the foundation and principles underlying the construction of the models, the way they are calibrated, the range of flow to which they may be applied and the limitations posed by particular closure forms. In the absence of such insight, models are often regarded as options in a computational selection panel, like sweets in a vending machine. Negative consequences of this superficial approach to turbulence modelling include the inappropriate application of models to challenging flow conditions; disappointment about poor predictions, coupled with lack of understanding for the reasons; and growing distrust of the usefulness of CFD in an industrial setting.

This book does not offer a recipe for choosing a particular model from the said 'vending machine' with its hundreds of compartments. Rather, its purpose is to put practitioners in a position to appreciate the rationale underpinning different model types and their relative merits, and to illuminate the route by which models are formulated and calibrated. While this is done by reference to specific models, these models are used as representative examples of their respective classes, and no attempt is made to provide an exhaustive account of the many models documented in hundreds of papers published in the technical literature over the past 50 years; there are literally hundreds of variants and variations, each involving specific assumptions, calibration and validation. Despite an all-out effort made in this book to bring out the rational elements underpinning the various turbulence models, it is an inescapable fact that the models are all amalgams of logic, intuition, 'curve-fitting' — or, put more charitably, 'calibration' by reference to experimental observations — and laborious trial-and-error-based corrections. This makes it impossible to justify, on rational ground, all aspects of modelling, at whatever level. It is this mix of ingredients that makes RANS modelling, and the secure application of models, much more challenging than scale-resolving simulations.

There are already a number of books on turbulence in the public domain. However, none provides the focus this book does on practical turbulence models — starting from the simplest concepts and forms, and ending in the most elaborate ones. In addition, the book deliberately adopts a highly accessible style, terminology and nomenclature, avoiding complex mathematical constructs, implicit manipulations and functionals that tend to obscure the implications. Turbulence does, inevitably, rely on mathematics to describe its characteristics — and, of course, models are ultimately mathematical constructs. However, care has been taken throughout the book to emphasise and bring out links to physical phenomena and implications. This reflects my observation that what students find hardest about turbulence is linking mathematical manipulations and results to the physics, a failure aggravated by insufficient preparatory material, which simply describes certain basic phenomena that require appreciation in relation to subsequent assumptions and simplifications.

It is my hope, above everything else, that this book will contribute to a more insightful engagement of CFD users in the subject of turbulence modelling, and will thus benefit the quality with which CFD is used in industry for predicting turbulent flows. This aspiration is not only confined to pure RANS schemes, but also extends to the increasingly popular category of hybrid large-eddy simulation-RANS (LES-RANS) formulations, in which the near-wall layer is approximated by a RANS model within a global LES method, as well as to currently emerging, more ambitious, embedding strategies, in which RANS patches are embedded within a larger LES domain, or vice versa.

Finally, I wish to express my gratitude to several colleagues who have assisted me, in one way or another, in the course of this book's highly intermittent and fragmented evolution. My thanks go, in particular, to Prof. Suad Jakirlić, TU Darmstadt; and Prof. Kazuhiko Suga, Osaka Prefecture University, for providing me with several figures; to Prof. Kemo Hanjalić, Technical University of Delft; Prof. Wolfgang Rodi, Karlsruhe Institute of Technology; and Dr. Stefan Wallin, FOI, for their insightful and (in parts) robust feedback on several contentious topics; to Prof. Dominique Laurence and Mr. Ryan Tunstall, University of Manchester, for their comments on, and corrections to, several chapters; and to Dr. Lionel Agostini and Dr. Sylvain Lardeau for their help towards improving the quality of some figures.

Contents

Foreword: What is this Book About, and for Whom? vii

1. Statistical Viewpoint of Turbulence — Motivation and Rationale **1**

2. What Makes Turbulence Tick? **7**

2.1 *Eddies, vortices and their scales* 7
2.2 *A semblance of order and organisation* 20
2.3 *Forcing turbulence — a first look* 25
2.4 *Summary and lessons* 30

3. Reynolds-Averaging **33**

3.1 *Decomposition and time-integration* 33
3.2 *The Reynolds-averaged Navier–Stokes (RANS) equations for steady flow* 37
3.3 *The Reynolds-averaged Navier–Stokes equations for unsteady flow (URANS)* 39
3.4 *Spatial averaging and statistical homogeneity* 45
3.5 *Summary and lessons* 46

4. Fundamentals of Stress/Strain Interactions **49**

4.1 *A rational framework for describing the Reynolds stresses* 50
4.2 *The case of simple shear* 54

4.3 *Manifestations of stress anisotropy* 57
4.4 *Summary and lessons* 61

5. Fundamentals of Near-Wall Interaction 65
5.1 *Turbulence in the viscous sublayer* 66
5.2 *Turbulence processes in the 'buffer layer'* 71
5.3 *The velocity distribution in the near-wall region* 73
5.4 *Summary and lessons* 76

6. Fundamentals of Scalar-Flux/Scalar-Gradient Interactions 79
6.1 *The exact equations for the turbulent fluxes* 80
6.2 *Some key interactions* 81
6.3 *Summary and lessons* 86

7. The Eddy Viscosity 87
7.1 *Conceptual foundation* 87
7.2 *Quantification of the eddy viscosity — a first attempt* 92
7.3 *The turbulent-velocity scale* 97
7.4 *The turbulent length scale* 102
7.5 *Length-scale transport* 110
7.6 *Summary and lessons* 113

8. One-Equation Eddy-Viscosity Models 115
8.1 *Introductory comments* 115
8.2 *Turbulence-energy-based models* 117
8.2.1 *The Wolfshtein (1969) model* 118
8.2.2 *The Norris–Reynolds (1975) model* 120
8.3 *Eddy-viscosity-transport models* 122
8.3.1 *The Spalart–Allmaras (1992) model* 123
8.3.2 *Models combining the k- and ε-equations* 128
8.4 *Summary and lessons* 132

9. Two-Equation Models **135**

9.1 ***Options for length-scale surrogates*** 135
9.2 ***The basic $k - \varepsilon$ model*** 137
9.3 ***Low-Reynolds-number $k - \varepsilon$-model extensions*** 149
9.3.1 *The Lam–Bremhorst (1981), Chien (1982) and Launder–Sharma (1974) models* 155
9.3.2 *The Lien–Leschziner model (1994, 1996)* 159
9.4 ***Alternative $k - \phi$ models*** 161
9.4.1 *The basic $k - \omega$ model* 164
9.4.2 *Low-Reynolds-number $k - \omega$-model extensions* 169
9.4.3 *Hybrid $k - \omega/k - \varepsilon$ modelling — the SST model* 172
9.5 ***Reductions of two-equation models to one-equation forms*** 176
9.6 ***Summary and lessons*** 179

10. Wall Functions for Linear Eddy-Viscosity Models **181**

10.1 ***The purpose of 'wall functions'*** 181
10.2 ***Log-law-based wall functions*** 182
10.3 ***Eddy-viscosity-based wall functions*** 189
10.4 ***Numerical wall functions*** 191
10.5 ***More general wall functions*** 193
10.6 ***Wall functions for heat transfer*** 195
10.7 ***Summary and lessons*** 196

11. Defects of Linear Eddy-Viscosity Models, Their Sources and (Imperfect) Corrections **199**

11.1 ***The need for corrections*** 199
11.2 ***Realisability*** 202
11.3 ***Curvature*** 208
11.4 ***Swirl*** 214
11.5 ***Rotation*** 217

11.6 *Body forces — buoyancy* 219
11.7 *Length-scale corrections* 223
11.8 *Summary and lessons* 225

12. Reynolds-Stress-Transport Modelling **227**
12.1 *Rationale and motivation* 227
12.2 *The exact Reynolds-stress equations* 231
12.3 *Closure — some basic rules* 234
12.4 *Realisability and its implications for modelling* 235
12.5 *Turbulent transport* 239
12.6 *Dissipation* 244
12.7 *Pressure-velocity interaction* 257
12.7.1 *Basic considerations* 257
12.7.2 *Modelling of the slow term* $\Phi_{ij,1}$ 262
12.7.3 *Modelling of the rapid term* $\Phi_{ij,2}$ 265
12.7.4 *Modelling* $\Phi_{ij,1} + \Phi_{ij,2}$ *collectively* 275
12.7.5 *Near-wall effects* 280
12.7.6 *Effects of body forces* 290
12.7.7 *Elliptic relaxation of pressure-strain correlation* 291
12.8 *Summary and lessons* 299

13. Scalar/Heat-Flux-Transport Modelling **303**
13.1 *The case for flux-transport closure* 303
13.2 *Closure of the flux-transport equations* 308
13.3 *Summary and lessons* 312

14. The $\overline{v^2} - f$ Model **315**
14.1 *Relationship to second-moment closure and elliptic relaxation* 315
14.2 $\overline{v^2} - f$ *model formulation and variants* 317
14.3 *Summary and lessons* 326

15. Algebraic Reynolds-Stress and Non-Linear Eddy-Viscosity Models **327**

15.1 ***Rationale and motivation*** 327

15.2 ***Explicit algebraic Reynolds-stress models (EARSMs)*** 332

15.2.1 *General formalism* 334

15.2.2 *The model of Pope (1975)* 338

15.2.3 *The model of Gatski and Speziale (1993)* 339

15.2.4 *The model of Wallin and Johansson (2000)* 342

15.2.5 *Near-wall behaviour* 348

15.2.6 *Scale-governing equations* 352

15.3 ***Approximations to rigorous EARSMs*** 353

15.3.1 *The Model of Abe et al. (1997, 2003)* 354

15.3.2 *The Model of Apsley and Leschziner (1998)* 359

15.4 ***Non-linear eddy-viscosity models (NLEVMs)*** 366

15.4.1 *The model of Shih et al. (1995)* 367

15.4.2 *The model of Craft et al. (1996)* 369

15.5 ***Summary and lessons*** 376

Appendix: Basic Tensor Algebra and Rules **379**

References **383**

Index **397**

Chapter One

1. Statistical Viewpoint of Turbulence — Motivation and Rationale

Turbulence is a manifestation of a fluid's 'unwillingness' to flow along a smooth, steady path when it is sheared and its speed is higher than a very modest value. Shear, in combination with vorticity, is typically imparted to the fluid by a solid boundary, and this tends to make the flow susceptible to instability if viscous damping is too weak and the flow is not highly constrained within a narrow passage. Turbulence may thus be thought of as an unstable reaction to 'forcing' — shear being one of several types — resulting in the vorticity being fragmented and redistributed in the form of a seemingly chaotic multiplicity of mutually interacting, three-dimensional, continuously evolving eddies, with a size range typically covering several orders of magnitude. In the absence of a better explanation, the chaotic nature of this process is said to reflect the '*non-linear character of the Navier–Stokes equations*' that govern all fluid flows. An inevitable consequence of turbulence is that the state of a turbulent flow cannot be described, experimentally or numerically, in any way other than as an infinite sequence of instantaneous, unique snapshots of the complete three-dimensional velocity field. This is, self-evidently, a formidable obstacle to a general description of turbulence and turbulent flows.

Scientists have spent decades trying to unravel the details of, and mechanics implicated in, the path along which turbulence evolves. While mathematical analysis of the linearised Navier–Stokes equations and numerical simulations with high-performance computers have shed light on some aspects of this evolution, no satisfactory explanation or description has emerged of the path leading to turbulence or, indeed, of the state of fully established turbulence. Our current picture of turbulence is thus derived from thousands of particular experiments, real or computational, for different flow conditions.

The engineer, who is concerned with designing equipment that relies on, or interacts with, fluid flow, is unlikely to be especially interested in the minutiae of turbulence mechanics or the details of the turbulence structure, but will be much more concerned with the practical manifestations of turbulence, and interested in pragmatic tools allowing the practical effects of turbulence to be described — surface pressure, frictional losses, mixing efficiency, species concentration and heat-transfer rate, to name but a few examples. This is not to say that the engineer can divorce himself/herself from the underlying physics of turbulence. Indeed, a firm grasp of the physics by which turbulence interacts with its environment is crucially important for understanding the engineering tools used to analyse turbulence. However, the physical processes at issue here are not at the unsteady, micro-structure level, but are macroscopic and integral in character, arising from a statistical viewpoint of turbulence. It is only via this statistical route that important phenomenological features of turbulent flows can be described in an accessible and meaningful form, and it is only via this route that the practical effects of turbulence can be quantified and predicted — albeit often with a fair margin of error.

Most computational fluid dynamics (CFD) practitioners will be aware of the impact turbulence and its modelling can have in a predictive computational framework. In contrast, novices may view turbulence as just one of many elements in the CFD mix, and may even regard it as peripheral. The fact is, however, that the representation of turbulence can have profound consequences to the usefulness of a numerical solution. To be clear: this is not merely a question of close fidelity and accuracy, but one that can involve errors of hundreds of percent, especially in attempts to predict heat transfer and scalar transport. Turbulence dictates the spreading rates of boundary layers, jets and wakes; the losses in a turbomachine; the level of wall friction; the ease with which, and the location at which, a boundary layer separates from a curved wall when subjected to an adverse pressure gradient; the size of the separated zone, and hence the pressure field; the viability of any combustion process; the effectiveness of wall cooling; the mixing of pollutants and heat in water and air flows; the transport of solid and liquid particles; vibrations; noise and many other engineering and natural phenomena. Therein lies the importance of the subject, in general, and the modelling of turbulence, in particular.

The starting point of any journey into turbulence, involving analysis, is the set of Navier–Stokes and mass-conservation equations that form a (presumed exact) closed system for the evolution of the velocity and pressure $\{\vec{U}(\vec{x},t), P(\vec{x},t)\}$, $\vec{x} \in \Omega$, $t > t_o$ in time t and the flow domain Ω. The only approach to obtaining a high-fidelity (though never exact) solution of the equations for non-trivial conditions — and then only for a particular set of initial and boundary conditions — is by a direct numerical simulation (DNS). However, this is an exceedingly expensive method for any but simple, low-speed flows, because the range of eddy size broadens in proportion to $Re^{3/4}$, Re being the Reynolds number, $U\Lambda/\nu$, based on a flow velocity U and a length scale Λ characteristic of the flow in question (ν is the kinematic viscosity). This implies a rise in mesh-resolution requirements in proportion to $Re^{9/4}$ and of CPU time in proportion to Re^3 (at best). The objective of such a simulation may be to gain insight into structural and temporal features of turbulence by visualising, for example, the evolution of vortices, deriving spectra from time-series and analysing the interaction between different eddy-size sub-ranges. However, more frequently, the desired outcome is in the form of statistical quantities that are derived from integrating the time-varying fields, in which case the temporal details, expensively resolved, are lost or deemed irrelevant.

Statistical approaches strive to reduce the complexity and cost of the task by focusing attention on the velocity and pressure at any one spatial location, averaged over some period of time, thus directly yielding 'filtered' or averaged (i.e. virtual) equivalents of the respective instantaneous variables. Thus, if the filtered velocity is denoted by $\tilde{\vec{U}}$ then the instantaneous velocity may be represented as $\vec{U} = \tilde{\vec{U}} + \vec{u}$, where the lower-case symbol denotes the velocity fluctuation filtered out by the time-integration process. In the extreme case of the integration being effected over a period much larger than that characterising the longest-lasting turbulent fluctuations, the result is $\vec{U} = \bar{\vec{U}} + \vec{u}$, where the overbar indicates the time-mean velocity — as contrasted with $\tilde{\vec{U}}$, which is time-varying, but at a rate lower than the actual turbulent value. The former decomposition may be appropriate, for example, in the case of a turbulent flow subjected to a sinusoidal actuation via a bounding wall or fluctuating inlet stream, in which case $\tilde{\vec{U}}$ would be the sinusoidally varying phase-averaged velocity, while $\vec{u}$ would be the turbulent fluctuation relative to $\tilde{\vec{U}}$. The latter decomposition of the velocity (and any other flow variable) into a time-averaged value and a turbulent fluctuation is the more common case and is the basis for all conventional Reynolds-averaged Navier–Stokes (RANS) methods.

To derive a mathematical framework for determining the time-mean velocity, the decompositions $\vec{U} = \bar{\vec{U}} + \vec{u}$ and $P = \bar{P} + p$ are inserted into the Navier–Stokes and mass-conservation equations, followed by time-averaging of the equations themselves — a process to which we shall turn in Chapter 3. In the context of this preliminary discussion, it is sufficient to note that the outcome of this process is the set of equations governing $\bar{\vec{U}}$ and $\bar{P}$, analogous to the parent Navier–Stokes equations, but which now contain additional terms in the form of the correlations $\{\overline{uu}, \overline{vv}, \overline{ww}, \overline{uv}, \overline{uw}, \overline{vw}\}$. These correlations constitute the 'penalty' for the simplification achieved by time-integrating the original Navier–Stokes equations — a 'penalty', because they are unknown quantities, introduced as a direct consequence of the decomposition and averaging process. Importantly, they are also the key quantities that represent turbulence within the statistical framework, and it is only through them that the time-mean flow is affected by turbulence. For example, turbulence is known to be extremely effective at mixing momentum, as well as other properties such as species concentration and energy. This mixing is due to the complex eddy structure within a turbulent flow, a process that is entirely assimilated in the above correlations, in the case of momentum. Hence, their accurate determination is crucially important to the prediction of the primary time-mean properties.

It is instructive to dwell briefly on the nature and meaning of the above correlations. Self-evidently, the auto-correlations $\{\overline{u^2}, \overline{v^2}, \overline{w^2}\}$ must be finite and positive. In fact, the sum $0.5(\overline{u^2} + \overline{v^2} + \overline{w^2})$ is referred to as the 'turbulence energy' (per unit mass) — the counterpart of the time-mean kinetic energy $0.5(\overline{U}^2 + \overline{V}^2 + \overline{W}^2)$. It is much less clear, however, whether the cross-correlations are finite and what sign they have. As will transpire later, finite values arise due to straining — by shear, for example — and this is connected to the distortion that straining causes to the eddy structure, especially of the large, energetic eddies that are most effective in the mixing process. A corollary is that the cross-correlations vanish in unstrained turbulence (albeit with a time lag following the removal of the strain). Viewed in probabilistic terms, a finite value of a cross-correlation implies that the probability of the two velocity fluctuations having the same sign is different to that of fluctuations having opposite signs. In contrast, a zero cross-correlation implies equal probabilities of sign combinations. This arises, for example, in a uniform, unobstructed stream in a constant-area wind tunnel, with turbulence initially imparted to the flow by a grid located at the inlet, and decaying downstream as a consequence of viscous dissipation — a process referred to as 'decaying isotropic, homogeneous turbulence'.

The fact that the auto-correlations can be viewed as components of the turbulence energy has already been noted above. If any of the correlations is multiplied by the fluid density, it is readily shown that the product has the dimensions of a force per unit area, i.e. a stress. This is the reason for the correlations being referred to as the 'Reynolds stresses'. Physically, the correlations play a role analogous to the viscous stresses: they redistribute momentum by a mechanism that is, phenomenologically, akin to diffusion: viscous diffusion is caused by Brownian motion, while turbulence produces a momentum redistribution by eddy-induced mixing. Importantly, the turbulence process is, typically, orders of magnitude more vigorous than its viscous counterpart, which is why an accurate determination of the correlations is the key to a realistic prediction of the interaction between turbulence and the time-mean flow.

Unfortunately, the derivation of RANS equation is only the start — indeed, the simplest step — of a long journey towards a closed framework for predicting turbulent flows. The difficult part of this journey is to derive a mathematical model for the above unknown correlations that is applicable to a wide range of flows. The problem is that the correlations react sensitively to variations in scenarios involving different strain types, boundary types and conditions, body forces, compressibility, density gradients and chemical reaction. Constructing a model that possesses a significant degree of generality is a formidable task, which has never been accomplished satisfactorily. As will emerge from this book, the conflict between the desire to achieve generality and the need to predict restricted classes of flows quickly and cheaply, without too much attention to fundamental fidelity, has spawned a bewildering array of models, each restricted in some respects, and each requiring a careful appreciation of its limitations. Unravelling some of the chaos, and untying some of the knots that have arisen from this highly unstructured and unfocused approach to modelling are the principal objectives of this book.

Chapter Two

2. What Makes Turbulence Tick?

The very first sentence in the Foreword of this book begins, "This book is **not** about *turbulence*, but emphatically about how the time-mean manifestations of turbulence are described by means of statistical turbulence models...". However, unsurprisingly, turbulence modelling cannot be entirely divorced from the fundamentals of turbulence mechanics, and it is necessary, as well as informative, to cover some *basic* physical concepts and phenomena that are pertinent to later modelling decisions. The material is covered in largely non-mathematical, descriptive terms, because the principal target is to convey insight into the *consequences* of turbulence mechanics on the mean-flow properties that are the outcome of Reynolds-averaged Navier–Stokes (RANS) solutions. We shall consider, broadly, three major issues: (i) turbulence as a collection of eddies or vortices; (ii) characteristics that might be regarded as reflecting a degree of organisation in the chaos; and (iii) the consequences of 'forcing' turbulence by applying some external action to the flow.

2.1 *Eddies, vortices and their scales*

Turbulence is often referred to as being 'random' or 'chaotic'. It is certainly not 'random', because randomness implies that events occur purely by chance, and that the probability of any one event occurring within a range of possible events is unrelated to any other event within that range. A powerful argument negating the concept of randomness is that the evolution of turbulence from some initial state is governed by the NS equations, presumed to be exact. Thus, given a precise description of some initial state of a flow system, and the conditions at its boundaries, the exact solution is deterministic and reproducible. Turbulence does, however, broadly comply with the definition of chaos, insofar as a turbulent flow is a complex 'dynamical system' with many degrees of freedom, and one that is highly sensitive to initial conditions — thus making the evolution of the system extremely difficult to predict (the 'butterfly effect'). Crucially, turbulence displays distinctive structural and statistical properties, some having a degree of universality, and the response of its statistical properties to forcing is predictable, if only in a restricted sense to be discussed below. Hence, it may be said that turbulence features some aspects of order and organisation.

A key property of turbulence is vorticity (i.e. rotationality). In simple terms, turbulence is a collection of vortices or eddies. Figure 2.1 shows cross-sections through two turbulent fields that arose from the agitation of a nominally stagnant fluid body in a box, that is now in the process of decaying.[1] Typically, neighbouring

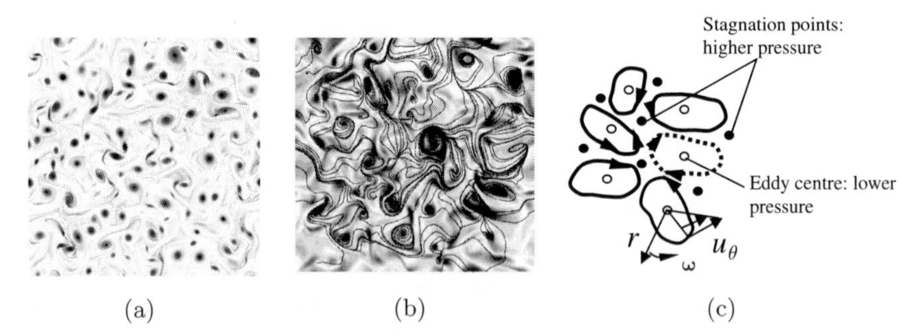

Fig. 2.1: Statistically homogeneous isotropic turbulence in a box. (a) two-dimensional box; blue and red eddies have opposite rotational sense (courtesy L. Rossini); (b) cut through a three-dimensional box (reproduced with permission of World Scientific Publishing from Leschziner *et al.* (2009)); (c) idealised two-dimensional representation of an eddy group.

[1] It is important to keep in mind that turbulence is three-dimensional and that the cross-section shown in Fig. 2.1(b) only gives a partial view of the turbulent field. Thus, the process in any one plane interacts strongly with processes normal to that plane.

eddies rotate in opposite directions, at least in two-dimensional turbulence (Figs. 2.1(a) and (c)), necessary to satisfy kinematic constraints — conservation of mass, in particular. This eddy structure — reflecting an important degree of 'organisation in the chaos' — is critical to the ability of turbulence to mix properties such as heat, scalar concentration and momentum. It is also one that is fundamentally different from Brownian motion, which is not rotational. Hence the oft-stated analogy between the two has very limited validity.

The formal definition of vorticity in a two-dimensional field $U(x, y)$, $V(x, y)$ is:

$$\boldsymbol{\Omega}_{xy} = \frac{1}{2}\left(\frac{\partial U}{\partial y} - \frac{\partial V}{\partial x}\right). \quad (2.1)$$

To recognise that this is simply the local rate of rotation of fluid elements, we may assume that the motion in the vicinity of the centre of an eddy conforms to solid body rotation, $u_\theta = \omega r$, as indicated in Fig. 2.1(c). The decomposition of velocity and radius into their respective $x - y$ components then leads to:

$$\frac{1}{2}\left(\frac{\partial u}{\partial y} - \frac{\partial v}{\partial x}\right) = -\omega. \quad (2.2)$$

Hence, Ω_{xy} represents the rotational property of the fluid elements in the clockwise direction.

Each eddy is (nominally) surrounded by an induced, radially varying irrotational velocity field $u_\theta(r) \propto 1/r$ that affects neighbouring eddies (referred to as the Biot–Savart law). Each eddy thus convects other eddies and is itself convected by neighbouring eddies. An important characteristic of turbulence is therefore 'self-convection'. In addition, each eddy deforms and 'stretches' other eddies. The latter includes stretching along the rotational axes of eddies, thus leading to a reduction in the diameter of eddies, and an increase in rotational speed, reflecting the conservation-of-angular-momentum principle.

As is evident from Fig. 2.1, eddies cover a range of sizes in a turbulent domain. In fact, in a highly turbulent flow, the eddy size

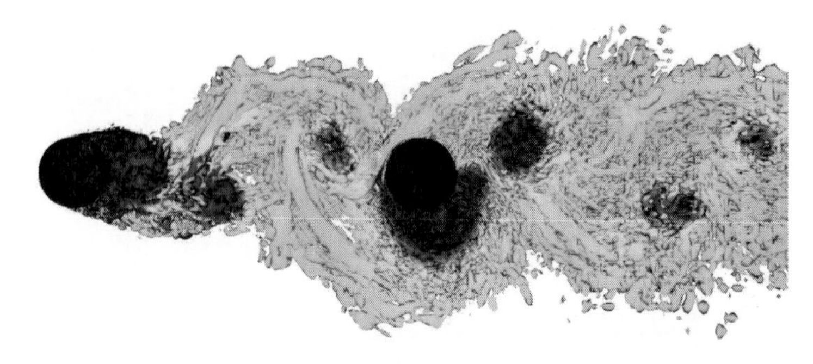

Fig. 2.2: Turbulence field in a flow around two circular cylinders in tandem; dark patches represent pressure troughs, identifying shed von Kármán vortices (reproduced with permission of World Scientific Publishing from Leschziner *et al.* (2009)).

extends over several decades. Turbulence is thus referred to as a 'multi-scale phenomenon'. This is well brought out in the simulation of vortex shedding from two cylinders in tandem, shown in Fig. 2.2. As an aside, this figure also serves to highlight the fact that turbulence often coexists with coherent (periodic) unsteady motion — in this case shed von Kármán vortices — that are not part of the turbulence field, but arise from a separate instability mechanism. This ambiguity poses an especially serious problem for turbulence models — an issue that will be discussed under the heading Unsteady RANS ("URANS") in Section 3.3.

Two properties that are extremely important in turbulence modelling are the turbulence energy and its rate of dissipation, both per unit mass, and conventionally denoted by k and ε, respectively. The former is the time-averaged value of the square of the magnitude of the velocity fluctuations, i.e. a statistical measure of the kinetic energy of the turbulent eddies.

In a typical turbulent flow, the velocity at any one location will vary greatly with time, as is clearly implied by Figs. 2.1 and 2.2. If we measure, within a Cartesian system, the x-component of the velocity fluctuation, $u(t) = U(t) - \overline{U}$, where $\overline{U}$ is the time-averaged value, and record the proportion of time, or the number of measured discrete events, of its value falling within a fixed increment (a 'pot'), we get a probability (-density) curve, such as that shown in Fig. 2.3.

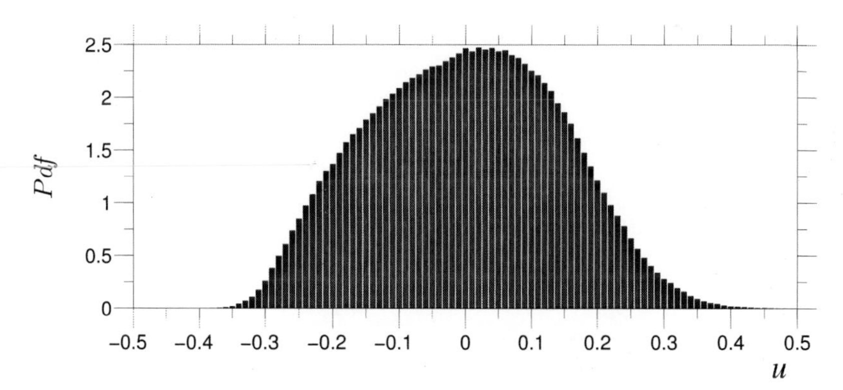

Fig. 2.3: Probability-density function of the streamwise-velocity fluctuations near the wall in a turbulent channel flow, evaluated from a direct numerical simulation.[2]

This distribution has the property $\int_{-\infty}^{+\infty} Pdf(u)du = 1$ — i.e. the probability of the velocity fluctuation assuming any value within the range $-\infty < u < \infty$ must be 1. The turbulence energy associated with the u' component then arises from:

$$\overline{u^2} = \int_{-\infty}^{+\infty} u^2 Pdf(u)du. \tag{2.3}$$

The above can be repeated for v and w, thus yielding $\overline{v^2}$ and $\overline{w^2}$. The turbulence energy is therefore:

$$k = 0.5\big(\overline{u^2} + \overline{v^2} + \overline{w^2}\big). \tag{2.4}$$

The turbulence energy may be used to define a turbulent velocity scale, $\sqrt{k}$, which may be taken to characterise, in an average sense, the energetic fluctuating motion. The turbulence energy is an amalgam of the kinetic energy of the eddies, the size of which typically covers a range of 4–5 orders of magnitude. If any eddy size is denoted by ℓ, it is conventional to characterise this by the associated 'wave number', which is defined by $\kappa = 2\pi/\ell$, essentially the inverse of the

[2]Note the non-Gaussian — 'skewed' — shape of the PDF. This is indicative of the non-randomness of turbulence.

eddy size. The 'energy density' of the field of eddies, $E(\kappa)$, is defined as the turbulent kinetic energy contained in any eddy-size increment divided by that increment, i.e. the integral of $E(\kappa)$ over κ is the turbulence energy itself:

$$k = \int_{\kappa_\Lambda}^{\kappa_\eta} E(\kappa) d\kappa, \tag{2.5}$$

where η is the smallest and Λ is the largest eddy size in the field. The former will be discussed later in this chapter, while the latter can safely be assumed to be of the order of, or somewhat smaller than, the dimension of the 'box' which confines the entire flow. Expression (2.5) is based on the concept that the field of velocity fluctuations can be Fourier-transformed and thus represented as a set of Fourier components, each component associated with a wave number κ. This also implies that the turbulence-energy contribution associated with κ can be determined, either from experiment or a DNS.

In a well-developed turbulent flow, at a sufficiently high Reynolds number, $E(\kappa)$ tends to display a near-universal behaviour, as is shown in Fig. 2.4(a). The range in this figure contains three sub-ranges: 'A' identifies large eddies, with a representative length scale Λ, which carry most of the energy; 'B' identifies the 'inertial sub-range'; and 'C' identifies the smallest eddies, of length scale η, which are most strongly affected by viscosity, to the extent that the formation of even smaller eddies is prevented by their energy being dissipated into heat.

The range 'B' — the 'inertial sub-range' — is especially interesting and important to modelling. In this range, the large eddies 'fragment' into smaller eddies by the action of 'vortex stretching' — a process by which eddies are stretched along their axes by other eddies and then fragment due to instability.[3] The variation in this range may be derived from dimensional-homogeneity considerations, based on the reasonable assumption that sub-range 'B' is unaffected

[3]It must be acknowledged, however, that we do not have an entirely clear view of the detailed mechanisms at play.

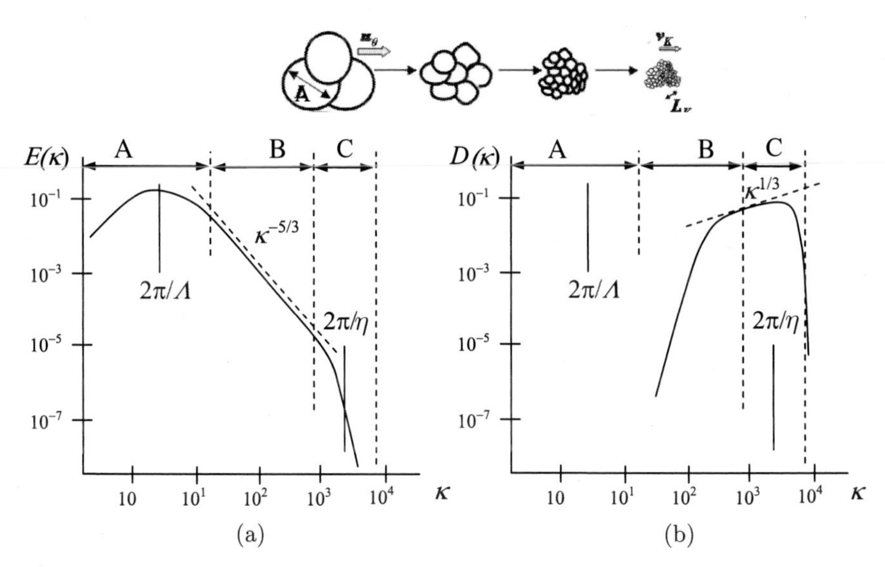

Fig. 2.4: Schematic of a typical spectrum of (a) the turbulence energy $E(\kappa)$, and (b) the turbulence dissipation $D(\kappa)$, in fully developed turbulence at high Reynolds number (see Eqs. (2.7) and (2.9)).

by the viscosity — the eddies being too large to be materially damped — and that the κ-wise energy density depends only on the rate at which energy is transferred from sub-range 'A' to sub-range 'C', i.e ε, the kinematic rate of turbulence-energy dissipation (again, per unit mass). Hence, dimensional reasoning implies:

$$E(\kappa) = f(\kappa, \varepsilon). \tag{2.6}$$

As $E(\kappa)$ is the kinematic energy per wave-number increment, its dimensions are $[L^3/T^2]$, where L and T are the dimensions of length and time, respectively. Moreover, the dimensions of the independent variables are $[\kappa] = [1/L]$, $[\varepsilon] = [L^2/T^3]$. Dimensional-consistency constraints then readily lead to:

$$E(\kappa) = c_k \varepsilon^{\frac{2}{3}} \kappa^{-\frac{5}{3}}. \tag{2.7}$$

This relationship was first presented by Kolmogorov (1941) who showed that, for the particular case of spatially homogeneous and directionally isotropic turbulence at sufficiently high Reynolds number, $c_k = 1.5$.

The energy transferred to progressively smaller eddies is eventually dissipated by friction in sub-range 'C'. Just as $E(\kappa)$ describes the wave-number dependence of the turbulence energy, another distribution, $D(\kappa)$, describes the dependence of the dissipation rate on the wave number in sub-range 'C', as shown in Fig. 2.4(b). Here, viscosity plays a central role, so the dimensional dependence must be of the form:

$$D(\kappa) = g(\nu, \kappa, \varepsilon), \tag{2.8}$$

where ν is the kinematic viscosity. Dimensional analysis then readily leads to:

$$D(\kappa) = c_\varepsilon \nu \varepsilon^{\frac{2}{3}} \kappa^{\frac{1}{3}}, \tag{2.9}$$

with $c_\varepsilon = 2c_k$, as derived by Kolmogorov. The true variation of $D(\kappa)$ is shown in Fig. 2.4(b), and this demonstrates that the dissipation actually starts rising in the inertial sub-range 'B', ahead of the $\kappa^{1/3}$ sub-range (note, however, the logarithmic scale), and then declines rapidly beyond it.

In summary, turbulent energy contained in sub-range 'A' — initially produced, say, by the flow passing through a physical mesh or past some obstacle — may be thought of as 'cascading down a pipeline' between sub-ranges 'A' and 'C'.[4] In other words, at any value κ within the inertial sub-range 'B', the rate of energy transfer is essentially constant and equal to the rate of energy dissipation ε in sub-range 'C'.[5] Importantly, as will be argued in Section 7.4, this allows the dissipative processes at the length scale η to be linked

[4]There is, 70 years after Kolmogorov (1941), still an ongoing debate about the precise mechanisms driving the 'cascade'. What is accepted, however, is that the net down-scale flux is a result of an imbalance between up-scale and down-scale energy fluxes. We shall ignore this debate and accept the cascade concept as stated.

[5]An evolution equation that describes how the energy density is connected to the production rate at low wave numbers, to the destruction rate at high wave numbers and to the energy transfer rate in the inertial sub-range can be found in several books on turbulence, including Hinze (1975), Pope (2000, p. 250), Davidson (2004, p. 475).

to processes at the length scale Λ, obviating the need to consider, explicitly, the statistics of the small-scale motions in formulating turbulence models.

The behaviour represented by Eq. (2.7) is also observed, qualitatively, in inhomogeneous, anisotropic turbulence typical of practical flows, provided attention focuses on volumes that are smaller than Λ^3 and the Reynolds number is sufficiently high. In fact, one of several quality indicators often used to demonstrate the adequacy of large eddy simulations (LES) for quite complex flow conditions is the presence of a $\kappa^{-5/3}$ range in spectra extracted from the simulations at various highly turbulent locations.

It has already been noted that Λ is of the order of, or somewhat smaller than, the size of the flow (e.g. the thickness of a shear layer or a jet, or smallest dimension of a cavity). Two questions that arise, therefore, in relation to Fig. 2.4 are: (a) what is the magnitude of η; and (b) what determines the width of the distribution, i.e. the ratio Λ/η? The first question can be answered upon noting the following:

(i) Dimensional reasoning dictates that the energy-transfer rate towards sub-range 'C' must be of order $u^2(u/\ell)$ — i.e. the kinematic energy per unit time — where u is the turbulent-velocity scale at the length scale ℓ (i.e $u \equiv u(\ell)$).
(ii) The energy-transfer rate is $u^2(u/\ell) = \varepsilon$.
(iii) As ℓ approaches η, viscous effects are of the same order as inertial effects, i.e. $Re_\eta = u_\eta \eta / \nu \approx 1$.

Thus, with $\ell = \eta$ and $u = u_\eta$ eliminated with the aid of ε, via condition (ii) above, the combination of (i) to (iii) (ignoring any constant multiplier) leads to:

$$\eta = \left(\frac{\nu^3}{\varepsilon}\right)^{\frac{1}{4}}. \tag{2.10}$$

For considerations pursued below, it is also useful to derive the velocity and time scales of the eddies at the length scale η. The condition under (iii) above immediately gives:

$$u_\eta = (\nu\varepsilon)^{\frac{1}{4}}, \tag{2.11}$$

and the time scale follows from:

$$\tau_\eta = \frac{\eta}{u_\eta} = \left(\frac{\nu}{\varepsilon}\right)^{\frac{1}{2}}. \tag{2.12}$$

Because of the condition under (ii) above, it follows that the velocity scale u decreases along with the length scale, although at a lower rate, and so does the time scale. Also, u/ℓ, the inverse of the time scale, may be regarded as a measure of the 'vorticity' of the turbulent eddies, and this increases steadily as the eddies decrease in size.

Next, we turn our attention to the ratio Λ/η. For the sake of clarity of the argument, we consider a simple plane shear layer of thickness δ created by a stream of velocity U flowing over a stagnant fluid body. With $\Lambda \approx \delta$ and a velocity scale of the largest eddies being u_Λ (typically 10% of U), the Reynolds number (Re) characterising the largest scales is:

$$Re_\Lambda = \frac{u_\Lambda \Lambda}{\nu}, \tag{2.13}$$

which we expect to be typically 10% of the mean Reynolds number. Consistent with item (ii) above, $u_\Lambda^2(u_\Lambda/\Lambda) \approx \varepsilon$, and substitution into Eq. (2.10) leads (within a constant multiplier) to:

$$\frac{\Lambda}{\eta} = Re_\Lambda^{\frac{3}{4}}. \tag{2.14}$$

Also, it is easy to show, using Eqs. (2.11) and (2.12), that:

$$\frac{u_\Lambda}{u_\eta} = Re_\Lambda^{\frac{1}{4}} \tag{2.15}$$

$$\frac{\tau_\Lambda}{\tau_\eta} = Re_\Lambda^{\frac{1}{2}}. \tag{2.16}$$

These are very important results, especially in the context of simulating turbulent flow. Equation (2.14) shows that the range of length scales present in a turbulent flow rises almost linearly with Re. Self-evidently, the range of the eddy-volume scale rises in proportion to $Re^{2.25}$. This dependence, and the added diminution of the time scale in proportion to $Re^{1/2}$, poses enormous challenges to the direct (scale-resolving) numerical simulation of turbulence at high Reynolds numbers. It is precisely this barrier that RANS

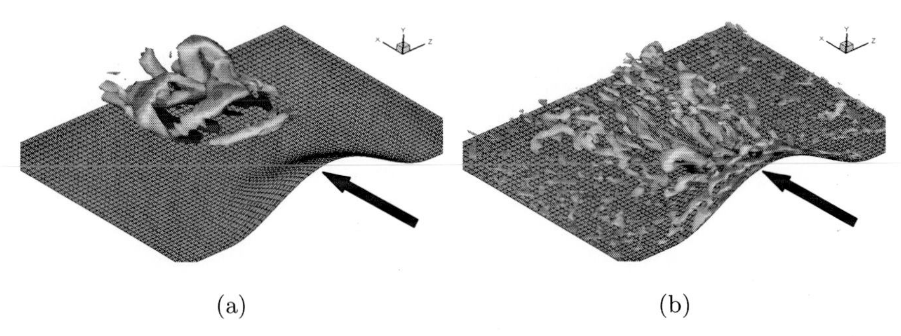

(a) (b)

Fig. 2.5: Snapshots of streamwise-velocity-fluctuation fields in a turbulent flow behind a three-dimensional hill on a flat wall at $Re = U_o H/\nu = 1300$ and 130,000, where U_o is the free-stream velocity above the hill, and H is the hill height (reproduced with permission of World Scientific Publishing from Leschziner *et al.* (2009)).

methods avoid — albeit at the penalty of the complexities that go with the construction of turbulence models — by assimilating the entire range of length and time scales into the statistical framework. The dependence of the eddy-size range on the Reynolds number is illustrated, qualitatively, in Fig. 2.5, which shows snapshots of streamwise-velocity fluctuations derived from two simulations for a flow past a three-dimensional hill-shaped body on a flat plate, with the flow separating from the leeward side of the hill.

An important implication of the substantial scale separation conveyed in Fig. 2.4 is that the major dynamics of a high-Reynolds-number shear flow do not depend directly on the viscosity, except close to walls where viscous mixing dominates over turbulent mixing. This is because the redistribution of momentum — the mixing — is primarily dictated by large-scale structures. The viscosity is important, but principally as a means of dissipating the turbulence energy 'cascading' down the length-scale range towards the small-scale eddies. These eddies are not effective 'mixers', however. Hence, two free turbulent jets, as shown in Fig. 2.6, one at a moderate and the other at a high Reynolds number, will spread at virtually the same angle and will have very similar (appropriately normalised) profiles of velocity and statistical turbulence variables that are associated with the energetic scales (such as the turbulence energy) and thus govern the mean-flow characteristics. The main difference will be that the

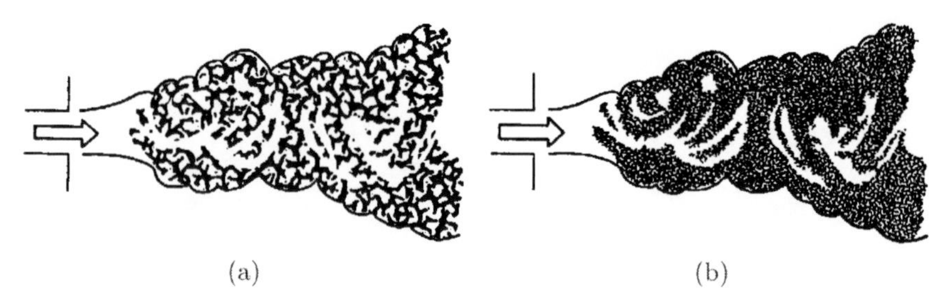

Fig. 2.6: Two turbulent free jets: (a) low Reynolds number and (b) high Reynolds number. The spreading rates are virtually unaffected, but the finest-scale structures are very different (reproduced with permission from MIT Press from Tennekes and Lumley (1972)).

high-Reynolds-number flow will feature a finer structure of eddies, as is readily seen from Eq. (2.14).

The situation is quite different near a wall, however, where the turbulent velocity fluctuations decline towards zero at the wall itself. Moreover, because of the kinematic constraints imposed by the wall (no-slip and impermeability conditions), the range of eddy size declines progressively, so that the largest eddies are not much larger than the smallest eddies associated with dissipation. In other words: the spectrum shown in Fig. 2.4 narrows progressively as the wall is approached. Hence, within a layer close to the wall — the 'viscosity-affected layer' or 'viscous sub-layer' — the flow is strongly and directly affected by the viscosity. Indeed, in an attached boundary layer, the velocity profile close to the wall is well described by the momentum equations pertaining to laminar flow. We shall return to the topic of near-wall turbulence in Chapter 5, and do so in some detail in view of the outstanding importance of near-wall turbulence.

To convey an impression of the actual size of the smallest eddies in a near-wall shear layer, we consider, in Table 2.1, the particular case of an air flow in a channel of height 200mm. To determine the Kolmogorov scale, it is necessary to know the dissipation rate (Eq. 2.10), and this is done here by resorting to results obtained from various DNS databases. As will be shown in Chapter 5, the dissipation rate varies significantly across any boundary layer, and the data given in Table 1 are the maximum values predicted by DNS

Table 2.1: Dependence of ratio of channel height to the Kolmogorov scale at the wall on the bulk Reynolds number in a channel of height 200 mm with air flow. The dissipation rate was extracted from various DNS data.

$Re = U_m 2h/\nu$	η_{wall} (mm)	$2h/\eta_{wall}$
5600	0.863	116
13750	0.371	270
31000	0.178	562
86700	0.061	1443

at the wall. Hence, the Kolmogorov scale is smallest at the wall. The striking message of Table 1 is that, even at practically modest Reynolds numbers, the smallest eddies are of order 10^{-3} of the boundary-layer thickness — the channel half-height, in this case — and this ratio can rise to 10^{-4} and higher in external aerodynamic applications. Thus, to perform a scale-resolving (DNS) simulation over a channel block of $(2h)^3$, down to the Kolmogorov scale, would require about 5 billion cells at $Re = 10^5$. Although a DNS need not resolve eddies quite as small as η, the discussion around Table 2.1 illustrates dramatically the enormous motivation for statistical treatments in a practical context.

So far, we have encountered two length scales, representing different ranges of eddy size, as shown in Fig. 2.4: Λ and η. There is a third length scale that is often mentioned in the turbulence literature: the 'Taylor microscale', λ. In essence, λ is an intermediate scale that represents relatively 'small' eddies that are, however, significantly larger than those associated with dissipation. This scale does not have as clear a physical foundation as do Λ and η, and it rarely features, explicitly, in RANS modelling. An intuitive argument is that this intermediate length scale should combine the Kolmogorov time scale (2.12) and the velocity scale $k^{1/2}$, rather than u_η in Eq. (2.11), which arises from imposing $Re_\eta = u\eta/\nu \approx 1$ and is thus inapplicable to medium-size eddies. In this case, $\lambda \propto k^{1/2}\tau_\eta = \sqrt{\frac{\nu k}{\varepsilon}}$. Readers interested in the formal derivation and relevance of λ are referred to Pope (2000, pp. 198–201). Suffice it to note here that the ratio λ/η, corresponding to Λ/η, (see Eq. (2.14)) is:

$$\frac{\Lambda}{\lambda} = \frac{Re_{\Lambda}^{\frac{1}{2}}}{\sqrt{10}}. \tag{2.17}$$

Thus, clearly, $\lambda > \eta$, and its separation from Λ is lower than that of η as the Reynolds number increases.

2.2 *A semblance of order and organisation*

One of the two images included in Fig. 2.1 is replicated in the upper part of Fig. 2.7 to convey important qualitative differences between unstrained 'box' turbulence and shear-strained turbulence, the latter in the lower part of Fig. 2.7 being ubiquitous in practical flows. The highly stylised, schematic representation of box turbulence on the left as a collection of 'spherical' eddies serves the purpose of contrasting

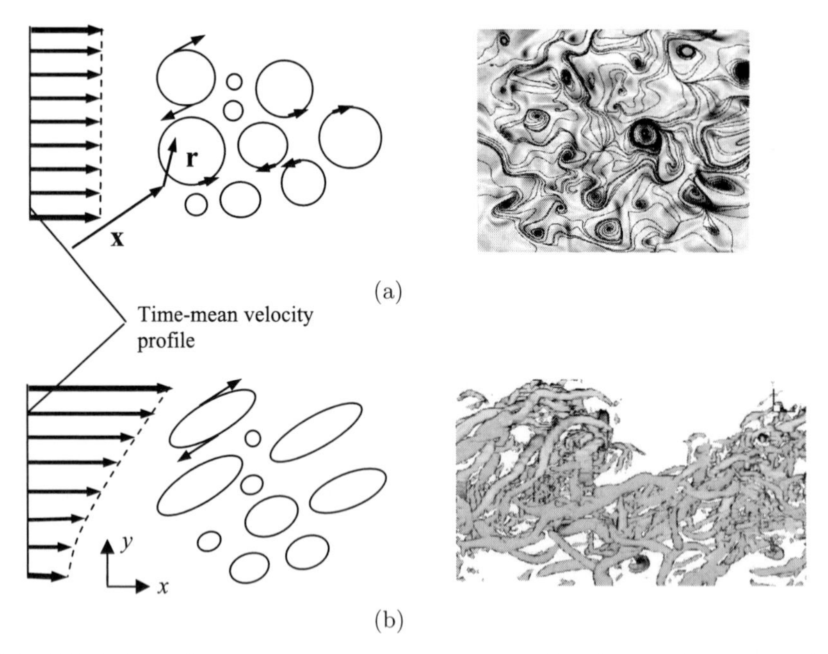

Fig. 2.7: Schematic representations and corresponding images from computer simulations of eddies in (a) unstrained isotropic and (b) sheared anisotropic turbulence (reproduced with permission of World Scientific Publishing from Leschziner *et al.* (2009)).

this simplest type of turbulence with that subjected to shear straining. Thus, when this field is sheared, the larger eddies deform in response to the shear, aligning themselves in a manner that reflects the shear motion. The small eddies, on the other hand, remain largely unaffected by the straining, because their length scale is far removed from the length scale characterising the mean shear — for example, the shear-layer thickness δ. This disparity in length scales hinders a direct interaction between the two — much like the fact that a mass-spring system would not be excited by an oscillation that is very different from the system's eigenfrequency. In fact, this insensitivity of the small-scale eddies to the mean strain, and implied universality, is an important pillar of the Kolmogorov theory mentioned in Section 2.1. The preferential orientation of eddies in shear flow has important consequences to the statistical properties of the associated turbulent flow, considered next.

One facet of the 'organisation' in turbulence, associated with its eddy structure, is that turbulent motions are statistically correlated in space — i.e. motions in one location are correlated with motions in neighbouring locations. Thus, if attention focuses on the larger eddies in Fig. 2.7, it should be obvious that velocity fluctuations (relative to the time-mean field) at any location $\mathbf{x}$ are correlated with those at location $(\mathbf{x}+\mathbf{r})$ for a restricted range of $\mathbf{r}$ of the order of the size of the large eddies. More generally, all components of the 'two-point correlation tensor',

$$R_{ij}(\mathbf{r}) \equiv \frac{\overline{u_i(\mathbf{x})u_j(\mathbf{x}+\mathbf{r})}}{\overline{u_i(\mathbf{x})u_j(\mathbf{x})}}, \tag{2.18}$$

in which u_i and u_j are the three Cartesian components of the fluctuating part of the velocity, decay gradually with $\mathbf{r}$, from unity at $\mathbf{r}=0$, thus indicating spatial correlation. This is illustrated, qualitatively, in Fig. 2.8 for one particular component, R_{11} and $\mathbf{r}=r_1$ (i.e. with $\mathbf{r}$ aligned with the x-axis), which characterises the correlation of two streamwise velocity fluctuations a distance r_1 apart. A clear implication of Fig. 2.8 is that turbulence is not a local process, and that the turbulent state at any one location cannot be expected to be captured by reference to conditions at this location only. This 'non-locality' is accentuated by the action of pressure

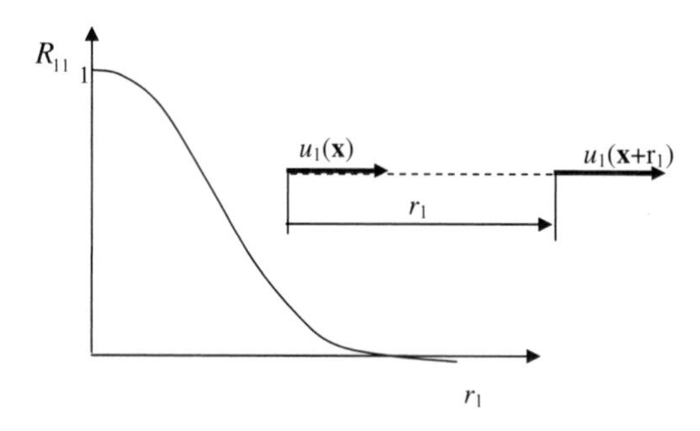

Fig. 2.8: Qualitative spatial variation of one component, R_{11}, of $R_{ij} \equiv \overline{u_i(\mathbf{x})u_j(\mathbf{x}+r_1)}/\overline{u_i(\mathbf{x})u_j(\mathbf{x})}$.

fluctuations which accompany the velocity fluctuations. Pressure is a highly 'elliptic' property, i.e. pressure fluctuations at any particular location tend to propagate in all spatial directions, are reflected — say, at solid boundaries — and then return to influence the condition at the position at which the original perturbation was generated. We shall discuss the specific subject of non-locality of pressure fluctuations in Section 12.7, when modelling correlations that involve these fluctuations.

Associated with the correlation of Fig. 2.8 is the length scale,

$$L_{11} = \int_0^\infty R_{11} dr_1. \tag{2.19}$$

This represents a 'macro-scale', that characterises the 'size' of the large eddies in the streamwise direction. The macro-scales associated with other components of R_{ij} can vary greatly, because eddies can be highly anisotropic. An extreme example is near-wall turbulence in a boundary layer or a channel, where turbulence is extremely 'streaky', as shown in Fig. 2.9, with L_{11} reaching several shear-layer thicknesses.

Another important feature of shear-strained turbulence, associated with its vortical character and negating the concept of random motion, is that the time-averaged cross-products of the turbulent velocity fluctuations are finite. In the particular case a flow that is sheared in the $x - y$ plane, such as that shown in Fig. 2.7(b),

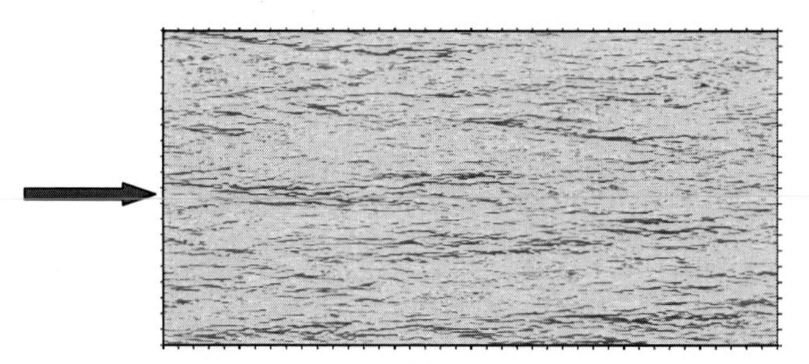

Fig. 2.9: Structure of a turbulent boundary layer in a plane parallel and close to the wall, derived from DNS: dark (blue) and light (green/red) regions are, respectively, high-velocity and low-velocity streaks. Note that the streamwise length of the structures is about 10–20 times longer than they are wide in the spanwise direction (courtesy L. Agostini).

$\overline{u_1 u_2} = \overline{uv}$ is far from zero. This is so, for example, in a jet or sheared boundary layer.

The non-zero value of this cross-correlation may be understood by noting, first, that, for the condition in Fig. 2.7(b), the rotational motion of the eddies tends to carry, on average, low (mean-) momentum fluid from lower to upper regions and high (mean-) momentum fluid from upper to lower regions — i.e. there is a redistribution or *mixing* of momentum. A negative (downward) velocity fluctuation, v, thus leads to a positive velocity fluctuation, u, in lower regions. Similarly, a positive v is associated with a negative u. Hence, in the case of Fig. 2.7(b), the correlation $\overline{uv}$ is finite and negative.

This argument is given a quantitative flavour in Fig. 2.10, derived from a DNS of a turbulent boundary layer. The figure shows contours of the probability of any streamwise fluctuation, u, occurring together with any other wall-normal fluctuation, v, in the sheared near-wall region, with the shear characterised by $\partial U/\partial y > 0$. A clear implication of Fig. 2.10 is that negative pairs of (u, v) are more likely than positive ones, so that the long-time average, $\overline{uv}$, must be negative. If the contours were symmetric relative to both axes, there would be no bias, and the implication would be $\overline{uv} = 0$. Correspondingly, $\partial U/\partial y < 0$ gives rise to $\overline{uv} > 0$.

A simple illustration of the above relationship is provided by a fully developed flow in a plane channel. Typical profiles of the velocity

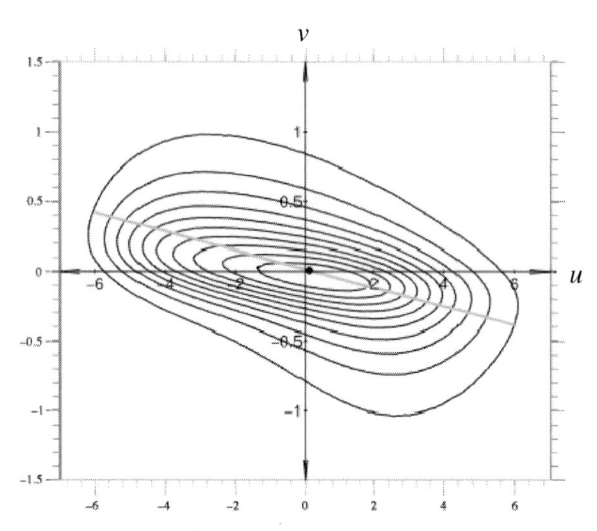

Fig. 2.10: Contours of probability of streamwise turbulent fluctuations, being correlated with wall-normal fluctuations in a boundary layer close to the wall, derived from DNS. The contour lines represent 10% (outermost contour), 30%, 50%, 70% and 90% (innermost contour) of the probability. Note the tendency of positive u fluctuations to be preferentially correlated with negative v fluctuations and vice versa (reproduced with permission of Cambridge University Press from Agostini *et al.* (2014)).

U and the correlation $\overline{uv}$, whether determined by measurement or from time-averaging DNS data, are shown in Fig. 2.11. Consistent with statements made above, the signs of the velocity gradient and of the correlation $\overline{uv}$ oppose each other, and the symmetry in U must mean that $\overline{uv}$ must be anti-symmetric with respect to the channel centre-line. The fact that the correlation declines to zero at the walls is an inevitable consequence of the zero-wall-velocity constraint, so that all fluctuations must vanish. As the fluctuations decline, the flow conditions are increasingly dictated by fluid viscosity. In fact, very close to the wall, the velocity profile arises from the solution of the U-momentum equation:

$$\mu \frac{\partial^2 U}{\partial y^2} = \frac{dP}{dx}, \tag{2.20}$$

in which the streamwise pressure gradient is virtually independent of y. The solution of Eq. (2.20) is an almost linear function of the

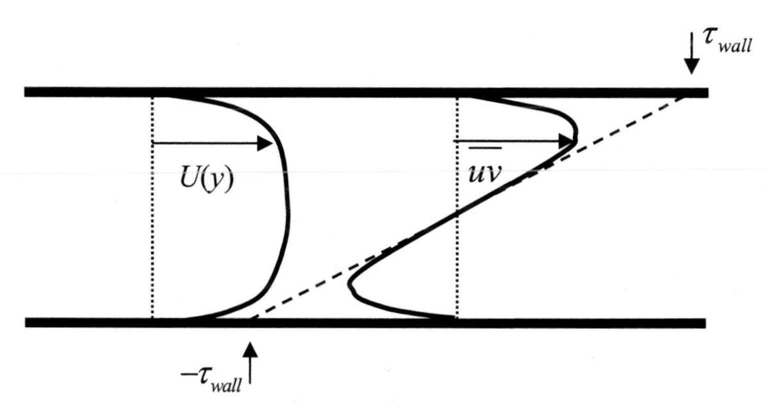

Fig. 2.11: Profiles of streamwise velocity and the correlation $\overline{uv}$ in a fully developed plane-channel flow (note that these profiles remain invariant in the streamwise direction, by the definition of fully developed flow).

wall distance as $y \to 0$:

$$U = \alpha y + \beta \frac{dP}{dx} y^2. \tag{2.21}$$

Away from the immediate vicinity of the wall, the velocity profile is dictated by momentum mixing caused by the turbulent eddies — in other words, the fluctuations u and v. In fact, we shall see in Chapter 3 that the decisive quantity is the correlation $\overline{uv}$, which represents a diffusion mechanism akin to, but much stronger than, the viscous stress $\mu \frac{\partial U}{\partial y}$.

One final point that is worth highlighting by reference to Fig. 2.10 is that streamwise fluctuations, u, tend to be substantially larger than wall-normal fluctuations, v (note the scales of the axes). This is a first indication of what is called 'anisotropy': the fact that the intensity of turbulent fluctuations tends to differ in different directions in the presence of straining — in this case, shear straining.

2.3 *Forcing turbulence — a first look*

'Box' turbulence, if left unstrained following its creation, decays as a consequence of energy being extracted from the turbulence by viscous dissipation. Turbulence can only be sustained if subjected to forcing — straining being the most frequent and effective form of such forcing. We have already encountered a classic type of forcing in the previous section — namely, the imposition of shear by the presence of a wall (Figs. 2.9–2.11). We shall see in Section 4.2, by reference to exact manipulations of the NS equations, that the coexistence of mean strain and turbulent fluctuations gives rise to a continuous generation of turbulence that compensates for viscous dissipation. In particular, given a shearing motion in a boundary layer or the channel shown in Fig. 2.11, we shall show that turbulence energy is generated (produced) at the rate:

$$P_{k,shear} = -\overline{uv}\frac{\partial U}{\partial y}. \tag{2.22}$$

If Eq. (2.22) is multiplied by the density, then the result can be interpreted as a stress, i.e. force per unit area in the $x - z$ plane, multiplied by a velocity difference per unit height, i.e. a force per unit volume multiplied by a velocity difference, and hence the power generated per unit volume. For the particular case of the channel flow, Fig. 2.11, the two terms on the right-hand side of Eq. (2.22) have been shown to be of opposite sign. Hence, the generation of turbulence energy is positive (but this is not always so, as will be explained in Chapter 11). Correspondingly, it can be readily shown that the production in Eq. (2.22) arises with the opposite (negative) sign in the kinetic-energy equation of the time-mean flow. In other words: this term represents the transfer of energy from the mean flow to the turbulence field. In fact, in 'simple' shear flow — a boundary layer, or a jet, or a mixing layer between two parallel streams moving at different velocities — the rate of turbulence generation is close to the rate of dissipation, a condition complying with the 'energy-cascade' concept introduced earlier in relation to Fig. 2.4 and referred to as 'local turbulence-energy equilibrium'.

Other types of kinematic forcing include compressive and extensive straining, but these are much less effective than shearing in generating turbulence energy. This difference will be demonstrated in Chapter 11, by reference to exact manipulations of the NS equations. Qualitatively, the effectiveness of shear can be appreciated by noting that shear is the primary source of instabilities in laminar

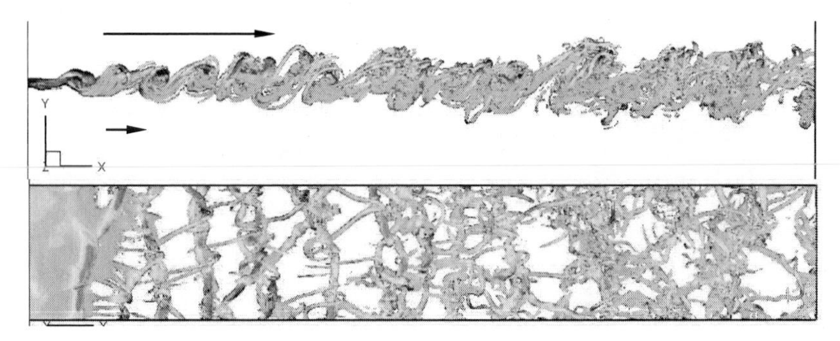

Fig. 2.12: Computer simulation of transition to turbulence in a free shear layer (side and top views), between two streams of different velocities, visualised via a vortex-identification criterion (reproduced with permission of World Scientific Publishing from Leschziner *et al.* (2009)).

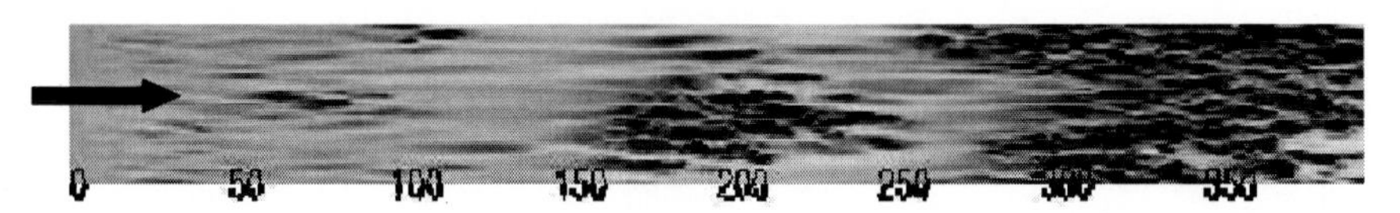

Fig. 2.13: Computer simulation of transition to turbulence in a boundary layer. Dark regions identify high velocity and bright regions low-velocity regions near the wall. Wavy instabilities are seen to precede a ‘turbulent spot’, followed by calming and then full transition in the last quarter of the streamwise stretch (reproduced with permission of ASME from Lardeau *et al.* (2007)).

flow, leading to transition and then fully established turbulence, as is illustrated in Figs. 2.12 and 2.13.

Forcing also arises from density stratification (buoyancy), streamline curvature, magnetic fields acting on electrically conducting liquids, and shocks (i.e. a strong normal pressure gradient) interacting with compressibility-associated density differences. An example for the first type, shown in Fig. 2.14, is turbulence created by heating a surface below an initially cold stagnant fluid body — say, the ground heated by the sun on a cold winter’s day. The initial consequence of such heating is the creation of regular (or cellular) updrafts, an instability referred to as ‘Rayleigh–Benard convection’. At sufficiently high heating rates (or temperature differences), these vertical updrafts become increasingly more convoluted and break up, by internal shearing, into turbulence. The equivalent

Fig. 2.14: Turbulence generated by heating a surface below a cooler fluid: (a) initial process of instability generating cellular structure; (b) ultimate turbulent state (courtesy J. Lülff, Institute for Theoretical Physics, University of Münster).

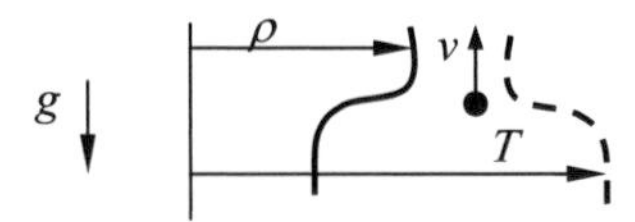

Fig. 2.15: Unstably stratified stagnant fluid body, used to interpret Eq. (2.23).

of shock-induced turbulence in a variable-density system is called 'Richtmyer–Meshkov instability'. In the case of Rayleigh–Benard convection, it can be shown that the rate of production of turbulence energy — corresponding to that given in Eq. (2.22) for shear-induced generation — is:

$$P_{k,buoyancy} = \rho g \beta \overline{T'v}, \tag{2.23}$$

where g is the gravitational acceleration in the y-direction, β is the volumetric thermal-expansion coefficient, and T' is the turbulent temperature fluctuation relative to the time-averaged temperature at the location in question. Here again, it is instructive to interpret Eq. (2.23) by reference to a simple physical example, apart from the fact that it can be shown, again by dimensional reasoning, to be power per unit volume.

Consider an unstably stratified stagnant fluid body, as shown in Fig. 2.15, the lower layer being warmer than the upper layer. Imagine a positive fluctuation v being provoked. This causes the transfer of warm, light fluid into a colder, heavier environment. The consequence is a positive temperature fluctuation in the upper layer. Hence the correlation $\overline{T'v}$ tends to be positive. In addition, when the lump of higher-temperature fluid finds itself in a colder environment, it has the tendency to be driven upwards by buoyancy, which is an unstable process enhancing the fluctuation and hence the turbulence activity. This enhancement is compensated for by a loss of potential energy that is associated with the mixing process and the erosion of the stratification. When stratification is stable — i.e. with the density declining with height — the correlation in Eq. (2.23) is negative, and this demonstrates the counter-intuitive situation in which generation can be negative. In such circumstances, turbulence activity tends to decline and eventually ceases — a situation that gives rise to smog formation in unfavourable weather conditions. Here, the reduction in turbulence activity is compensated for by a gain in potential energy that arises from cross-stratification motion being suppressed.

A simple example illustrating the role of curvature-related body forcing is shown in Fig. 2.16: a curved shear layer with two alternative velocity profiles. Without curvature, $U_\theta = U$, and the shear-produced turbulence generation would simply be that given by Eq. (2.22). In

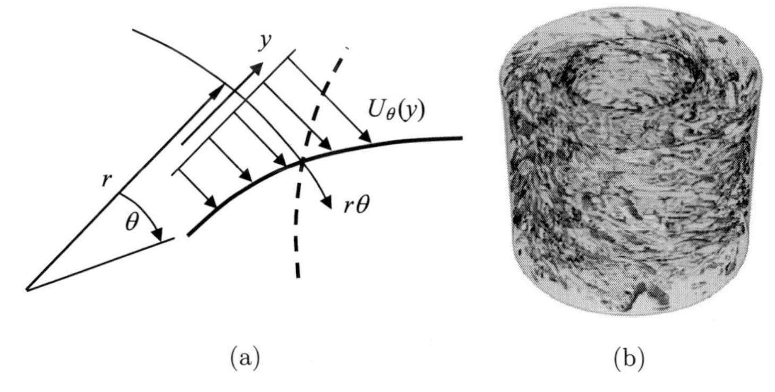

Fig. 2.16: Shear layer subjected to curvature (a) schematic used to interpret Eq. (2.24), solid and dashed lines indicate alternative velocity profiles causing opposite effects on turbulence production; (b) DNS-computed turbulence in Couette–Taylor flow between two cylinders, with inner cylinder rotating (courtesy S. Dong — see also Dong (2007)).

contrast, with curvature included, turbulence is produced at the rate:

$$P_k = -\overline{vu_\theta}\left(\frac{\partial U_\theta}{\partial y} - \frac{U_\theta}{r}\right). \tag{2.24}$$

Hence, the curvature-related forcing is:

$$P_{k,curvature} = \overline{vu_\theta}\frac{U_\theta}{r}, \tag{2.25}$$

and this opposes the principal part of the production. Hence, curvature of the type shown in Fig. 2.16(a), in combination with the solid velocity profile, tends to depress turbulence. In the particular case of solid-body motion $U_\theta = \omega r$, the production in Eq. (2.24) is zero. In contrast, if the velocity gradient reverses its sign, curvature tends to amplify turbulence. A simple example is the free-vortex motion $U_\theta = c/r$ — the dashed profile in Fig. 2.16(a) — in which case, the two strain components in Eq. (2.24) are additive. A practical example for the latter case is the amplification of turbulence in a Taylor–Couette flow in the gap separating a rotating inner cylinder from an outer stationary cylinder — akin to a journal bearing with a very thick fluid film. At relatively low rotation rate, the flow is initially characterised by an axially ordered array of Taylor vortices spanning the gap, a reflection of an instability similar to that of Rayleigh–Benard convection. However, as the rotation rate increases, the flow becomes turbulent due to the combination of shear and curvature-induced amplification.

2.4 *Summary and lessons*

We return to the question of 'what makes turbulence tick?' This chapter presented some simple examples, which illustrated, largely in statistical and descriptive terms, that turbulence needs to be driven by input of energy, either taken from the mean kinetic energy, as is the case in strained flows, or from the potential energy, as is the case in buoyancy-driven flows. Other forcing scenarios involve spatially varying body forces, due to magnetic fields or rotation, and shock waves. Following the onset of instability and transition to turbulence, or forced agitation, this energy feeds primarily into the large eddies, which interact most readily with the mean flow, because the length scale of the large eddies is of the same order as that associated with the mean flow — the thickness of shear layer or the thickness of the density layer, for example. These eddies then interact mutually, leading to vortex stretching, eddy breakup and a reduction in eddy size. The overall process is thus one of turbulence energy being produced at the largest eddy-size scale, and then 'cascading' down the scale range towards the smallest eddies that cannot survive the destructive influence of viscosity. It has been shown that the width of the cascade, in terms of the eddy-size range that separates the largest scales, associated with generation, and the smallest scales, associated with viscous destruction, rises steeply with increasing Reynolds numbers.

Turbulence not subjected to forcing — say, upon the removal of straining or agitation — progressively loses its energy: as the rate of energy transfer declines, the small scales die first, the scale range narrows, the turbulent eddies become progressively more isotropic, and eventually turbulence dies altogether. Importantly, the characteristics of the large, energetic turbulent eddies are sensitive to the structure of the mean flow, the type of forcing and the proximity of the turbulence to walls or any near-discontinuity within the fluid. In the presence of shear — the most frequent type of forcing — the eddies are elongated, and the intensity of the fluctuations can be very different in different directions, turbulence close to a wall being an extreme example. Why this is so, and why this is important, will be explored in Chapter 4.

Chapter Three

3. Reynolds-Averaging

3.1 *Decomposition and time-integration*

In a paper presented to the Royal Society of London in 1895, marking the beginning of statistical turbulence modelling, Osborne Reynolds set out the foundation of a method, later referred to as the 'Reynolds-averaged Navier–Stokes (RANS) approach'. Starting with the (Reynolds) velocity decomposition[1]:

$$u = \bar{u} + u', \tag{3.1}$$

with $\bar{u}$ interpreted as the ensemble-averaged, density-weighted velocity at any location within a turbulent flow:

$$\bar{u} = \frac{1}{N\bar{\rho}} \sum_{n=1}^{N} \rho u, \tag{3.2}$$

where N is assumed to be a large number of realisations over a long sampling period. Reynolds then inserted into the Navier–Stokes (NS) equations his decomposition for all three velocity components and for

[1]Reynolds' own nomenclature and presentation style are intentionally used in Eqs. (3.1)–(3.4), in particular, u, $\bar{u}$, u' for the velocity, its ensemble average and its fluctuating component, respectively.

the pressure. From these, he derived the 'equations of mean-motion':

$$\begin{aligned}
\rho\frac{d\bar{u}}{dt} &= -\frac{d}{dx}\left(\overline{p_{xx}} + \rho\bar{u}\bar{u} + \rho\overline{u'u'}\right)\\
&\quad -\frac{d}{dy}\left(\overline{p_{yx}} + \rho\bar{u}\bar{v} + \rho\overline{u'v'}\right)\\
&\quad -\frac{d}{dz}\left(\overline{p_{zx}} + \rho\bar{u}\bar{w} + \rho\overline{u'w'}\right) \qquad (3.3)\\
\&c. &= \qquad\qquad \&c.\\
\&c. &= \qquad\qquad \&c.,
\end{aligned}$$

in which $\overline{p_{xx}}$ etc. are the 'force components per unit of area', i.e. arising from pressure and viscous stresses. He followed this by the derivation of the equations of 'relative mean-motions' (i.e. the velocity fluctuations) and the equation for the 'energy of the relative mean-motion', defined as:

$$2\overline{E'} = \rho\left(\overline{u'^2} + \overline{v'^2} + \overline{w'^2}\right), \qquad (3.4)$$

which is twice the (dynamic) turbulence energy, conventionally denoted by ρk in modern literature. The implied retention of the time derivative in the substantive derivative of Eq. (3.3) is intriguing, and hints at Reynolds perceiving the concept of 'scale separation' that underlies the present use of RANS methods for some unsteady problems, i.e. those driven by unsteady boundary conditions, with a time scale much larger than that of the turbulent eddies.

To this day, the equations underpinning all RANS methods are, essentially, those presented by Reynolds, although they are derived via time averaging, rather than ensemble-averaging, and normally without density weighting. They are also written slightly differently, in a more compact form, usually in Cartesian-tensor notation. In this chapter, we will derive the relevant equations in terms of this notation, with its associated simple set of rules stated in the Appendix of this book (for a much broader account, see Jeffreys (1969)). In contrast to Reynolds' notation, we use capital letters to denote the

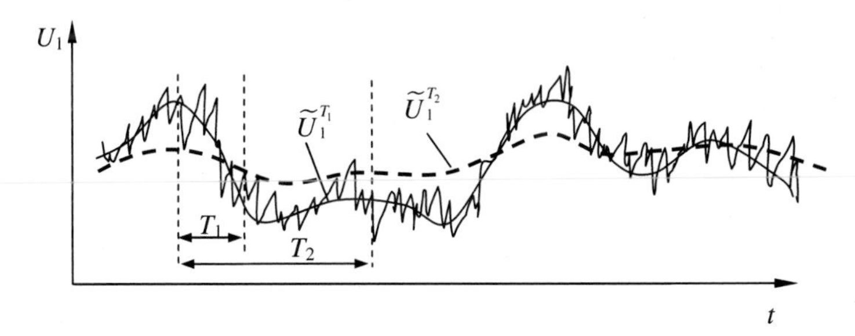

Fig. 3.1: Schematic of a turbulent-velocity signal $U_1(t)$ and two integrated (partially averaged or 'time-filtered') signals $\tilde{U}_1(t)$.

velocity or its averaged value, and lower-case letters to indicate velocity fluctuations.

Rather than starting with conventional time-averaging over a (nominally) infinitely long period of time, we will first consider the more general case of averaging over a finite interval T, a case in which only the higher-frequency range of the unsteady variables is assimilated into the average. We do so primarily to address the question of whether a time-varying turbulent flow may be represented by the RANS equations.

We propose the decomposition:

$$U_i = \tilde{U}_i + u_i, \quad P = \tilde{P} + p, \tag{3.5}$$

in which $i = 1, 2, 3$ denotes the Cartesian directions x, y, z, and the tilde denotes that the respective velocity and the pressure are averaged over the chosen period T,

$$\tilde{U}_i(t) = \frac{1}{T} \int_{\frac{t-T}{2}}^{\frac{t+T}{2}} U_i \, dt. \tag{3.6}$$

As is evident from Fig. 3.1, the average value $(\widetilde{\ldots})$ changes with time if the integration period is shorter than the longest time scale of the unsteady motion, in which the low-frequency unsteadiness may be part of the turbulence, or may be due to some imposed unsteady disturbance. Moreover, $(\widetilde{\ldots})$ is clearly dependent on the averaging interval. Substitution of Eq. (3.5) into the NS equations, applicable to incompressible flow of a Newtonian fluid without body forces (say, due to buoyancy or electro-magnetic forcing), then gives:

$$\frac{\partial \rho(\tilde{U}_i + u_i)}{\partial t} + \frac{\partial \rho(\tilde{U}_i + u_i)(\tilde{U}_j + u_j)}{\partial x_j}$$
$$= -\frac{\partial(\tilde{P} + p)}{\partial x_i} + \frac{\partial}{\partial x_j}\mu\left(\frac{\partial(\tilde{U}_i + u_i)}{\partial x_j} + \frac{\partial(\tilde{U}_j + u_j)}{\partial x_i}\right). \quad (3.7)$$

Next, the above equation, as a whole (i.e. all terms therein), is averaged over the same period. Subject to the approximations, $\tilde{u}_i = 0$, $\tilde{p} = 0$[2] and hence $\widetilde{\widetilde{U}} = \widetilde{U}$, $\widetilde{\widetilde{P}} = \widetilde{P}$, the result is:

$$\frac{\partial \rho\tilde{U}_i}{\partial t} + \frac{\partial \rho\tilde{U}_i\tilde{U}_j}{\partial x_j} = -\frac{\partial \tilde{P}}{\partial x_i} + \frac{\partial}{\partial x_j}\mu\left(\frac{\partial \tilde{U}_i}{\partial x_j} + \frac{\partial \tilde{U}_j}{\partial x_i}\right)$$
$$- \frac{\partial}{\partial x_j}\left(\rho\widetilde{\tilde{U}_i\tilde{U}_j} - \rho\tilde{U}_i\tilde{U}_j\right) - \frac{\partial}{\partial x_j}\left(\rho\widetilde{\tilde{U}_i u_j} + \rho\widetilde{\tilde{U}_j u_i}\right)$$
$$- \frac{\partial}{\partial x_j}\rho\widetilde{u_i u_j}. \quad (3.8)$$

In principle, this is an equation for the unsteady, 'filtered', velocity $\tilde{U}_i$. However, it is not a useful foundation for a practical predictive framework. First, as already noted by reference to Fig. 3.1, the division between $\tilde{U}_i$ and u_i clearly depends on the (arbitrary) choice of the integration interval T. Second, the solution of the equation would require the three sets of unknown correlations $\widetilde{\tilde{U}_i\tilde{U}_j}$, $\widetilde{\tilde{U}_i u_j}$ and $\widetilde{u_i u_j}$ to be approximated by respective models that link all the correlations to the velocity $\tilde{U}_i$, so as to close the equations. Finally, even if we were able to devise such models, they would need to be tied to the chosen (unphysical) time scale T. Indeed, the nature and details of the models would, collectively, dictate the distinction between $\tilde{U}_i$ and

[2] This approximation relies on the filtered distribution being fairly smooth relative to the original fully turbulent signal, implying that the integration period T should span over a fair proportion of the high-frequency range of the turbulent signal.

the fluctuations relative to it, and would thus introduce a substantial level of ambiguity into the physical meaning of $\tilde{U}_i$. Does this mean that we cannot compute unsteady flows with time-averaged equations? This is a question that will be discussed in Section 3.3.

3.2 *The Reynolds-averaged Navier–Stokes (RANS) equations for steady flow*

In the limiting case of the averaging period being much longer than the largest time scale of the turbulent motion, we may set $\tilde{U}_i = \overline{U}_i$, with the overbar denoting the true long-time-average value. In this case, $\rho\widetilde{\tilde{U}_i u_j} + \rho\widetilde{\tilde{U}_j u_i} = 0$ and $\rho\widetilde{\tilde{U}_i\tilde{U}_j} - \rho\tilde{U}_i\tilde{U}_j = 0$, because $\tilde{U}_i$ is no longer time-dependent and can, therefore, be taken outside the tilde spanning the products. The former sum is then $\rho\overline{\overline{U}}_i\bar{u}_j + \rho\overline{\overline{U}}_j\bar{u}_i$, which is, self-evidently, zero, because the average of the fluctuations must be zero. Eq. (3.8) thus reduces to:

$$\frac{\partial \rho\overline{U}_i\overline{U}_j}{\partial x_j} = -\frac{\partial \overline{P}}{\partial x_i} + \frac{\partial}{\partial x_j}\mu\left(\frac{\partial \overline{U}_i}{\partial x_j} + \frac{\partial \overline{U}_j}{\partial x_i}\right) - \frac{\partial}{\partial x_j}\rho\overline{u_i u_j}, \tag{3.9}$$

in which $\rho\overline{u_i u_j}$ are the *six* independent 'Reynolds-stresses'. The reference to 'stress' is rooted in two facts. First, the dimensions of $\rho\overline{u_i u_j}$ is of force per unit area. Second, upon grouping together the last two terms in Eq. (3.9), $-\rho\overline{u_i u_j}$ can be interpreted as the turbulence-equivalent of the viscous term $\mu\left(\frac{\partial \overline{U}_i}{\partial x_j} + \frac{\partial \overline{U}_j}{\partial x_i}\right)$, which is the set of viscous stresses in a Newtonian fluid. The emphasis on *six* stresses is intended to convey the fact that the three off-diagonal components of the stress tensor,

$$\overline{u_i u_j} \equiv \begin{pmatrix} \overline{u_1^2} & \overline{u_1 u_2} & \overline{u_1 u_3} \\ \overline{u_2 u_1} & \overline{u_2^2} & \overline{u_2 u_3} \\ \overline{u_3 u_1} & \overline{u_3 u_2} & \overline{u_3^2} \end{pmatrix}, \tag{3.10}$$

are identical to their respective symmetric counterparts, i.e. $\overline{u_2 u_1} = \overline{u_1 u_2}$; $\overline{u_1 u_3} = \overline{u_3 u_1}$; $\overline{u_3 u_2} = \overline{u_2 u_3}$. Note that if one mean-velocity component is zero throughout, say $\overline{U}_3 = 0$, so that the flow is statistically two-dimensional in (x_1, x_2), the shear stresses associated with

x_3 must vanish, i.e. $\overline{u_1u_3} = \overline{u_3u_1} = 0$; $\overline{u_3u_2} = \overline{u_2u_3} = 0$. However, importantly, $\overline{u_3^2} \neq 0$, because any realistic turbulence field, even one subject to zero mean-motion in one direction (x_3), must have three fluctuation components.[3] In the more general case of statistical two-dimensionality in (x_1, x_2), $\overline{U}_3$ may be finite, but the flow must be statistically homogeneous in this direction, i.e. $\partial\overline{U}_i/\partial x_3 = 0$, in which case all stress components are, in general, finite. An example for this set of conditions is an axisymmetric (i.e. two-dimensional) swirling flow, in which the azimuthal (swirl) component, $\overline{U}_3 = \overline{U}_\theta$, does not vary with θ.

Equation (3.9) is complemented by the Reynolds-averaged mass-conservation equation:

$$\frac{\partial \rho \overline{U}_j}{\partial x_j} = 0, \tag{3.11}$$

which arises, trivially, from the parent equation governing the unsteady velocity vector U_i and having the same form as Eq. (3.11), except for the overbar and the time derivative of the density, which vanishes upon time averaging.

It is now the task of a turbulence model to relate the stresses to the time-averaged velocities and any other known or determinable quantities; this is the subject of much of the material to follow in this book.

In cases where the transport of scalar quantities is of interest — say, mass-specific enthalpy or species concentration — there is also a need to derive a Reynolds-averaged conservation equation for the time-averaged scalar property in question. If this property is denoted by the general symbol Φ, considerations entirely analogous to those above, with $\Phi = \overline{\Phi} + \phi$, lead to:

$$\frac{\partial \rho \overline{U}_j \overline{\Phi}}{\partial x_j} = \frac{\partial}{\partial x_j} \Gamma_\Phi \left(\frac{\partial \overline{\Phi}}{\partial x_j} \right) - \frac{\partial}{\partial x_j} \rho \overline{u_j \phi} + \rho \overline{S}_\Phi, \tag{3.12}$$

[3] A (theoretical) exception arises at liquid/gas interfaces, subject to the assumption that the interface is rigid. In such a case, interface-parallel fluctuations are finite, while interface-normal fluctuations vanish.

where Γ_Φ is the relevant fluid conductivity, and $\overline{S}_\Phi$ is any net source (per unit volume) that may be relevant — say, due a chemical reaction that generates $\overline{\Phi}$. The inclusion of this equation introduces the need to determine, in addition to the Reynolds stresses, the correlations $\rho\overline{u_j\phi}$ from an appropriate turbulence model. Upon grouping together the second and third terms in Eq. (3.12), as argued in relation to the RANS equations (3.9), namely:

$$\frac{\partial \rho \overline{U}_j \overline{\Phi}}{\partial x_j} = \frac{\partial}{\partial x_j}\left[\Gamma_\Phi \left(\frac{\partial \overline{\Phi}}{\partial x_j}\right) - \rho\overline{u_j\phi}\right] + \rho \overline{S}_\Phi, \tag{3.13}$$

we may interpret these correlations as being the components of a flux vector 'diffusing' the property $\overline{\Phi}$. Hence, the turbulent correlations constitute a vector analogous to the vector of viscous fluxes, although virtually never having the same direction and being, typically, two to three orders of magnitude larger than its viscous counterpart.

3.3 *The Reynolds-averaged Navier–Stokes equations for unsteady flow (URANS)*

There are many circumstances in fluid mechanics where the flow contains a mixture of stochastic turbulence and 'coherent' unsteadiness, the latter associated either with periodic external disturbances or with internal-instability mechanisms. These coherent motions can have a time scale that overlaps with that of the low-frequency turbulent motions. In this case, the question arises whether the RANS strategy may be used to compute the unsteady 'mean' (non-turbulent) field. Unfortunately, this is a question that cannot be easily answered.

A case in point is vortex shedding behind a cylinder. As shown in Fig. 3.2, this flow is characterised by a combination of turbulent (stochastic) unsteadiness and a periodic, non-turbulent, component. The latter is associated with the periodic shedding of vortices (a 'von Kármán vortex street'), i.e. a non-turbulent process that reflects an absolute instability, forms of which are almost always observed in bluff-body flows. The periodic component is especially prominent in the lift coefficient in Fig. 3.2(c), and this also illustrates the presence

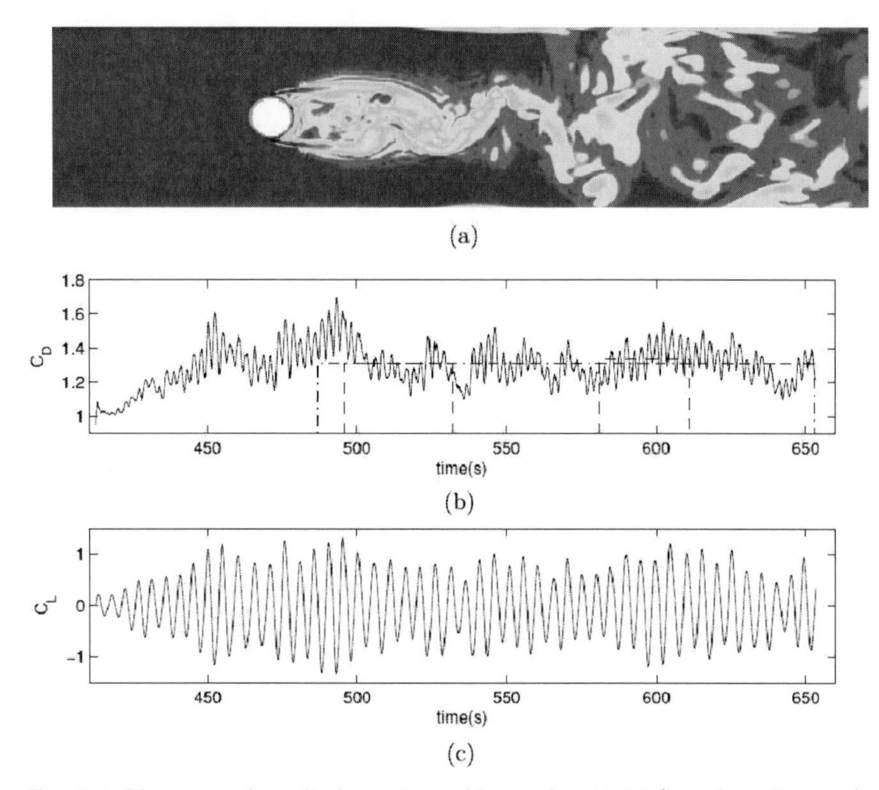

Fig. 3.2: Flow around a cylinder at Reynolds-number 10000 (based on diameter): (a) snapshot from a simulation; (b) history of drag coefficient; (c) history of lift coefficient; derived from LES data (see Leschziner *et al.* (2009)).

of a third time scale, associated with the slow modulation over a time interval much longer than the shedding period.

Long-time averaging over this flow would give an integral view — say in the form $\overline{U}_i(x, y)$ — that includes *all* unsteady processes, at *all* time scales. If we now imagine that we derive this view by solving the RANS equations (3.9), the steady-state result would only be correct if we provide the equations with the correct level of the correlations $\overline{u_i u_j}$. However, these correlations would not represent the turbulence activity alone. Rather, it would, necessarily, represent an amalgam of both turbulence and unsteady contributions unrelated to turbulence.

Consider, as an alternative to the above, the proposition of adding to Eq. (3.9) the temporal gradient of $\overline{U}$, reflecting the intention to resolve the unsteady 'mean-motion'. As the solution is unsteady, it is more appropriate to use $(\widetilde{\ldots})$ rather than $(\overline{\ldots})$, so that Eq. (3.9) now becomes:

$$\frac{\partial \rho \tilde{U}_i}{\partial t} + \frac{\partial \rho \tilde{U}_i \tilde{U}_j}{\partial x_j} = -\frac{\partial \tilde{P}}{\partial x_i} + \frac{\partial}{\partial x_j} \mu \left(\frac{\partial \tilde{U}_i}{\partial x_j} + \frac{\partial \tilde{U}_j}{\partial x_i} \right) - \frac{\partial}{\partial x_j} \rho \widetilde{u_i u_j}. \tag{3.14}$$

The implication of Eq. (3.14) — correct or not — is that the averaging is effected over a time interval that includes all turbulent scales, but excludes the coherent unsteadiness. A comparison of Eq. (3.14) with Eq. (3.8) shows that a number of correlations appearing in the latter are neglected in the former. In particular, the omission of terms such as $\widetilde{\tilde{U}_i u_j}$ implies the assumption that the low-frequency fluctuations are uncorrelated with the higher-frequency ones. Moreover, the omission of the contribution $\widetilde{\tilde{U}_i \tilde{U}_j} - \tilde{U}_i \tilde{U}_j$ implies slow temporal variations in the 'mean' velocity. Whether this simplification is adequate depends greatly on the particular flow circumstances encountered — especially on the question of whether the spectrum of turbulent motions overlaps with the coherent unsteady components. In general, a condition that needs to be satisfied is 'scale separation' — i.e. a spectral gap between the turbulent and the coherent motions.

Assuming that the simplified form (3.14) is adequate, there remains the question of whether its solution will, in fact, be unsteady for the flow in Fig. 3.2, and if so, whether the unsteady motion will be represented realistically. This depends greatly on the properties of the turbulence model used to represent the Reynolds-stresses $\widetilde{u_i u_j}$ (i.e. that same model that normally gives $\overline{u_i u_j}$). If the model tends to return, by virtue of its construction, relatively low stress levels — lower than is appropriate to represent the combined stochastic and unsteady motions — then the solution is likely to be periodic (but not necessarily with the correct period). With other models, and in different flows, the periodic motion may well be suppressed.

A pertinent issue that most readers may well not fully appreciate ahead of Chapters 7–12 is that turbulence models are designed and, above all, calibrated by reference to, and with input data from, steady canonical flows — for example, simple shear layers and decaying grid turbulence. It follows, therefore, that any such model is not applicable, in principle, to unsteady flows, especially ones in which the time scale of the unsteady flow overlaps with the time scale of the most energetic, large-scale turbulent motions. Hence, it is entirely possible that a model returning accurate solutions for steady flows will fail to do so in unsteady conditions, even if the periodic motion *is* resolved with that model.

Despite the above limitations, it is equation (3.14) that is often solved in practice when periodic flows are computed, in which the periodicity is induced by a strong internal instability (e.g. vortex shedding, Rayleigh–Bénard convection; and Taylor vortices in rotating flow), or enforced by unsteady boundary conditions (e.g. IC-engine flows and stator-rotor interactions in turbomachinery). This is what is generally understood as constituting 'Unsteady RANS' or 'URANS'.

There are a number of examples in the literature which illustrate that the use of conventional models to approximate $\widetilde{u_i u_j}$ can provide entirely adequate solutions — e.g. Johansson *et al.* (1993), and Hanjalić and Kenjeres (2001). Both these studies show that the sum $\overline{(\widetilde{U} - \overline{U})^2} + \overline{u_i u_j}$ — the former term derived from *a-posteriori* averaging of the unsteady solution itself, and the latter from a conventional model — agrees well with measurements or a direct numerical simulation. Nevertheless, it is important to appreciate the fundamental limitations and dangers inherent in URANS when this approach is applied to flows in which the instability mechanisms that give rise to the periodic motion is not as strong as it is in some judiciously chosen laboratory flows, such as that shown in Fig. 3.2. Examples where significant uncertainties arise include unsteady separation from the suction side of a stalled aerofoil, the wake of a streamlined car body, the flow behind a turbomachine blade with rounded trailing edge and a short cylinder or smooth hill-like obstacle on a flat plate.

In principle, a fundamentally superior alternative to the above framework involves a decomposition that distinguishes explicitly between the stochastic and periodic components — referred to as the 'triple decomposition':

$$U = \underbrace{\underbrace{\overline{U}}_{\text{Mean}} + \underbrace{\widetilde{u}}_{\text{Periodic}}}_{\text{Phase-average } \langle U \rangle} + \underbrace{u'}_{\text{Stochastic}} . \tag{3.15}$$

If this is inserted into the Navier–Stokes equations, followed by time-averaging, the result (subject to the assumption that u' and $\widetilde{u}$ are not correlated) is:

$$\frac{\partial \rho \overline{U}_j \overline{U}_i}{\partial x_j} = -\frac{\partial \overline{P}}{\partial x_i} + \mu \frac{\partial^2 \overline{U}_i}{\partial x_j \partial x_j} + \frac{\partial}{\partial x_j} \rho \left(-\overline{\tilde{u}_i \tilde{u}_j} - \overline{u_i' u_j'} \right). \tag{3.16}$$

This equation requires, in principle, models for both $\overline{u_i' u_j'}$ and $\overline{\tilde{u}_i \tilde{u}_j}$. As the former term has, essentially, the same meaning as the turbulent Reynolds stresses $\overline{u_i u_j}$ in Eq. (3.9), this may be obtained, validly, from any conventional turbulence model of the type discussed later in the book, starting with Chapter 7. The correlations $\overline{\tilde{u}_i \tilde{u}_j}$ are more problematic, however. Even if we had a model for them, Eq. (3.16) would only give the steady-state solution. Obtaining the periodic solution itself requires the derivation of an equation for $\tilde{u}_i$. We shall not do this here, but refer the interested reader to Hussain (1983), and we only give the result below:

$$\frac{\partial \rho \tilde{u}_i}{\partial t} + \frac{\partial \rho \tilde{u}_i \tilde{u}_j}{\partial x_j} = -\frac{\partial \tilde{p}}{\partial x_i} + \mu \frac{\partial^2 \tilde{u}_i}{\partial x_j \partial x_j} + \frac{\partial}{\partial x_j} \rho \left(\overline{\tilde{u}_i \tilde{u}_j} + \overline{u_i' u_j'} - \langle u_i' u_j' \rangle \right)$$
$$- \frac{\partial}{\partial x_j} \rho \left(\overline{U}_i \tilde{u}_j + \overline{U}_j \tilde{u}_i \right), \tag{3.17}$$

in which $\langle u_i' u_j' \rangle$ is the phase average of $u_i' u_j'$. The meaning of phase averaging is explained by reference to the sketch in Fig. 3.3.

Given a phase ϕ within any period T, the signal is sampled repeatedly at intervals of T, at identical locations ϕ. With the fluctuation relative to the periodic mean denoted by u', the phase average is defined as:

$$\langle u' u' \rangle = \frac{1}{N} \sum_{n=0}^{N-1} (u' u')|_{(\phi + nT)}. \tag{3.18}$$

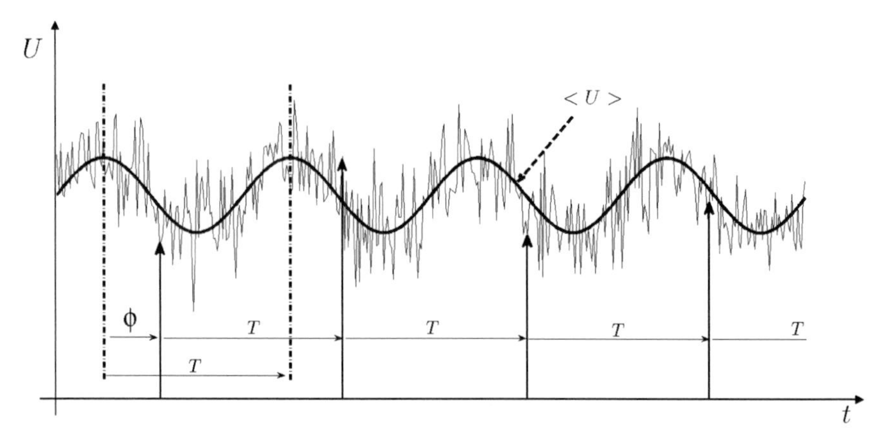

Fig. 3.3: Schematic used to explain phase averaging (see Eqs. (3.15) and (3.18)).

Hence, to solve the sets (3.17) and (3.18), we now require models for three sets of correlations, namely $\overline{u_i'u_j'}$, $\overline{\tilde{u}_i\tilde{u}_j}$ and $\langle u_i'u_j'\rangle$. This is not, however, a practically tenable proposition, and this route, although fundamentally correct, has never been adopted in practice. Therefore, Eq. (3.15) remains the only practical path, but we need to be very clear about the dangers it harbours.

An illustration of the uncertainty that can arise with URANS is given in Fig. 3.4. This shows the turbulence field generated in free (Rayleigh–Bénard) convection due to lower-wall heating of a fluid confined between two walls. The flow is observed to contain a mixture of quasi-organized cells and turbulence, the former due to buoyancy-induced (Rayleigh–Bénard) instability. The direct numerical simulation (DNS) result shows a relatively fine-scale structure, although the Rayleigh number is relatively modest, so that the range of turbulent eddies is not very broad. The URANS solution was obtained with an elaborate turbulence model — a second-moment closure of the type covered in Chapter 12. The model is seen to return a much smoother field, because the small-scale part of the eddy-size spectrum has been assimilated within the turbulence model, which returns correspondingly elevated stresses $\widetilde{u_iu_j}$ (and also heat fluxes $\widetilde{u_i\phi}$). The remaining unsteady field (Fig. 3.4(b)), although relatively smooth, is clearly turbulent, and cannot be said to qualify as an

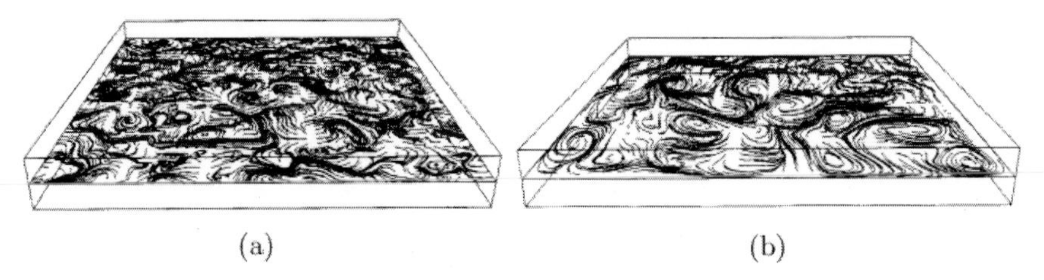

Fig. 3.4: Vortical structure in Rayleigh–Bénard convection (lower wall heated) at Rayleigh number 6.5×10^5: (a) DNS; (b) URANS (reproduced with permission of Springer from Hanjalić and Kenjeres (2001)).

unsteady (periodic) 'mean' flow. The actual time-averaged stresses arise upon performing the averaging $(\overline{\widetilde{u_i u_j}} + \widetilde{U}_i\widetilde{U}_j - \overline{U_i}\,\overline{U_j})$ over the time-dependent solution. Whether or not the result correctly represents the true stress levels depends greatly on the model used. Any other model will almost certainly give a different split between the modelled and resolved portions of the spectrum, and thus a different solution, possibly one that suppresses all unsteadiness.

3.4 *Spatial averaging and statistical homogeneity*

A flow in which the unsteadiness is purely a consequence of true turbulence may be said to be statistically homogeneous in time. In other words, its average is not a function of the integration period once that period is significantly larger that the turbulent time scales. A more frequent use of the term 'statistically homogeneous' is encountered in the technical literature in relation to spatial averaging. Returning to Fig. 3.2, which relates to a (nominally) infinitely long cylinder, we may derive a spanwise-averaged, spatially two-dimensional representation of the flow by performing the integration:

$$\overline{U}_i^z(x, y, t) = \frac{1}{Z} \int_{z=0}^{Z} U_i \, dz, \tag{3.19}$$

where Z is a spanwise distance significantly larger than all of the turbulent length scales. This averaging is only appropriate, however, if the spanwise boundary conditions do not introduce spanwise distortions to the mean flow (say, by being periodic in z or in the form of zero-spanwise gradients). In terms of the RANS equations, spanwise homogeneity means that we may restrict the solution of Eq. (3.9) to the two components U and V and also ignore the shear stresses $\overline{uw}$ and $\overline{vw}$ (but not $\overline{w^2}$!). In the case of a periodic flow, for which we might choose to use the URANS equations (3.14), the two-dimensional restriction is especially risky, because periodic motions are very rarely spanwise homogeneous. Hence, this restriction prevents energy from being transferred to spanwise motions, and this can aggravate the misrepresentation of the physical processes caused by inherent fundamental flaws in the URANS framework.

The ultimate homogeneous flow is 'box turbulence', shown in Fig. 2.1(b). In this case, statistical homogeneity extends to all three directions, and the spatial integral of the velocity over the box volume will obviously be zero. However, importantly, the correlations $\overline{u^2}, \overline{v^2}, \overline{w^2}$ will be finite, but directionally invariant, and will decay (slowly) in time as the turbulence energy is dissipated into heat. Also, if the early readjustment of the turbulence following the initial turbulence-generation process is ignored, this flow is very close to being statistically isotropic, i.e. $\overline{u^2} = \overline{v^2} = \overline{w^2}$. While this flow is not practically interesting, it is an important one from a fundamental perspective, as it forms a test bed for investigating basic fundamental questions of turbulence mechanics, the cascade process in particular.

3.5 *Summary and lessons*

The aim of deriving equations for time-averaged properties is to obviate the need to resolve, explicitly, the highly unsteady, small-scale turbulent fluctuations. We have seen that this can be done in several ways. The simplest and most secure route is to adopt the Reynolds (double) decomposition, and to time average the Navier–Stokes equations and the transport equation for any other scalar property. This results in a set of equations for the time-mean properties, from which the time derivative is naturally omitted. However, the penalty for this simplification is the need to provide a closure model for the unknown correlations of the turbulent fluctuations.

In the case of periodic flows, in which periodicity is self-induced by an internal instability or imposed by external (boundary) conditions, the steady-state framework requires the correlations to express the combined effects of the stochastic turbulence and the periodic motions. However, existing turbulence models are not suited to this task, as they are explicitly designed for, and calibrated by reference to, statistically steady flows that do not contain periodic (coherent) motions. While it is possible to solve unsteady versions of the RANS equations for periodic flows by adding the time derivative, this can result in unrealistic solutions, because the nature of the solution depends significantly on the characteristics of the turbulence model used to represent the turbulent correlations. Indeed, depending on the turbulence model used, the solution may turn out to be either steady or unsteady. Therefore, if this route is adopted, great care is needed when judging the physical significance of the statement it yields.

A fundamentally superior framework appears to be offered by the use of triple decomposition, which distinguishes, explicitly, between the stochastic and periodic motions. However, while this approach yields equations for the time-averaged velocity and its periodic component, it requires models for three sets of correlations, and it does not, therefore, constitute a practically tenable route.

Before we start on the long journey of modelling, it will be instructive to discuss a number of qualitative relationships that can be gleaned from simple rational considerations. This discussion, pursued in the next chapter, provides valuable indicators on what to expect from different modelling approaches. At the most basic level, a turbulence model has to relate the unknown correlations $\overline{u_i u_j}$ to the mean-velocity components which, apart from the pressure, are the only quantities determinable from the RANS equations. In fact, Galilean invariance dictates that the relationship must *not* be to the velocity components themselves, but to their derivatives, whether of order one or higher. It is this linkage we shall now explore.

Chapter Four

4. Fundamentals of Stress/Strain Interactions

The primary task of a turbulence model is to provide the Reynolds-averaged Navier–Stokes (RANS) equations with closure relations for the Reynolds-stresses $\rho\overline{u_i u_j} \equiv \left\{\rho\overline{u^2}, \rho\overline{v^2}, \rho\overline{w^2}, \rho\overline{uv}, \rho\overline{uw}, \rho\overline{vw}\right\}$ and possibly the scalar fluxes $\rho\overline{u_i\phi} \equiv \left\{\rho\overline{u\phi}, \rho\overline{v\phi}, \rho\overline{w\phi}\right\}$. 'Closure' is a term that signifies the process by which the stresses and fluxes are related to known or determinable quantities, i.e. geometric parameters, flow scales, strains (but not the velocity itself!) and the mean-scalar gradients in question. When facing the task of closure, we need to develop a basic understanding of what 'drives' the stresses, for these drivers must then clearly be included as major elements of any closure based on a rational foundation.

In Section 2.3 we observed, essentially in qualitative terms, that turbulence is driven by generation processes that involve strains, stresses and body forces. Now we shall pursue these types of interaction more rigorously, by exploiting exact equations and manipulating them in a manner that brings out the sources of important properties of turbulence observed in real flows. This chapter restricts itself to simple shear, i.e. situations in which the strain field is dominated by the shear strain $\frac{\partial \overline{U}}{\partial y}$. This is the case, for example, in plane-channel and duct flows and in a boundary layer over a flat plate. More complex interactions, involving compressive strain, curvature and rotation, will be discussed in Chapters 10 and 12.

4.1 *A rational framework for describing the Reynolds stresses*

An excellent basis for illuminating some key physical interactions between stresses and strains is the set of *exact* equations governing the evolution of the Reynolds stresses. This is also the best starting point for constructing turbulence models which hold some promise of generality — a subject considered in detail in Chapter 12. While the general set of equations for all Reynolds-stresses will be discussed therein, it is instructive at this point to consider a simplified framework. Without significant loss of generality, at this initial stage of discussing the subject, we restrict ourselves to two-dimensional mean flow $\{\overline{U}(x,y),\overline{V}(x,y)\}$ and the Reynolds-stress components[1] $\left\{\rho\overline{u^2},\rho\overline{v^2},\rho\overline{w^2},\rho\overline{uv}\right\}$. In this case, all z-wise derivatives of time-averaged quantities (including time-averaged correlations of turbulence quantities) vanish. Moreover, we restrict the detailed derivation to the equation governing the *Reynolds shear stress.* Equations for the normal stresses arise along an entirely analogous route. A general derivation will be pursued Chapter 12, in the context of modelling turbulence within the framework of second-moment closure.

The equations for the stresses can be derived by a somewhat laborious, but otherwise straightforward, combination of the Navier–Stokes equations and their Reynolds-averaged forms. If the former are denoted, symbolically, by $NS(\overline{U}_i + u_i) = 0$,[2] then the manipulations that lead to the requisite set of equations can be written:

$$\overline{u_i NS(\overline{U}_j + u_j) + u_j NS(\overline{U}_i + u_i)} = 0. \tag{4.1}$$

In the following exposition, the manipulations are performed in detail for the specific case of the shear stress $\overline{u_1u_2} = \overline{uv}$. Purely for the sake of algebraic simplicity, and without significant loss of generality, the

[1]In reality, turbulence is always three-dimensional. What is meant here is that we consider variations in flow and turbulence properties in a two-dimensional domain, the third dimension being homogeneous.

[2]*NS* should be understood as Eq. (3.7), with the right-hand-side terms transferred to the left, and the tilde replaced by overbar.

derivation will be subject to two constraints:

- two-dimensionality of the time-averaged flow — $\overline{W} = 0$, $\frac{\partial \overline{\{...\}}}{\partial z} = 0$ — and
- a state of two-component turbulence — $w = 0$, $\frac{\partial \{...\}}{\partial z} = 0$, in which the braces indicate any turbulent-velocity component.

The former is entirely realistic, the latter only in extreme circumstances, such as in very shallow flows, close to an interface of two immiscible liquids and very close to a wall normal to z (except for the vanishing z-wise gradient). The single term that is absent from the final equation, as a consequence of the latter simplification, will be added and highlighted at the end.

The process expressed by Eq. (4.1) involves four major steps:

1. Introduce the Reynolds decomposition into the NS equations, subtract from these the associated RANS equations, and then multiply each equation by the turbulent-velocity fluctuation in the other direction. The result is:

$$v\frac{\partial u}{\partial t} + v\overline{U}\frac{\partial u}{\partial x} + v\overline{V}\frac{\partial u}{\partial y} + vu\frac{\partial \overline{U}}{\partial x} + vv\frac{\partial \overline{U}}{\partial y} + v\frac{\partial uu}{\partial x} + v\frac{\partial uv}{\partial y}$$

$$= -v\frac{1}{\rho}\frac{\partial p}{\partial x} + 2\frac{\mu}{\rho}v\frac{\partial^2 u}{\partial x^2} + \frac{\mu}{\rho}v\frac{\partial}{\partial y}\left(\frac{\partial u}{\partial y} + \frac{\partial v}{\partial x}\right) + v\frac{\partial \overline{u^2}}{\partial x} + v\frac{\partial \overline{uv}}{\partial y}$$

$$u\frac{\partial v}{\partial t} + u\overline{U}\frac{\partial v}{\partial x} + u\overline{V}\frac{\partial v}{\partial y} + uu\frac{\partial V}{\partial x} + uv\frac{\partial \overline{V}}{\partial y} + u\frac{\partial uv}{\partial x} + u\frac{\partial vv}{\partial y}$$

$$= -u\frac{1}{\rho}\frac{\partial p}{\partial y} + 2\frac{\mu}{\rho}u\frac{\partial^2 v}{\partial y^2} + \frac{\mu}{\rho}u\frac{\partial}{\partial x}\left(\frac{\partial u}{\partial y} + \frac{\partial v}{\partial x}\right) + u\frac{\partial \overline{uv}}{\partial x} + u\frac{\partial \overline{v^2}}{\partial y}. \quad (4.2)$$

2. Add the two equations:

$$\left\{v\frac{\partial u}{\partial t} + u\frac{\partial v}{\partial t}\right\} + \left\{v\overline{U}\frac{\partial u}{\partial x} + u\overline{U}\frac{\partial v}{\partial x}\right\} + \left\{v\overline{V}\frac{\partial u}{\partial y} + u\overline{V}\frac{\partial v}{\partial y}\right\}$$

$$= -\left\{vu\frac{\partial \overline{U}}{\partial x} + vv\frac{\partial \overline{U}}{\partial y} + uu\frac{\partial \overline{V}}{\partial x} + uv\frac{\partial \overline{V}}{\partial y}\right\}$$

$$-\frac{1}{\rho}\left\{v\frac{\partial p}{\partial x} + u\frac{\partial p}{\partial y}\right\} + \frac{1}{\rho}\left\{2\mu v\frac{\partial^2 u}{\partial x^2} + \mu v\frac{\partial}{\partial y}\left(\frac{\partial u}{\partial y} + \frac{\partial v}{\partial x}\right)\right.$$

$$+ 2\mu u\frac{\partial^2 v}{\partial y^2} + \mu u\frac{\partial}{\partial x}\left(\frac{\partial u}{\partial y} + \frac{\partial v}{\partial x}\right)\Bigg\}$$

$$-\left\{v\frac{\partial\left(uu - \overline{u^2}\right)}{\partial x} + v\frac{\partial\left(uv - \overline{uv}\right)}{\partial y} + u\frac{\partial\left(uv - \overline{uv}\right)}{\partial x}\right.$$

$$\left. + u\frac{\partial\left(vv - \overline{v^2}\right)}{\partial y}\right\}. \tag{4.3}$$

3. Collect terms and impose the mass-conservation constraint $\frac{\partial\overline{U}}{\partial x} + \frac{\partial\overline{V}}{\partial y} = 0$, $\frac{\partial u}{\partial x} + \frac{\partial v}{\partial y} = 0$:

$$\frac{\partial uv}{\partial t} + \left\{\overline{U}\frac{\partial uv}{\partial x} + \overline{V}\frac{\partial uv}{\partial y}\right\}$$

$$= -\left\{vu\frac{\partial\overline{U}}{\partial x} + vv\frac{\partial\overline{U}}{\partial y} + uu\frac{\partial\overline{V}}{\partial x} + uv\frac{\partial\overline{V}}{\partial y}\right\} - \frac{1}{\rho}\left\{v\frac{\partial p}{\partial x} + u\frac{\partial p}{\partial y}\right\}$$

$$+ \left\{\frac{\mu}{\rho}\frac{\partial^2 uv}{\partial x^2} + \frac{\mu}{\rho}\frac{\partial^2 uv}{\partial y^2} - 2\frac{\mu}{\rho}\frac{\partial u}{\partial x}\frac{\partial v}{\partial x} - 2\frac{\mu}{\rho}\frac{\partial u}{\partial y}\frac{\partial v}{\partial y}\right\}$$

$$- \left\{\frac{\partial\left(vuv - v\overline{uv}\right)}{\partial y} + \frac{\partial\left(uuv - u\overline{uv}\right)}{\partial x}\right\}$$

$$- \left\{v\frac{\partial\left(uu - \overline{u^2}\right)}{\partial x} + u\frac{\partial\left(vv - \overline{v^2}\right)}{\partial y}\right\}. \tag{4.4}$$

4. Time-average the equation, noting that terms such as $u\frac{\partial\overline{v^2}}{\partial y}$ and $v\frac{\partial\overline{u^2}}{\partial x}$ vanish when averaged, and trivially reorder the terms:

$$\underbrace{\left\{\overline{U}\frac{\partial\overline{uv}}{\partial x} + \overline{V}\frac{\partial\overline{uv}}{\partial y}\right\}}_{\text{Convection}}$$

$$= \underbrace{\frac{\mu}{\rho}\left\{\frac{\partial^2\overline{uv}}{\partial x^2} + \frac{\partial^2\overline{uv}}{\partial y^2}\right\}}_{\text{Viscous diffusion}} - \underbrace{\left\{\overline{vu}\frac{\partial\overline{U}}{\partial x} + \overline{vv}\frac{\partial\overline{U}}{\partial y} + \overline{uu}\frac{\partial\overline{V}}{\partial x} + \overline{uv}\frac{\partial\overline{V}}{\partial y}\right\}}_{\text{Generation}}$$

$$
\underbrace{-\left\{\frac{\partial(\overline{uuv})}{\partial x}+\frac{\partial(\overline{vuv})}{\partial y}\right\}}_{\text{Turbulent diffusion}}-\underbrace{\frac{1}{\rho}\left\{\overline{v\frac{\partial p}{\partial x}}+\overline{u\frac{\partial p}{\partial y}}\right\}}_{\text{Pressure-velocity interaction}}
$$

$$
\underbrace{-2\frac{\mu}{\rho}\left\{\overline{\frac{\partial u}{\partial x}\frac{\partial v}{\partial x}}+\overline{\frac{\partial u}{\partial y}\frac{\partial v}{\partial y}}+\boxed{\overline{\frac{\partial u}{\partial z}\frac{\partial v}{\partial z}}}\right\}}_{\text{Viscous dissipation}}, \tag{4.5}
$$

in which the last term within the square frame arises as a consequence of the two-component turbulence constraint being removed.[3]

Equation (4.5) is a revealing result, for it shows that the shear-stress is dictated by processes that are analogous to those describing familiar flow properties, including energy, species concentration and momentum. We shall dwell on the physical interpretation of the terms, identified below the under-braces in some detail in Section 4.3. Here, we provide a preliminary summary first.

The first two terms should be obvious: they describe, respectively, convection and viscous diffusion of the stress (the latter along the two gradients of the stress). The third term represents the most influential process, namely the production (or generation) of the stress. This is recognised upon noting that $-\rho\overline{vv}\frac{\partial \overline{U}}{\partial y}$, for example, is a force per unit area multiplied by a velocity difference per unit length — i.e. power per unit volume (see Eq. (2.22)).

Mathematically, the fourth term is of the same type as viscous diffusion: both terms integrate to zero over a flow domain at the boundary of which the boundary-normal gradient of the shear stress is zero. The difference is that the fourth term represents 'diffusion' of the stress $\overline{uv}$ by *turbulent mixing* — i.e. the fluctuations u and v in the two gradient terms, respectively.

[3]Note that in Eq. (4.5) and following, the use of the term 'diffusion' is rather loose. Any term that is included under this heading has the property that its space integral is zero if Neumann (zero-gradient) conditions are imposed at all domain boundaries. In this case, the term is spatially distributive, i.e. does not generate or destroy the property in question.

The physical interpretation of the fifth term, identified as the 'pressure-velocity interaction', is left to future considerations; suffice to say here that this term causes a reduction in the shear stress by promoting the de-correlation between u and v via 'pressure scrambling'.

The last term is especially important to appreciate. Apart from viscous diffusion (the second term), this is the only one associated with fluid viscosity. It does not represent a diffusion process, however, because its form is not compatible with the interpretation of diffusion as described above — i.e. it is not distributive and does not integrate to zero over any volume in any circumstances. In fact, this is a term that is responsible for attenuating, by viscous damping, the motions that cause the shear stress; in other words, it represents a dissipative or destructive term. We have already encountered this type of process in Section 2.1, denoting it by ε. However, while ε is the dissipation-rate of the turbulence energy, $0.5\left(\overline{u^2}+\overline{v^2}+\overline{w^2}\right)$, the dissipation rate at issue here pertains to the shear stress.

4.2 *The case of simple shear*

To simplify matters further, we will now consider a simple shear layer, such as a channel flow, a boundary layer or a jet, in which case the dominant strain is $\frac{\partial \overline{U}}{\partial y}$ — indeed, this is the only strain in a fully-developed channel flow. Given that this type of flow evolves only slowly (if at all) in the x-direction, all x-wise gradients of time-averaged terms (but not gradients of turbulent fluctuations that contribute to time-averaged correlations) may be neglected. The equations governing the stress components $\left\{\overline{uv},\overline{u^2},\overline{v^2},\overline{w^2}\right\}$ are then:

$$\frac{D\overline{uv}}{Dt} = -\overline{v^2}\frac{\partial \overline{U}}{\partial y} + \overline{\frac{p}{\rho}\left(\frac{\partial u}{\partial y}+\frac{\partial v}{\partial x}\right)} - \frac{\partial}{\partial y}\left(\overline{uv^2}+\frac{\overline{pu}}{\rho}\right) + \frac{\mu}{\rho}\frac{\partial^2\overline{uv}}{\partial y^2}$$
$$-2\frac{\mu}{\rho}\left\{\overline{\frac{\partial u}{\partial x}\frac{\partial v}{\partial x}}+\overline{\frac{\partial u}{\partial y}\frac{\partial v}{\partial y}}+\overline{\frac{\partial u}{\partial z}\frac{\partial v}{\partial z}}\right\}$$

$$\frac{D\overline{u^2}}{Dt} = -2\overline{uv}\frac{\partial U}{\partial y} + 2\overline{\frac{p}{\rho}\frac{\partial u}{\partial x}} - \frac{\partial}{\partial y}\left(\overline{u^2 v}\right) + 2\frac{\mu}{\rho}\frac{\partial^2 \overline{u^2}}{\partial y^2}$$
$$- 2\frac{\mu}{\rho}\left\{\overline{\frac{\partial u}{\partial x}\frac{\partial u}{\partial x}} + \overline{\frac{\partial u}{\partial y}\frac{\partial u}{\partial y}} + \overline{\frac{\partial u}{\partial z}\frac{\partial u}{\partial z}}\right\}$$
$$\frac{D\overline{v^2}}{Dt} = 0 + 2\overline{\frac{p}{\rho}\frac{\partial v}{\partial y}} - \frac{\partial}{\partial y}\left(\overline{v^3} + \frac{2\overline{pv}}{\rho}\right) + 2\frac{\mu}{\rho}\frac{\partial^2 \overline{v^2}}{\partial y^2}$$
$$- 2\frac{\mu}{\rho}\left\{\overline{\frac{\partial v}{\partial x}\frac{\partial v}{\partial x}} + \overline{\frac{\partial v}{\partial y}\frac{\partial v}{\partial y}} + \overline{\frac{\partial v}{\partial z}\frac{\partial v}{\partial z}}\right\}$$
$$\frac{D\overline{w^2}}{Dt} = 0 + 2\overline{\frac{p}{\rho}\frac{\partial w}{\partial z}} - \frac{\partial}{\partial y}\left(\overline{vw^2}\right) + 2\frac{\mu}{\rho}\frac{\partial^2 \overline{w^2}}{\partial y^2}$$
$$- 2\frac{\mu}{\rho}\left\{\overline{\frac{\partial w}{\partial x}\frac{\partial w}{\partial x}} + \overline{\frac{\partial w}{\partial y}\frac{\partial w}{\partial y}} + \overline{\frac{\partial w}{\partial z}\frac{\partial w}{\partial z}}\right\}. \tag{4.6}$$

Note that we have retained fragments such as $\overline{\frac{\partial u}{\partial x}\frac{\partial v}{\partial x}}$ in the dissipation terms (within braces), because the gradients of any fluctuation, in any direction, are comparable in magnitude; it is only x-wise gradients of time-averaged quantities that we may neglect.

A minor point to bring out here is that, in the derivation of Eq. (4.6) from Eq. (4.5), the following split has been introduced:

$$\underbrace{\frac{1}{\rho}\left(\overline{v\frac{\partial p}{\partial x}} + \overline{u\frac{\partial p}{\partial y}}\right)}_{\text{Pressure-velocity interaction}} = \underbrace{\frac{1}{\rho}\left(\overline{\frac{\partial vp}{\partial x}} + \overline{\frac{\partial up}{\partial y}}\right)}_{\text{Pressure diffusion}} - \underbrace{\overline{\frac{p}{\rho}\left(\frac{\partial v}{\partial x} + \frac{\partial u}{\partial y}\right)}}_{\text{Pressure-strain interaction}} . \tag{4.7}$$

The designation 'pressure diffusion' reflects the fact that this term, like turbulent diffusion in Eq. (4.5), vanishes when integrated across a domain subject to zero-derivative conditions at the boundaries. In other words, this term may be interpreted as one that supplements turbulent-diffusion, spatially redistributing the shear stress across the flow domain. The term 'pressure-strain interaction' simply identifies this as a product of pressure fluctuations and strain fluctuations.

Finally, the equation for the turbulence energy arises as half of the sum of the last three equations in Eq. (4.6) for the normal-stress components[4]:

$$\underbrace{\frac{Dk}{Dt}}_{\text{Convection}} = \underbrace{\frac{\partial}{\partial y}\left(\frac{\rho}{\rho}\frac{\partial k}{\partial y}\right)}_{\text{Viscous diffusion}} - \underbrace{\overline{uv}\frac{\partial \overline{U}}{\partial y}}_{\text{Generation}} - \underbrace{\frac{\partial}{\partial y}\left(\overline{vk} + \frac{2\overline{pv}}{\rho}\right)}_{\text{Turbulent diffusion}}$$

$$\underbrace{-\frac{\mu}{\rho}\left\{\begin{array}{c}\overline{\frac{\partial u}{\partial x}\frac{\partial u}{\partial x}} + \overline{\frac{\partial u}{\partial y}\frac{\partial u}{\partial y}} + \overline{\frac{\partial u}{\partial z}\frac{\partial u}{\partial z}} + \overline{\frac{\partial v}{\partial x}\frac{\partial v}{\partial x}} + \overline{\frac{\partial v}{\partial y}\frac{\partial v}{\partial y}} \\ + \overline{\frac{\partial v}{\partial z}\frac{\partial v}{\partial z}} + \overline{\frac{\partial w}{\partial x}\frac{\partial w}{\partial x}} + \overline{\frac{\partial w}{\partial y}\frac{\partial w}{\partial y}} + \overline{\frac{\partial w}{\partial z}\frac{\partial w}{\partial z}}\end{array}\right\}}_{\text{Dissipation}}. \tag{4.8}$$

It is instructive to first extend the interpretations given above for the shear-stress equation (4.5) to Eq. (4.8), and then derive, in the next section, some important qualitative features from Eq. (4.6).

As in the case of the shear stress, the first two terms in equation (4.8) represent, self-evidently, convection and viscous diffusion. The third term represents the creation of turbulence energy by the interaction between the shear stress and the shear strain (confirming Eq. (2.22)). A practical example is wind shear across a runway: ground friction creates a boundary layer in which the flow is sheared, and the presence of turbulent motions gives rise to a finite value of the product of shear stress and shear strain, enhancing the level of turbulence, and thus creating increasingly uncomfortable conditions for landing aircraft. The fourth term represents diffusion of k by the cross-flow mixing action of v, and this is augmented by a cross-flow 'mixing' of pressure fluctuations, which are akin to velocity-squared fluctuations. The fifth and last term is, again, representative

[4]The dissipation rate in this equation, as stated, is only the exact transfer of energy to heat in homogeneous turbulence in which the Reynolds-stresses do not vary in space. In non-homogeneous turbulence, the total energy transfer to heat by friction involves the additional contributions $\frac{\mu}{\rho}\left\{\frac{\partial^2\overline{u^2}}{\partial x^2} + \frac{\partial^2\overline{uv}}{\partial x\partial y} + \cdots + \frac{\partial^2\overline{w^2}}{\partial z^2}\right\}$, but this is minor.

of destruction by viscous action. This term is the only one that counteracts the rising intensity of turbulence due to generation. It transfers turbulence energy into heat by the viscosity interacting with the smallest-scale eddies in the spectrum shown in Fig. 2.4. As the turbulent eddies become smaller, due to the 'cascade' process (range B in Fig. 2.4), the derivatives of the fluctuations become larger and larger, causing intense shearing between the eddies, thus promoting viscous destruction. Importantly, the pressure-strain terms in Eqs. (4.6) and (4.7) do not appear in Eq. (4.8), because the sum of the pressure-strain terms in the $\overline{u^2}$, $\overline{v^2}$ and $\overline{w^2}$ equations reduces to zero, as a consequence of the mass-conservation principle (for incompressible conditions). This signifies that the pressure-strain term does not enhance or reduce the turbulence energy, but merely acts to redistribute it among the normal stresses.

4.3 *Manifestations of stress anisotropy*

We return to the equation set (4.6) to highlight a few further important interactions among the stresses. First, we note that only the $\overline{u^2}$ equation has a non-zero production term. We therefore expect the streamwise stress to be larger than the other components. Indeed, this is what is observed experimentally and in full turbulence simulations — a state referred to as *normal-stress anisotropy*. We further note that the production of the shear stress — the dynamically most important stress in a shear layer — is driven directly by the normal stress $\overline{v^2}$. This observation suggests, on its own, that it is of major importance to correctly predict the anisotropy of the normal stresses.

Three illustrations of the state of anisotropy are given in Figs. 4.1 to 4.3, the first for a plane wall jet, the second for a flat-plate boundary layer in zero pressure gradient, and the third for a fully developed channel flow. The plane wall jet may be thought of as consisting of a free outer shear layer, interacting with a boundary layer developing on the lower wall. In the free shear layer, the normal-stress peaks coincide with the maximum production of the streamwise stress, where the velocity gradient (the shear strain) is highest. The anisotropy $\overline{u^2}/\overline{v^2}$ is around 1.7, which is not very high,

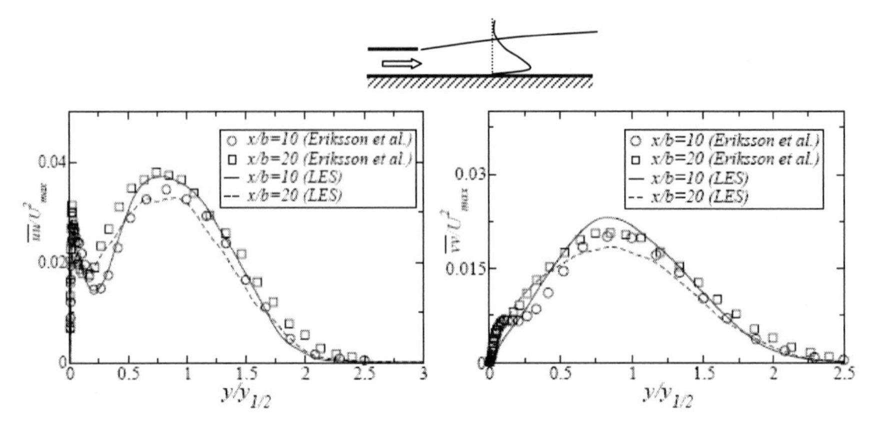

Fig. 4.1: Streamwise and wall normal Reynolds-stresses in a wall jet. The profiles at $x/b = 20$ are essentially self-similar. Symbols indicate experimental data, lines simulations; derived from LES data (reproduced with permission of AIP Publishing LLC from Dejoan and Leschziner (2005)).

but is nevertheless indicative of the important role played by generation in relation to the anisotropy.

In the boundary layer and the channel flow, however, the anisotropy is clearly far higher, and it would thus appear that a near-wall flow has characteristics which differ substantially from those in a free shear flow. In particular, at $y^+ \approx 15$ (see the definition of this normalised wall distance in the caption of Fig. 4.2 and footnote[5]), where the streamwise stress peaks, the ratio $\overline{u^2}/\overline{v^2}$ is around 30 (note that Fig. 4.2 shows root-mean-square (RMS) profiles for the normal stresses).

The results for the channel flow in Fig. 4.3 have been normalised and plotted in two ways — with wall scales (see caption of Fig. 4.2) and outer scales (channel half-height and bulk velocity) — to show that both can be made to yield identical distributions if the scales are adjusted judiciously. There is an important fundamental difference, however: wall scaling yields distributions which are virtually universal — i.e. are hardly dependent on the Reynolds number (except for

[5]The importance of this normalisation by the wall shear-stress will transpire in Chapter 5. In essence, it allows the near-wall velocity and stresses to be represented as universal distributions, independent of the Reynolds number of the flow.

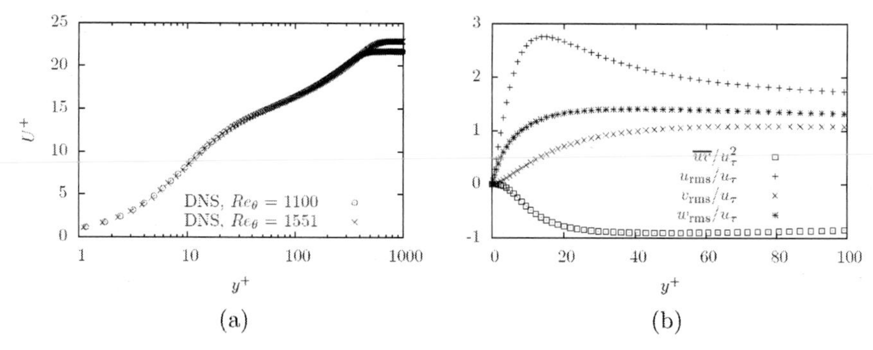

Fig. 4.2: Velocity distribution (note the log scale) and RMS of stresses (except shear-stress) in a boundary layer predicted by DNS. All quantities are normalised using wall scaling — for example, $U^+ = \overline{U}/u_\tau$; $y^+ = yu_\tau/\nu$; $u_\tau \equiv \sqrt{\tau_{wall}/\rho}$, where τ_{wall} is the wall shear-stress. Derived from DNS data (reproduced with permission of Taylor & Francis Ltd., from Bentaleb, Lardeau and Leschziner (2012)).

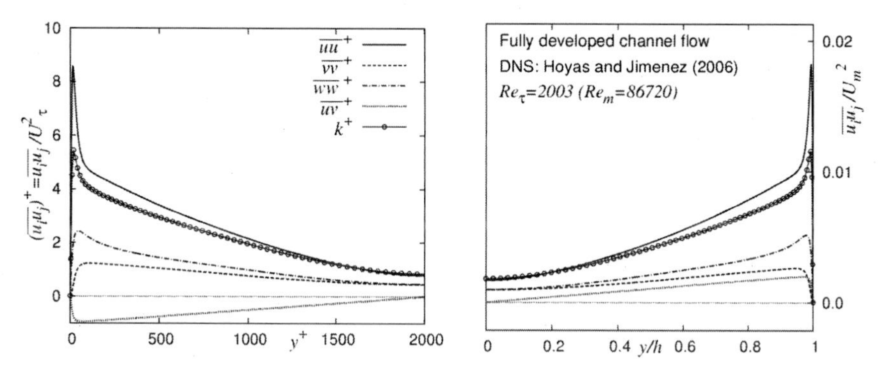

Fig. 4.3: Reynolds-stresses in fully developed channel flow at a bulk Reynolds-number of 86700 (based on channel height), depicted with wall scaling (left hand side) and outer scaling (right hand side). Note the identical shape of the variations when linear scaling of y^+ is adopted and the scaling of the stress magnitude is matched. Derived from DNS data by Hoyas and Jimenez (2006) (Courtesy S. Jakirlić).

very low values) — while outer scaling does not. For example, with outer scaling, the stress maxima, normalised by the square of the bulk velocity, change with Reynolds number. As a rough guide, for reasons explained in Section 5.2, the peak turbulence energy occurs at $y^+ \approx 15$ (but varies greatly in terms of y/h!), and is around $4-5u_\tau^2$ or 1%-2% U_m^2. Hence (again as a rough guide), $u_\tau \approx 0.05U_m$. We shall pursue further the subject of near-wall turbulence in Chapter 5.

The question might reasonably be posed as to why $\overline{v^2}$ and $\overline{w^2}$ are finite at all, since they are not generated (see Eq. 4.6). The answer is that the pressure-strain-interaction terms act to redistribute the turbulence energy among the normal stresses, tending to steer turbulence towards an isotropic state. The redistributive nature of these terms is most clearly reflected by the fact that their sum across the $\overline{u^2}$, $\overline{v^2}$ and $\overline{w^2}$ equations is zero, as already noted by reference to the discussion of Eq. (4.6). Hence, in this case, the pressure-strain process takes away energy from $\overline{u^2}$ and feeds this energy to the stresses $\overline{v^2}$ and $\overline{w^2}$. As is suggested by Figs. 4.1 and 4.2, the redistribution process is far more effective in a free shear layer than in a near-wall layer, for reasons explained in Chapter 5.

Unlike stress transport (diffusion), which is observed to be modest (in relative terms), except very close to the wall and in the vicinity of velocity maxima (where generation is locally zero), dissipation is a very important process in Eq. (4.6). This is quite evidently so from Eq (4.8), because in that equation the dissipation is the only process that destroys the turbulence energy being continuously created by the strain-driven production terms. An example in which stress transport *is* important is given in Fig. 4.4. This shows the Reynolds stresses (RMS values, in the case of the normal stresses) in a self-similar (fully established) round jet spreading in stagnant surroundings. Here again, stress anisotropy is evident, although it is even more modest than in the wall jet in Fig. 4.1. Crucially, the level of the turbulent fluctuations at the (statistical) centre-line of the jet is almost as high as it is in the highly strained region. Because generation must be zero at the centre-line, it follows that the normal stresses (and turbulence energy) are *diffused* towards the centre-line from outer jet regions, thus maintaining a high level of turbulence activity. In fact, and more precisely, Eq. (4.8) tells us that the state at the centre-line reflects a balance between cross-flow diffusive transport, convective transport from upstream, and dissipation by viscous action, the first two elevating the turbulence intensity, and the last diminishing it.

Although we cannot recognise from Eq. (4.6) how dissipation is distributed across the normal stresses, we have good reason to

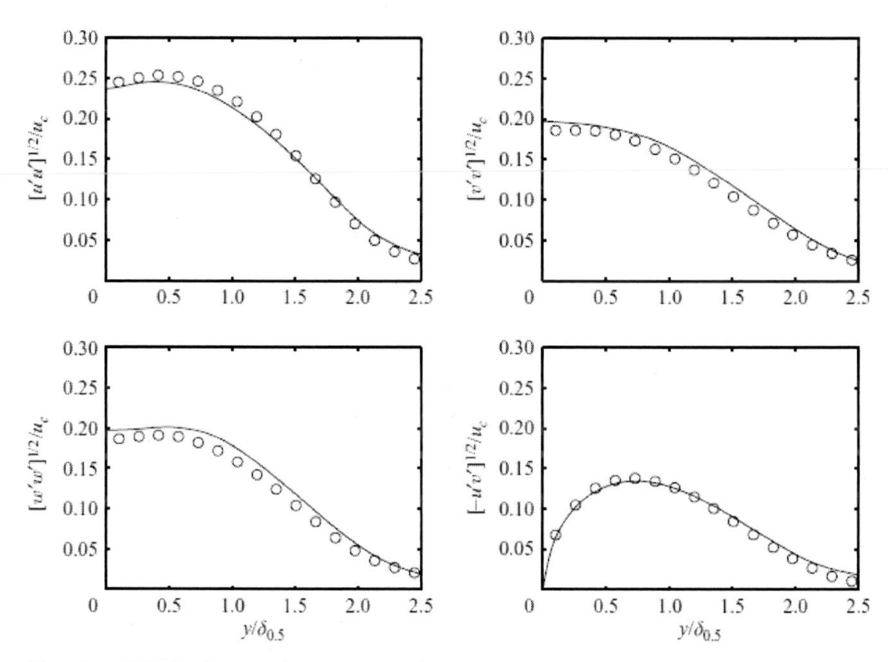

Fig. 4.4: RMS of normal stresses and shear-stress in a self-preserving round jet at $Re_d = 11000$, experimental and LES-computed distributions (reproduced with permission of Cambridge University Press from Bogey and Bailly (2009)).

suspect that the dissipation rates do not differ greatly from one another, as dissipation is associated with the smallest eddies of turbulence — the portion C in the spectrum of Fig. 2.4. These smallest eddies are remote (in terms of both their length and time scales) from the respective scales associated with the mean-flow strain and turbulence generation, the latter being within range A in Fig. 2.4. This 'disconnect' implies that the dissipative eddies are close to being isotropic, and this suggests, in turn, that the rate of dissipation is fairly uniformly distributed across the normal stresses. As will be explained in Chapter 5, however, this argument does not hold near a wall.

4.4 *Summary and lessons*

Based on the discussion so far, we may construct a closed cycle that conveys the essential interactions among the stresses, as gleaned from

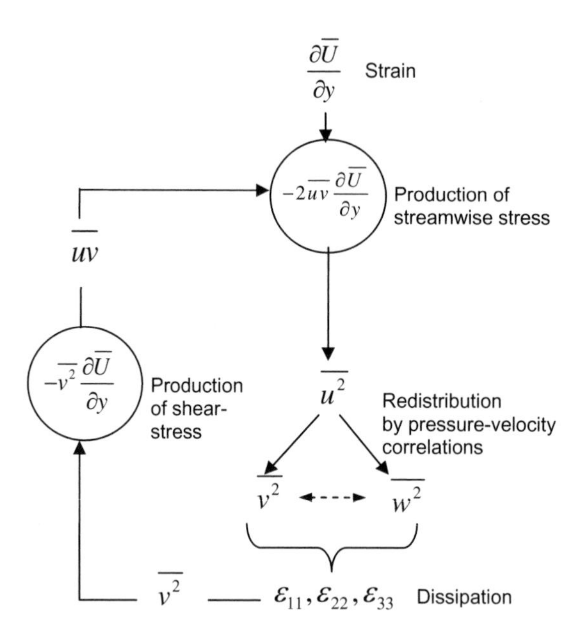

Fig. 4.5: The 'cycle of turbulence' in a simple shear flow. Note the focus on the productions which feed the streamwise stress and the shear-stress.

their respective equations. This cycle is shown in Fig. 4.5, which also conveys, graphically, how turbulence is sustained in a shear flow.

We have seen that the stresses (and the turbulence energy) may be regarded as transported quantities: they can be convected, diffused, generated and dissipated. On the assumption that *generation* is a dominant (or at least very important) process, we have concluded that any one stress is, in general, sensitively linked to all other stresses and strain components in a non-trivial manner. For the case of simple shear flow, we have observed that the only normal-stress component being generated is that in the streamwise direction. The sustenance of three-dimensional turbulence then hinges on a redistribution mechanism, which transfers energy from the streamwise direction to the other directions. This is a redistributive mechanism that tends to drive turbulence towards a state of isotropy, which is, however, only reached after a long period of time, and then only if generation ceases. Importantly, without this transfer, the shear-stress generation is disabled. This process is also the key to maintaining a finite level of turbulence energy, because this energy is only generated in the presence of a finite shear stress. Finally, the dissipation mechanism is the process that prevents turbulence rising indefinitely, for it acts against the production, by transferring turbulence energy into heat by viscous friction, due to the interaction between the viscosity and the intense strain fluctuations within the small-scale eddies.

Chapter Five

5. Fundamentals of Near-Wall Interaction

The reason for devoting this chapter to the near-wall state is two-fold. First, from a practical perspective, near-wall turbulence is of major interest, for it dictates the wall shear stress, which is responsible for friction drag, and the wall heat transfer, which is the key to cooling and heating. Second, as is evident from Figs. 4.1–4.3, there are some peculiar interactions at the wall that require particular attention, both from a fundamental perspective and, more importantly, for their relevance to modelling.

Figure 5.1(a) is intended to indicate, in highly idealized terms, what we expect to find near the wall: eddies that diminish in size and become increasingly elongated and 'flat' due to the blocking action of the wall. Unless we examine length scales, such as $\Lambda/\eta = Re_\Lambda^{3/4}$ (Eq. (2.14)), and others extracted from two-point correlations (e.g. Eq. (2.19)), we cannot really give quantitative substance to this notional picture — although visualisations, such as that in Fig. 5.1(b), provide some useful evidence. What we *can* do more easily, and more productively, however, is to examine how the components of the turbulent fluctuations — and hence the stresses — vary as the wall is approached. We can also derive useful information on the turbulence properties in the layers further removed from the wall, and hence use this information to describe the wall-normal velocity variation. This is the route taken in the subsections below.

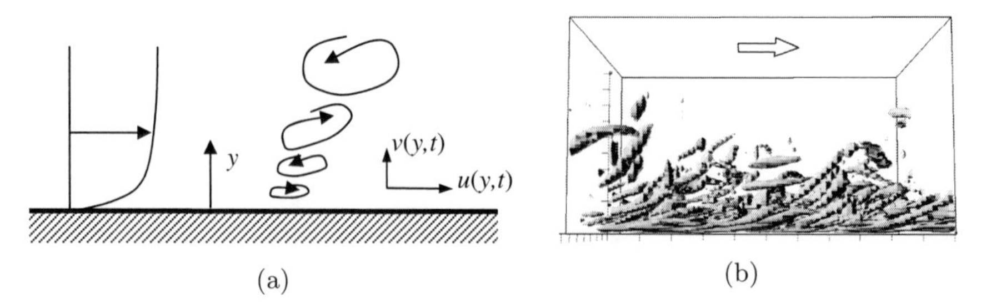

Fig. 5.1: Near-wall eddy structure (a) idealized schematic; (b) visualisation derived from DNS data (reproduced with permission of Cambridge University Press from Blackburn, Mansour and Cantwell (1996)).

5.1 *Turbulence in the viscous sublayer*

Very close to the wall, the wall-normal variation of the turbulent fluctuations may be written in the general form of the following Taylor-series expansions:

$$\begin{aligned} u(t,y) &= a_1(t) + b_1(t)y + c_1(t)y^2 + \cdots . \\ v(t,y) &= a_2(t) + b_2(t)y + c_2(t)y^2 + d_2(t)y^3 + \cdots . \\ w(t,y) &= a_3(t) + b_3(t)y + c_3(t)y^2 + \cdots . \end{aligned} \tag{5.1}$$

At the wall itself, $y = 0$, and all fluctuations vanish, so that $a_1 = a_2 = a_3 = 0$. Also, at the wall, $\partial u/\partial x|_0 = 0$, $\partial w/\partial z|_0 = 0$ and, therefore, because of mass conservation (in incompressible conditions), it follows that $\partial v/\partial y|_0 = 0$ and $b_2(t) = 0$. With the relevant terms omitted from Eq. (5.1), multiplication of the first two equations in this set gives:

$$u(t,y)v(t,y) = b_1(t)c_2(t)y^3 + [c_1(t)c_2(t) + b_1(t)d_2(t)]y^4 + \cdots , \tag{5.2}$$

Time-averaging Eq. (5.2) then gives:

$$\overline{uv}(y) = \alpha_1 y^3 + \beta_1 y^4 + \cdots , \tag{5.3}$$

where the coefficients α_1, β_1 are simply time-averaged forms of the associated coefficients in Eq. (5.2). It is, in fact, easy to show, by an entirely equivalent route, that:

$$\overline{u^2}(y) = \alpha_2 y^2 + \beta_2 y^3 + \cdots,$$
$$\overline{w^2}(y) = \alpha_3 y^2 + \beta_3 y^3 + \cdots, \qquad (5.4)$$
$$\overline{v^2}(y) = \alpha_4 y^4 + \beta_2 y^5 + \cdots.$$

From this we deduce four important facts:

(i) the wall-parallel normal stresses and the turbulence energy decay quadratically as the wall is approached;
(ii) the wall-normal stress decays by two orders of magnitude faster than the other two normal stresses;
(iii) it follows from (i) and (ii) that turbulence approaches a 'two-component state' at the wall, expressible as:

$$\frac{\overline{u^2}}{k} \rightarrow C_1; \quad \frac{\overline{w^2}}{k} \rightarrow C_2; \quad \frac{\overline{v^2}}{k} \rightarrow 0; \qquad (5.5)$$

(iv) the shear stress decays cubically — a result that will be used later in examining the behaviour of near-wall models.

Figure 5.2 provides a computational verification of the asymptotic behaviour implied by Eqs. (5.3) and (5.4) for the case of a fully developed channel flow, by way of results derived from a direct numerical simulation. The figures show that, while the asymptotic rates derived from the kinematic considerations are reproduced, the range in which this behaviour applies is quite restricted, typically within $y^+ = 5 - 10$, i.e. well within the viscous sublayer. Yet, Figs. 4.2 and 4.3 suggest that the wall exerts a strong influence on the anisotropy far beyond the viscous sublayer. For example, Fig. 4.3 shows that significant differences between $\overline{v^2}$ and $\overline{w^2}$ persist beyond $y^+ = 500$. Hence, importantly, the present asymptotic analysis exposes only a part of the mechanism by which the wall influences the anisotropy.

The result of Eq. (5.5) is verified in Fig. 5.3, again for a fully-developed channel flow, albeit at a higher Reynolds number than the case in Fig. 5.2. As in Fig. 5.2, the stresses and the wall distance have been normalised with the wall scale $u_\tau \equiv \sqrt{\tau_{wall}/\rho}$. Note the

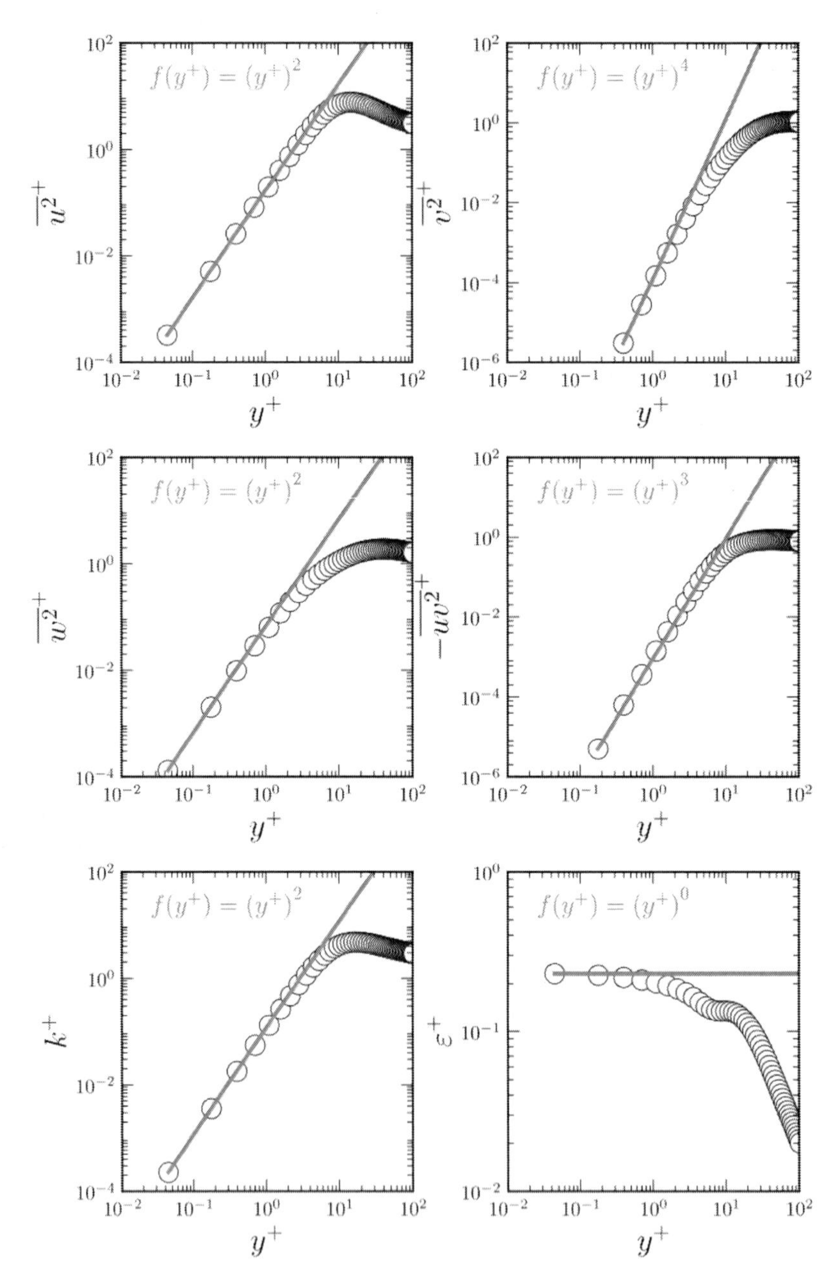

Fig. 5.2: Near-wall variations of the Reynolds stresses, the turbulence energy and the dissipation rate for the case of a channel flow at $Re_\tau = u_\tau h/\nu = 512$, $u_\tau = \sqrt{\tau_{wall}/\rho}$, where h is the channel half-height and the superscript '+' indicates normalisation with u_τ; derived from DNS data (reproduced with permission from F. Billard (2011)).

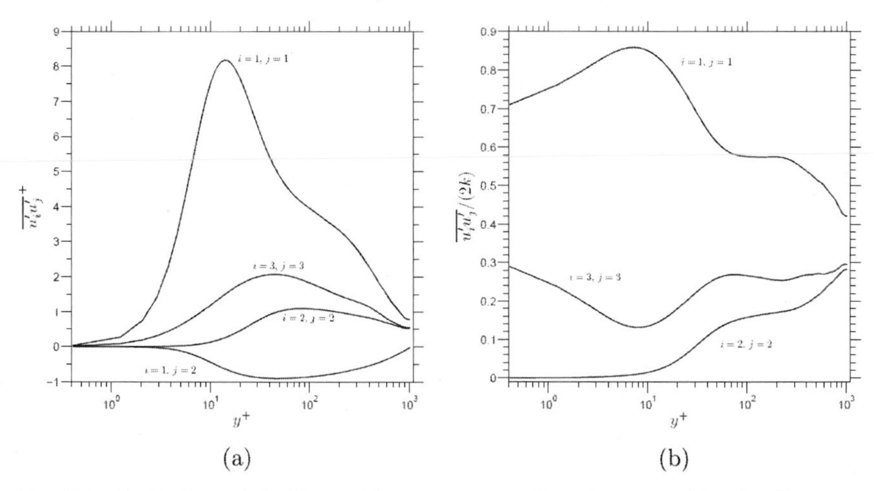

Fig. 5.3: Variations of the Reynolds stresses and the anisotropy (Eq. (5.5)) across a channel flow at $Re_\tau = 1000$, corresponding to a bulk Reynolds-number of about 40000; derived from DNS data (reproduced with permission of AIP Publishing LLC from Agostini and Leschziner (2014)).

very high level of anisotropy in the buffer region, $y^+ \approx 15$, also conveyed by Fig. 4.2. This reflects the high rate of generation of $\overline{u^2}$ in the region in which the viscous sublayer ($y^+ < 10$) merges into the turbulent layer above it ($y^+ > 30$), and the fact that a wall inhibits, preferentially, wall-normal fluctuations by an inviscid blocking process. We shall demonstrate the validity of the former in Section 5.2. As seen in Fig. 5.3(b), the anisotropy components $\overline{u^2}/k$ and $\overline{w^2}/k$ indeed approach constant values — around 0.7 and 0.3, respectively.

Another relevant deduction, made possible by reference to the turbulence-energy-transport equation, Eq. (4.8), relates to the near-wall variation of the dissipation rate. Very close to the wall, all fluctuations decay to an insignificant level, and only the viscous terms remain — i.e.:

$$\underbrace{\frac{\partial}{\partial y}\left(\frac{\mu}{\rho}\frac{\partial k}{\partial y}\right)}_{\text{Viscous diffusion}}\Bigg|_{y\to 0} = \underbrace{\varepsilon}_{\text{Dissipation}} . \tag{5.6}$$

Because k varies quadratically as the wall is approached, the left-hand side must be of the form:

$$\varepsilon = \nu(\alpha_2 + \alpha_3), \tag{5.7}$$

i.e. the constant value $2\nu \frac{k}{y^2}\Big|_{y \to 0}$, as emerges upon the insertion of the quadratic variation of k into Eq. (5.6). The deduction expressed by Eq. (5.7) is verified by Fig. 5.2. Thus, the near-wall variation $k^+ = f(y^+)$ suggests $\frac{k^+}{y^{+2}} \left(= \frac{\nu^2}{u_\tau^4} \frac{k}{y^2}\right) \approx 0.12$. As the wall-scaled dissipation rate is $\varepsilon^+ = \frac{\nu}{u_\tau^4}\varepsilon$, it follows that $\varepsilon^+ = 2\frac{k^+}{y^{+2}} \approx 0.24$, which agrees well with the asymptotic level shown in Fig. 5.2.

Equation (4.6) for the stress components in simple channel flow also shows that, at the wall, the dissipation rates of the individual stresses have to decay at the same rate as the second derivatives of stresses themselves. This then demonstrates, unambiguously, the validity of the assumption that the near-wall dissipation is highly anisotropic, as are the stresses. In particular, $\varepsilon_{11}/\varepsilon$ and $\varepsilon_{33}/\varepsilon$ approach constant values, which are the same as $\overline{u^2}/k$ and $\overline{w^2}/k$, respectively, while ε_{22} decays quadratically towards zero.

The validity of Eq. (5.6) is demonstrated in Fig. 5.4. This shows the normalized wall-normal distributions of the contributions to the turbulence-energy Eq. (4.8) for a flat-plate boundary layer, derived

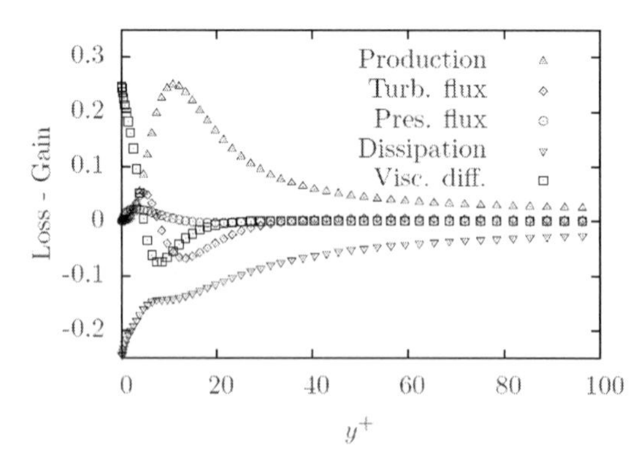

Fig. 5.4: Wall-normal profiles of the contributions to the turbulence-energy equation (4.8) for a boundary layer on a flat plate at a momentum-thickness Reynolds number of 1100; derived from DNS data (reproduced with permission of Taylor & Francis Ltd., from Bentaleb, Lardeau and Leschziner (2012)).

from direct numerical simulation (DNS). As the wall is approached, the dissipation (lowest curve) and viscous diffusion (square symbols) are finite at the wall and balance each other.

Another feature that is useful to highlight, ahead of the discussion in Section 5.2, is that the maximum in turbulence-energy production occurs at around $y^+ \approx 15$. In fact, here, this production is simply half of that of the streamwise normal stress $\overline{u^2}$ (compare Eqs. (4.6) and (4.8)), which is consistent with the peak of $\overline{u^2}$ itself at the same position, as shown in Fig. 5.3.

5.2 *Turbulence processes in the 'buffer layer'*

Next, we seek insight into why the turbulence energy reaches a maximum in a region that is beyond the immediate vicinity of the wall, at around $y^+ \approx 15$. As noted already, Fig. 5.4 shows the production (generation) of turbulence energy to reach a maximum at this height, but why? Here again, we can learn a great deal from examining a particular stress-strain interaction. We note first, from Eq. (4.8), that the production of turbulence energy is:

$$P_k = -\overline{uv}\frac{\partial \overline{U}}{\partial y}. \tag{5.8}$$

Hence, in general, the turbulence energy will be especially large where the product of shear stress and shear strain is a maximum. In the outer region of the boundary layer, the strain approaches zero, and so does the production. Close to the wall, the shear strain is high, but the turbulent fluctuations decay, eventually all reaching zero at the wall. The asymptotic near-wall variation of P_k follows readily from Eq. (5.3), and the observation that the velocity profile in the viscous sublayer is close to linear, unless the streamwise pressure gradient is very high. Hence, the asymptotic decay of the production follows the cubic variation:

$$P_k = \alpha_1 \left.\frac{\partial \overline{U}}{\partial y}\right|_{y \to 0} y^3. \tag{5.9}$$

In between the wall and the outer stream, there is a position at which the product in Eq. (5.8) reaches a maximum. To determine where this maximum is, we differentiate Eq. (5.8) and set the result equal to zero, i.e.:

$$-\frac{\partial \overline{uv}}{\partial y}\frac{\partial \overline{U}}{\partial y} - \overline{uv}\frac{\partial}{\partial y}\frac{\partial \overline{U}}{\partial y} = 0. \tag{5.10}$$

In the particular case of a flat-plate boundary layer in zero pressure gradient, it is trivial to show, by reference to the U-momentum equation, that the total shear stress is very close to constant and equal to the wall shear stress,[1] i.e.:

$$\mu\frac{\partial \overline{U}}{\partial y} - \rho\overline{uv} = \tau_{wall}, \tag{5.11}$$

which then leads to:

$$-\frac{\partial \overline{uv}}{\partial y} = \nu\frac{\partial}{\partial y}\frac{\partial \overline{U}}{\partial y}. \tag{5.12}$$

We note, in passing, that Eq. (5.11) implies a very rapid decrease in the velocity gradient, because Eq. (5.3) dictates that the shear stress increases cubically away from the wall (though, strictly, only within $y^+ \approx 10$ — see Fig. 5.2). Hence, the viscous contribution to the total stress must decline correspondingly. In other words, the curvature in the velocity profile is especially high in the buffer region.

Substitution of Eq. (5.12) into Eq. (5.11) gives:

$$\left(\nu\frac{\partial^2 \overline{U}}{\partial y^2}\right)\frac{\partial \overline{U}}{\partial y} - \overline{uv}\frac{\partial^2 \overline{U}}{\partial y^2} = 0, \tag{5.13}$$

or:

$$\nu\frac{\partial \overline{U}}{\partial y} = -\overline{uv}. \tag{5.14}$$

This is consistent with the observation, upon following the lowest curve in Fig. 4.2(b), that the turbulent shear stress declines to about

[1] This also applies to channel flow close to the wall, because the contribution of the streamwise pressure gradient to the streamwise-momentum balance, expressed by the momentum equation, decays with y^2 as the wall is approached, thus causing only minor deviations from Eq. (5.10).

half of its maximum at around $y^+ = 10 - 15$, and a similar value pertains to the shear-stress distribution in a channel flow, Fig. 5.3(a). Fig. 4.2(a) shows that this distance also corresponds to the (virtual) intersection of the outer logarithmic layer, in which the turbulent shear stress dominates, and the inner viscosity-dominated layer.[2] This is given expression by the designation 'buffer layer' to the region $y^+ \approx 10 - 30$.

To gain an appreciation of how physically close to the wall $y^+ = 10$ is, we consider a boundary layer of thickness δ_{99} (at the location 99% of the free-stream velocity) having a Reynolds number $Re = \frac{U_\infty \delta_{99}}{\nu} = 10{,}000$, which is a fairly low value by engineering standards. This corresponds, roughly, to $Re_\tau = \delta_{99}^+ = \frac{u_\tau \delta_{99}}{\nu} = 500$ (Re_τ is referred to as the 'friction Reynolds number'). Hence, $y^+ = 10$ corresponds to $10/500 \times \delta_{99}^+$, i.e. a distance from the wall of around 2% of the boundary-layer thickness.

5.3 *The velocity distribution in the near-wall region*

The result in Eq. (5.3), namely that the shear stress declines rapidly as the wall is approached, allows us to examine the mean-velocity distribution near the wall. The velocity profile across a boundary layer has already been shown in Fig. 4.2(a), in terms of what is called the 'universal law of the wall' — $U/u_\tau = f(yu_\tau/\nu)$ — where $u_\tau \equiv \sqrt{\tau_{wall}/\rho}$ is the 'friction velocity'. This result has been obtained from DNS, but measurements show an identical behaviour.

We will take advantage of Eq. (3.9), and simplify it to apply to a zero-pressure-gradient boundary layer, i.e.:

$$\frac{\partial \overline{U}^2}{\partial x} + \frac{\partial \overline{V}\,\overline{U}}{\partial y} = \frac{\partial}{\partial y}\left(\nu \frac{\partial \overline{U}}{\partial y} - \overline{uv}\right). \tag{5.15}$$

As an aside, although an important one, it is informative to highlight that, while the only dynamically active stress in a thin shear layer is the shear-stress (as is seen in Eq. (5.15)), the other stress components are, nevertheless, extremely important. In particular, reference to

[2]It will be shown in Section 5.3 that the intersection is close to $y^+ = 11$.

Fig. 4.5 (and Eq. (4.6)) shows that the shear stress is driven by the normal stress $\overline{v^2}$. This component is driven upwards by the transfer of energy from the vigorously generated streamwise stress $\overline{u^2}$. Hence, a single-track focus on the shear stress alone is a recipe for trouble.

Very close to the wall, advection is negligible and may be removed from Eq. (5.15). The turbulent shear stress declines as $O(y^3)$, while the velocity gradient varies as $O(1)$. Hence the solution of Eq. (5.15) is trivial, and the result, following a double integration and the insertion of the zero-velocity condition at the wall, is:

$$\overline{U} = \frac{\tau_{wall}}{\mu} y, \tag{5.16}$$

or, after dividing both sides by $u_\tau = (\tau_{wall}/\rho)^{1/2}$:

$$\overline{U}^+ = y^+. \tag{5.17}$$

Next, we turn our attention to the upper layer — that lying above the buffer layer. On the assumption that this layer is dominated by turbulent mixing, and that advection is still weak, Eq. (5.15) reduces to:

$$\frac{\partial \overline{uv}}{\partial y} = 0 \quad \text{or} \quad \overline{uv} = u_\tau^2. \tag{5.18}$$

To proceed further, we need to relate the shear-stress to the velocity. However, this is precisely what is understood by 'turbulence modelling', a step we are not ready to take at this stage. As an alternative, we resort to dimensional analysis.

A critical point to appreciate first is that, although the velocity itself is a function of the viscosity, $U = f(y, \mu \ldots)$ — because it necessarily links to the velocity in the viscous near-wall flow, Eq. (5.17) — the *gradient* of the velocity is dictated by local turbulent mixing, and this is *not* a function of the viscosity; i.e. $\frac{\partial U}{\partial y} \neq f(\mu \ldots)$. We can now examine the implication of this constraint, using the observation that $U^+ = f(y^+)$ is a universal near-wall description, unless the boundary layer is subjected to a strong pressure gradient or some other strong forcing. Thus:

$$\frac{\partial \overline{U}}{\partial y} = u_\tau \frac{\partial f}{\partial y^+}\frac{\partial y^+}{\partial y} = \frac{\rho u_\tau^2}{\mu}\frac{\partial f}{\partial y^+}. \tag{5.19}$$

If this gradient is to be independent of the viscosity, this must mean:

$$\frac{\partial f}{\partial y^+} = \frac{\partial \overline{U}^+}{\partial y^+} \propto \mu. \tag{5.20}$$

However, because the left-hand side is obviously dimensionless, the right hand side must likewise be so, and the only non-dimensional representation of the viscosity is, here, the inverse of the Reynolds-number $(yu_\tau/\mu)^{-1} = 1/y^+$. Hence,

$$\frac{\partial \overline{U}^+}{\partial y^+} \propto \frac{1}{y^+}, \tag{5.21}$$

which can be readily integrated to give:

$$\overline{U}^+ = A \ln y^+ + B. \tag{5.22}$$

Measurements (and simulations) show:

$$\overline{U}^+ = \frac{1}{0.4}\ln y^+ + 5.45, \tag{5.23}$$

where the value 0.4 is referred to as the 'von Kármán constant[3]' (usually denoted by κ).

Finally, the intercept of the logarithmic profile (Eq. (5.23)) with the viscous-layer profile (Eq. (5.17)) can easily be shown to occur at $y^+ \approx 11$. There is no such intercept in reality, of course. Rather, as noted earlier, a semi-viscous 'buffer layer', extending to $y^+ \approx 30$, connects the two, as is seen in Fig. 4.2(a) for a boundary layer, and in Fig. 5.5 for channel flow at different Reynolds-number values.

[3]There is no full consensus on the value of κ. Different studies at different Reynolds-numbers report values between 0.39 and 0.42. The most recent experimental investigations in high-Reynolds-number boundary layers suggest a value of 0.4.

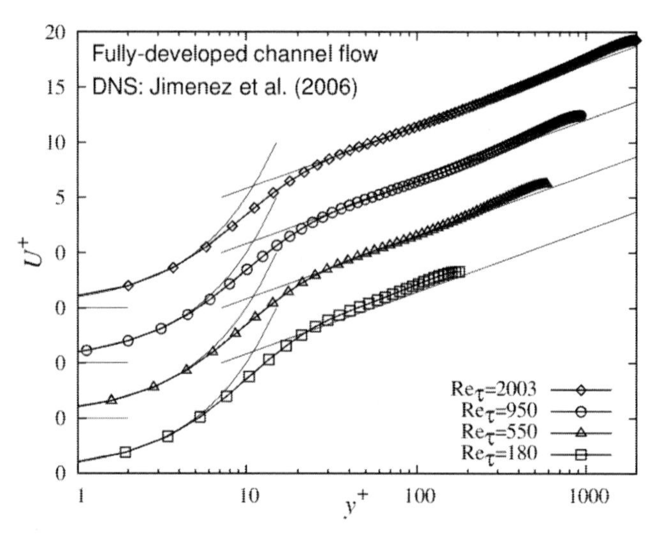

Fig. 5.5: Universal velocity profiles in fully developed channel flow at different Reynolds numbers (bulk Reynolds-number up to 42000), normalised with u_τ and plotted in semi-logarithmic coordinates; derived from DNS data by Hoyas and Jimenez (2006) and del Alamo and Jimenez (2003, 2004) (courtesy S. Jakirlić).

5.4 *Summary and lessons*

We have seen that a wall exerts a profound influence on the turbulence properties close to it. Crucially, a wall tends to block (or disproportionately inhibit) the wall-normal velocity fluctuations, thus causing very high levels of anisotropy and inhibiting cross-flow mixing. Indeed, at the wall itself, turbulence assumes a two-component state: while all turbulent stresses decay to zero, the wall-parallel stresses, when normalised by the turbulence energy, remain finite. Very close to the wall, within $y^+ \approx 10$, we have shown that the wall-parallel stresses and turbulence energy decay quadratically; the wall-normal stress decays with y^4; the shear stress decays with y^3; and the dissipation rate approaches a maximum and asymptotically constant level, balanced by viscous diffusion of turbulence energy towards the wall.

The 'buffer layer', connecting the viscous sublayer to the turbulent upper layer, is characterised by very high curvature in the velocity profile, as the linear profile within $y^+ \approx 10$ merges into the logarithmic profile in the fully turbulent layer at $y^+ > 30$.

In the buffer layer, the turbulent shear stress rises rapidly with wall distance, and the viscous shear stress (i.e. the velocity gradient) declines in harmony, so as to maintain (in zero or low pressure gradient) a constant total shear stress in the near-wall region. Around the middle of the buffer layer, $y^+ \approx 15$, where the viscous and turbulent shear stresses have the same value, the production of the turbulence energy reaches a maximum, giving rise to high levels of streamwise stress and turbulence energy. Around this maximum, the ratio $\overline{u^2}/\overline{v^2}$ reaches values in excess of 20, and this suggests that it may be crucially important to distinguish carefully between the wall-normal turbulence intensity and other components, if near-wall transport of mean-flow properties is to be modelled realistically.

Finally, in a statistically two-dimensional boundary layer, the fully turbulent upper layer is characterised by a universal logarithmic profile of the mean velocity, at least for sufficiently low streamwise pressure gradient. This result, here derived from dimensional analysis, is confirmed by numerous measurements and simulations, and it is often regarded, alongside the Kolmogorov theory for isotropic homogeneous turbulence (see Section 2), as one of very few 'fix points' in the subject of turbulence and turbulent flows. Indeed, as will be shown in Chapters 8 and 9, both are central to the calibration of turbulence-model constants.

Chapter Six

6. Fundamentals of Scalar-Flux/ Scalar-Gradient Interactions

In many engineering applications, the primary objective is not to describe flow as such — although this *has* to be determined first, in all circumstances — but rather to quantify the distribution of temperature and, hence, of heat transfer across a boundary, usually a wall. When chemical reactions are involved, as is the case in combustion, the task extends to determining the distribution of chemical species of various kinds.

We know already that the Reynolds-averaged transport equation for an intensive scalar property, Φ, is Eq. (3.12). In the case of heat transfer, this quantity is the enthalpy per unit mass, and this is related to the temperature via $h = c_p T$, where c_p is the specific heat at constant pressure. To solve the transport equation, we need to determine the flux vector $\overline{u_j \phi}$, equivalent to the stress tensor. In the case of buoyant flows, at least one component of this flux vector is a source of turbulence energy (see Eq. (2.23)), and the flux vector also affects the Reynolds-stress components, as will transpire in Section 11.6 and Chapter 12.

A rational basis for determining the flux vector is offered by an exact set of transport equations for the flux components, analogous to the transport equations for the stresses. As in the case of the latter set, the flux equations reveal highly pertinent relationships between the fluxes and the processes driving them, allowing important qualitative lessons to be learnt from them in relation to the likely validity of different modelling proposals. Learning these lessons is the primary purpose of the present chapter.

6.1 *The exact equations for the turbulent fluxes*

We start the derivation of the flux-transport equations with the equation for the instantaneous property denoted by Φ, into which we can insert, trivially, the decompositions:

$$\begin{aligned} \Phi &= \overline{\Phi} + \phi \\ U_j &= \overline{U}_j + u_j. \end{aligned} \tag{6.1}$$

As before, the right-hand side fragments are, respectively, the time-mean and the fluctuating values. In the absence of a source, the result is:

$$\begin{aligned} &\frac{\partial \overline{\Phi}}{\partial t} + \frac{\partial \phi}{\partial t} + \left(\overline{U}_j + u_j\right)\frac{\partial \overline{\Phi}}{\partial x_j} + \left(\overline{U}_j + u_j\right)\frac{\partial \phi}{\partial x_j} \\ &\quad = \frac{\partial}{\partial x_j}\left(\alpha \frac{\partial \overline{\Phi}}{\partial x_j}\right) + \frac{\partial}{\partial x_j}\left(\alpha \frac{\partial \phi}{\partial x_j}\right), \end{aligned} \tag{6.2}$$

In which $\alpha \equiv \Gamma_\Phi/\rho$ (cf. Eq. (3.12) for the mean scalar $\overline{\Phi}$). Next, subtracting Eq. (3.12) from the above gives:

$$\underbrace{\frac{\partial \phi}{\partial t} + \overline{U}_j \frac{\partial \phi}{\partial x_j}}_{D\phi/Dt} = \frac{\partial}{\partial x_j}\left(\alpha \frac{\partial \phi}{\partial x_j}\right) + \frac{\partial}{\partial x_j}\left(\overline{u_j \phi} - u_j \phi\right) - u_j \frac{\partial \overline{\Phi}}{\partial x_j}. \tag{6.3}$$

The scalar-flux equations may now be obtained from the manipulations:

$$\frac{D\overline{u_j \phi}}{Dt} = \overline{u_j \frac{D\phi}{Dt}} + \overline{\phi \frac{Du_j}{Dt}}, \tag{6.4}$$

where D/Dt denotes the substantive derivative identified in Eq. (6.3). The right-hand side of Eq. (6.4) entails a multiplication of Eq. (6.3) by u_j, and the equation governing the velocity fluctuation u_j by ϕ, followed by time averaging the sum of the two. We do not have, at this stage, an equation for Du_j/Dt, but this can easily be derived upon subtracting the Reynolds-averaged Navier–Stokes (RANS) equations from the parent Navier–Stokes (NS) equations, i.e. Eq. (3.7) from Eq. (3.8), with the finite-interval time averaging $(\widetilde{\ldots})$ replaced by conventional, long-time-averaging $(\overline{\ldots})$. The result of this process is:

$$\frac{\partial u_i}{\partial t} + \overline{U}_j \frac{\partial u_i}{\partial x_j} = -\frac{1}{\rho}\frac{\partial p}{\partial x_i} + \frac{\partial}{\partial x_j}\nu\left(\frac{\partial u_i}{\partial x_j} + \frac{\partial u_j}{\partial x_i}\right) + \frac{\partial}{\partial x_j}\left(\overline{u_i u_j} - u_i u_j\right) - u_j \frac{\partial \overline{U}_i}{\partial x_j}. \tag{6.5}$$

Straightforward manipulations in Eq. (6.4) then yield:

$$\underbrace{\frac{\partial \overline{u_i\phi}}{\partial t} + \overline{U}_j \frac{\partial \overline{u_i\phi}}{\partial x_j}}_{\text{Convection}} = \underbrace{-\overline{u_j\phi}\frac{\partial \overline{U}_i}{\partial x_j} - \overline{u_i u_j}\frac{\partial \overline{\Phi}}{\partial x_j}}_{\text{Production}} + \underbrace{\overline{\frac{p}{\rho}\frac{\partial \phi}{\partial x_i}}}_{\text{Pressure-scalar interaction}} \underbrace{-(\nu+\alpha)\overline{\frac{\partial \phi}{\partial x_j}\frac{\partial u_i}{\partial x_j}}}_{\text{Dissipation}} \underbrace{-\frac{\partial}{\partial x_j}\left(\overline{u_i u_j \phi} + \frac{\overline{p\phi}}{\rho}\delta_{ij} + \alpha\overline{u_i\frac{\partial \phi}{\partial x_j}} + \nu\overline{\phi\frac{\partial u_i}{\partial x_j}}\right)}_{\text{Diffusion}}, \tag{6.6}$$

where δ_{ij} is the unit tensor (equivalent to the unit matrix $\{\mathbf{I}\}$), with diagonal terms '1' and all other terms '0'. This equation (with some important qualifications) is analogous to Eq. (4.5) for the shear stress, except that it governs the three components of the flux vector, namely $\overline{u_j\phi} = \{\overline{u\phi}, \overline{v\phi}, \overline{w\phi}\}$.

6.2 *Some key interactions*

As in the case of the stresses in the RANS equations, the terms in Eq. (6.6) may be interpreted as representing distinct physical processes — some quite obvious, if only by analogy, to the terms in the equations for the Reynolds stresses and the turbulence energy. The analogy is not entirely valid, however, and this calls for care in the interpretation. For example, the pressure-scalar-interaction term cannot be referred to as 'redistributive' in the sense of the pressure-strain process in the Reynolds-stress equations (4.6), if only because the flux components do not correspond to the normal Reynolds stresses, and there is no equation equivalent to Eq. (4.8) for the turbulence energy.[1] Also, it is not obvious that the term identified as 'dissipation' is a sink, i.e. that it always has a negative value. The last term represents, unequivocally, diffusion. This is easily recognised upon assuming $\alpha = \nu$, applicable when the Prandtl–Schmidt number — the ratio of the viscosity to the diffusivity — is unity, in which case the last two terms combine to become:

$$\alpha \overline{u_i \frac{\partial \phi}{\partial x_j}} + \nu \overline{\phi \frac{\partial u_i}{\partial x_j}} = \nu \frac{\partial \overline{u_i \phi}}{\partial x_j}, \tag{6.7}$$

which is obviously a term representing viscous diffusion of the fluxes.

There are reasons for expecting the terms representing production to be especially influential — although it has to be said, here again, that these terms are neither obviously nor necessarily always of a sign that results in an increase in the magnitude of the fluxes. Notwithstanding this question mark, the form of these production terms leads to some interesting conclusions.

We observe, first, that any flux component is linked to all other flux components, to all Reynolds-stress components, to the mean scalar gradients and to the strain components. Hence, if the scalar under consideration is the enthalpy (or temperature), we cannot expect the flux in any one direction to be driven (solely) by the temperature gradient in that direction, as we might expect from ordinary conduction considerations in solids and laminar flow. Indeed, somewhat counter-intuitively, we must expect the fluxes to be partly driven by the velocity gradients.

An instructive example to consider is a two-dimensional, thin, horizontal mixing/thermal layer, $\overline{U}(y)$, $\overline{T}(y)$, in which the only significant non-zero mean-property gradients are $\partial\overline{U}/\partial y$ and $\partial\overline{T}/\partial y$.

[1] It is easily possible, however, to derive an equation for the variance of the temperature fluctuations, $\overline{\phi^2}$, which is the closest equivalence to k.

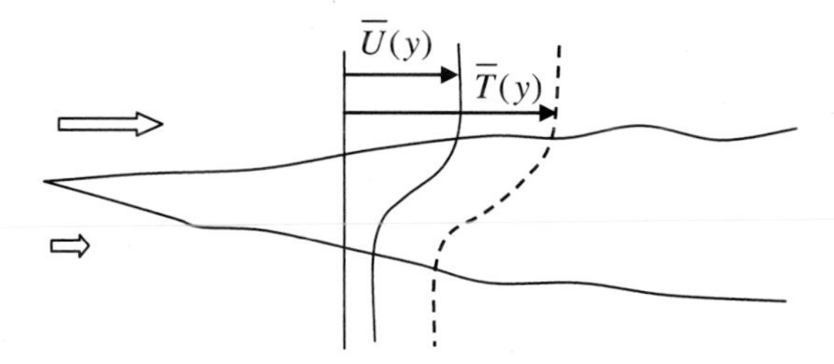

Fig. 6.1: Heated mixing layer (relating to Eq. (6.8)).

In this case, shown in Fig. 6.1, the productions of the two relevant heat fluxes reduce to:

$$
\begin{aligned}
P_{\overline{vT'}} &= -\overline{v^2}\frac{\partial \overline{T}}{\partial y} \\
P_{\overline{uT'}} &= -\overline{vT'}\frac{\partial \overline{U}}{\partial y} - \overline{uv}\frac{\partial \overline{T}}{\partial y}.
\end{aligned}
\tag{6.8}
$$

As expected, the cross-flow flux is driven by the cross-flow temperature gradient. However, an entirely counter-intuitive result is that the streamwise flux, which we may expect to be negligible, may in fact be considerably larger than the cross-flow flux, as it is driven by both the vertical temperature *and* velocity gradients. However, because this flow evolves only slowly in the streamwise direction, the streamwise *gradient* of this flux tends to be quite small, and this is the reason why this flux does not contribute materially to the behaviour of the temperature, as determined from the solution of mean-scalar-transport Equation (3.12).

It is instructive to give quantitative evidence for some of the qualitative arguments made above. To this end, the results of direct numerical simulation (DNS) heat-transfer simulations of Kawamura, Abe and Matsuo (1999) for channel flow are included in Figs. 6.2 and 6.3. The Prandtl number, 0.71, is that for air, while lower values relate to liquid metals (mercury, for instance, has a value of 0.015). The higher Reynolds-number ($Re_\tau = 395$), corresponds to a bulk Reynolds-number of around 7,000 (based on the channel half height). Heating was imposed by prescribing a heat flux at the wall. In Fig. 6.2, the heat flux components normal to the wall and in the

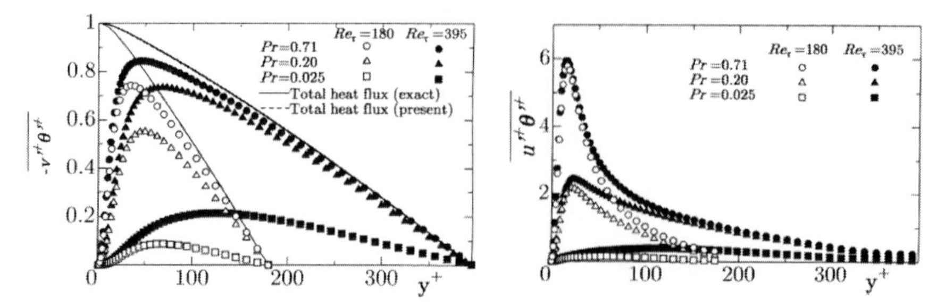

Fig. 6.2: Wall-normal and streamwise heat-flux components in channel flow at different Reynolds-numbers and Prandtl numbers; open circles relate to air at Prandtl number of 0.71; derived from DNS data (reproduced with permission of ASME from Kawamura, Abe and Matsuo (1999)).

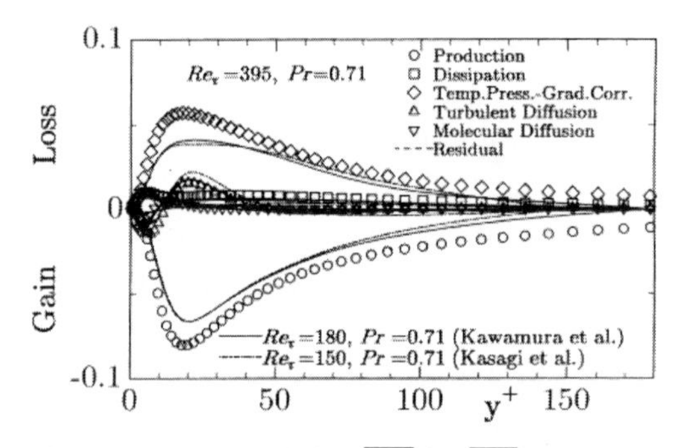

Fig. 6.3: Budgets for the wall-normal flux $\overline{v'\theta'}$ (or $\overline{vT'}$ in the present terminology) in channel flow, with walls heated at constant total heat flux; derived from DNS data (reproduced with permission of ASME from Kawamura, Abe and Matsuo (1999)).

streamwise direction are denoted by $\overline{v'\theta'}$ and $\overline{u'\theta'}$, respectively, with θ' being the temperature fluctuation. The wall-normalisation is such that the total wall-normal heat flux becomes '1' at the wall, in the same way as the total shear stress divided by the wall shear-stress becomes '1'. The two main features to highlight are, first, that the streamwise flux is substantially larger than the wall-normal flux; and second, that the maximum fluxes occur close to the region in which the maximum of turbulence energy occurs.

Figure 6.3 shows the contribution of different physical processes corresponding to terms in Eq. (6.6) for the wall-normal flux $\overline{vT'}$.

The dominant term is the production, which reaches a maximum at around $y^+ = 18$, corresponding to the thickness of the thermal viscous layer.[2] Reference to the production term in Eq. (6.6) shows that the only active term (with $i = 2$) is $-\overline{v^2}\frac{\partial \overline{T}}{\partial y}$, in agreement with the first equation in the set of Eq. (6.8). This is the term equivalent to $-\overline{v^2}\frac{\partial \overline{U}}{\partial y}$, which generates the shear stress (see first equation in the set of Eq. (4.6) for the shear stress, and Fig. 4.5). It is not surprising, therefore, that an *analogy* is observed between the shear stress and the wall-normal heat flux in simple shear flow.

As noted already, there is no such obvious analogy in the case of $\overline{uT'}$, to which Fig. 6.4 relates. This figure shows that $\overline{uT'}$ is driven by a production term about 5 times higher than $-\overline{v^2}\frac{\partial \overline{U}}{\partial y}$, reflecting the fact that $\overline{uv}$ is much larger than $\overline{v^2}$ in the first term of the production $P_{\overline{uT'}}$ of Eq. (6.8), and the additional influence of the strain in the second term of $P_{\overline{uT'}}$. Interestingly, the budget in Fig. 6.4 bears a fairly close resemblance to the turbulence-energy budget in Fig. 5.4. In both cases, the production is balanced mainly by dissipation.

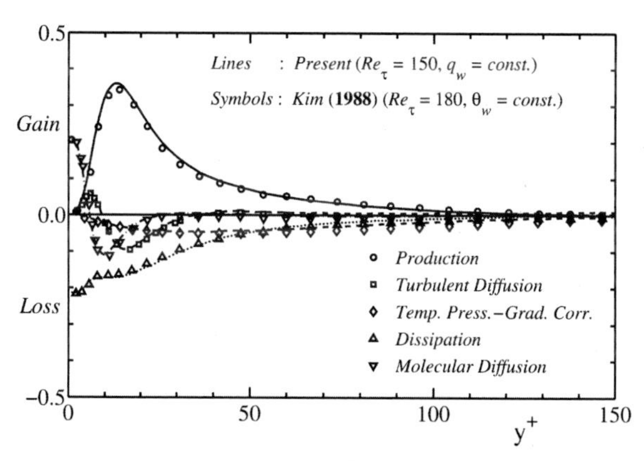

Fig. 6.4: Budgets for the streamwise flux $\overline{uT'}$ in channel flow, with walls heated at constant total heat flux; derived from DNS data (reproduced with permission of ASME from Kasagi, Tomita and Kuroda (1992)).

[2] The dynamic viscous layer is within $y^+ = 11$, but the Prandtl number of 0.71 leads to a thermal diffusivity which is larger than the momentum diffusivity (i.e. the viscosity). Hence, the thermal viscous sublayer is roughly 1.4 times thicker than the corresponding velocity layer.

As the turbulence-energy budget is dominated by contributions from the $\overline{u^2}$-transport equation, the equivalence is, in fact, between $\overline{uT'}$ and $\overline{u^2}$, and this reflects a high correlation between the fluctuations u and T'.

6.3 *Summary and lessons*

Yet again, we are faced with the importance of stress anisotropy — here, in dictating the transport of any scalar flux across a shear layer, via the cross-flow normal stress appearing as a multiplier in the production of the cross-flow heat flux. Another important conclusion emerging from the exact flux equations is that the assumption of a Fourier–Fick law (i.e. a diffusity $\times$ scalar-gradient law) for the turbulent scalar fluxes lacks a physical foundation. Rather, any heat-flux component is driven, in general, by scalar gradients in all directions, as well as the strain rates. Thus, in the particular case of a shear layer, with the streamwise scalar gradient being zero, the streamwise flux is driven by both the cross-flow scalar gradient and the shear strain, yielding, counter-intuitively, a streamwise flux that is much larger than the cross-flow component. This is not practically important in thin shear layers, in which the streamwise gradient of the streamwise flux is small, but can be extremely influential in more complex conditions — for example, in recirculating flows.

Chapter Seven

7. The Eddy Viscosity

7.1 *Conceptual foundation*

In this chapter, we turn to the key question of how to determine the unknown turbulence correlations — the Reynolds stresses — in the Reynolds-averaged Navier–Stokes (RANS) equations, i.e. Eq. (3.9). The discerning reader may well already recognise that the most rational and straightforward approach to this question would be to attempt to solve more general forms of equations of the type Eq. (4.6), which govern the evolution of the Reynolds stresses. Indeed, this is a possible route that will be discussed in detail in Chapter 12. However, it is also a highly challenging route, because many terms in these equations are unknown, and their modelling is very difficult. Even after closure is achieved, there remains the far from trivial task of solving, numerically, the highly coupled, non-linear set of equations for the Reynolds stresses — a task that is computationally often more challenging than solving the RANS equations themselves. This is certainly not the route that was adopted, therefore, in early efforts to model the correlations and solve the RANS equations numerically or analytically — the latter only possible for a few especially simple shear flows.

An extremely influential idea, introduced by Joseph Valentin Boussinesq in *Essai sur la théorie des eaux courantes* (1877), is the eddy-viscosity concept. This is based on the notion that the action of turbulence is analogous to the Brownian motion that is responsible for fluid viscosity — an analogy that is, ultimately, flawed, for it overlooks the spatial coherence of turbulent structures (two-point correlations and vortical motions, in particular). Thus, the proposal simply states that, similar to fluid viscosity in laminar flows, a flow-properties-dependent turbulent viscosity may be added to the molecular agitation to represent turbulent mixing or diffusion. In the case of the shear-stress in a simple shear layer, this may simply be expressed as:

$$\mu \frac{d\overline{U}}{dy} - \rho\overline{uv} = (\mu + \mu_t) \frac{d\overline{U}}{dy}, \tag{7.1}$$

where μ_t is some unknown function of the flow properties, and hence is dependent on the spatial location at which the turbulent stress is to be determined. This proposal thus expresses the concept of (turbulent) diffusion being driven by the gradient of the property being diffused.

Even if the shear stress is the only stress sought, substitution (7.1) into the relevant RANS equation does not, in itself, constitute a 'closure', because the task of determining the shear stress is simply pushed sideways to one of determining the unknown eddy viscosity. Despite this, the proposal, when extended to other stress components, does allow a substantial simplification (albeit at a cost in fidelity) of the task of determining the set of active Reynolds stresses in more complex circumstances than in the case of a simple shear layer.

Conceptual support for Eq. (7.1) is offered by the following slight rearrangement of Eq. (3.9):

$$\frac{\partial \rho \overline{U}_i \overline{U}_j}{\partial x_j} = -\frac{\partial \overline{P}}{\partial x_i} + \frac{\partial}{\partial x_j} \left\{ \mu \left(\frac{\partial \overline{U}_i}{\partial x_j} + \frac{\partial \overline{U}_j}{\partial x_i} \right) - \rho \overline{u_i u_j} \right\}. \tag{7.2}$$

This rearrangement makes clear, if only on dimensional grounds, that the Reynolds stresses play a similar role to that of the viscous stresses. Hence, if this analogy is accepted, it is reasonable to suppose that Eq. (7.1) could be generalised as[1]:

$$-\rho\overline{u_i u_j} = \mu_t\left(\frac{\partial\overline{U}_i}{\partial x_j} + \frac{\partial\overline{U}_j}{\partial x_i}\right). \tag{7.3}$$

However, there is an important and fundamental difference between Eqs. (7.1) and (7.3): while the former can be satisfied exactly by an inspired choice of μ_t, so as to fit the required variation of $\overline{uv}$, a single function for the scalar μ_t *cannot* possibly be derived such that it satisfies all components $\overline{u_i u_j}$. Hence, no matter how μ_t is determined, Eq. (7.3) will necessarily be an approximation — and more likely than not, a crude one.

Regardless of how μ_t is determined, a correction is required to Eq. (7.3), which is not initially obvious from the generalisation of Eq. (7.1). Clearly, in the case of strain-free flow, Eq. (7.3) returns a zero value for all stress components. However, this is not a physically correct result — for example, in the case of a uniform (strain-free) flow that contains turbulence as a consequence of upstream agitation and subsequent convection, as shown in Fig. 7.1. In this case, the wakes of the mesh wires cause strong local shear and separation, and thus generate turbulence, which is then convected forward in the wind tunnel. While the turbulence intensity declines as a consequence

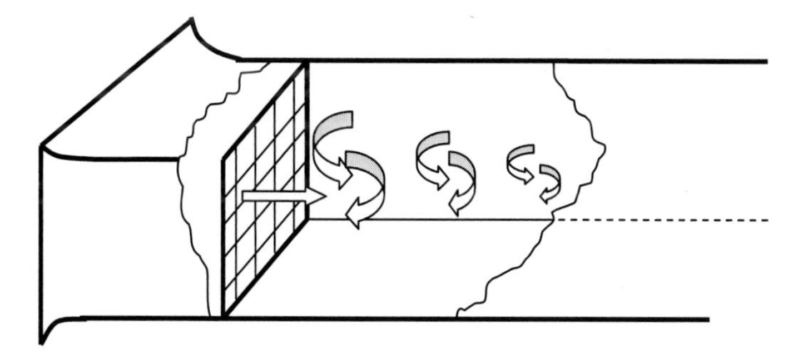

Fig. 7.1: Decaying turbulence in an empty wind tunnel with an upstream mesh.

[1]We shall demonstrate shortly that this form violates a constraint, which then requires the augmentation of the equation by a term containing the turbulence energy.

of viscous dissipation, this process is relatively slow, and the flow thus contains turbulence energy. But this energy is, by definition:

$$k = \frac{1}{2}(\overline{u_1^2} + \overline{u_2^2} + \overline{u_3^2}), \tag{7.4}$$

which thus clearly contradicts Eq. (7.3). To re-establish compatibility, Eq. (7.3) is augmented as follows:

$$-\rho\overline{u_i u_j} = \mu_t \left(\frac{\partial \overline{U}_i}{\partial x_j} + \frac{\partial \overline{U}_j}{\partial x_i} \right) - \frac{2}{3}\delta_{ij}\rho k. \tag{7.5}$$

This augmentation has two consequences. First, it implies that, in the absence of normal straining (i.e. $\partial\overline{U}/\partial x = \partial\overline{V}/\partial y = \partial\overline{W}/\partial z = 0$), all the normal-stress components are the same:

$$\overline{u_1^2} = \overline{u_2^2} = \overline{u_3^2} = \frac{2}{3}k. \tag{7.6}$$

This is a reasonable proposition in the situation depicted in Fig. 7.1 — at least far downstream, because the absence of forcing (straining) allows turbulence to 'relax' towards a state of isotropy. Second, Eq. (7.5) implies that the eddy-viscosity terms only account for *deviations* of the normal stresses from their average $2/3k$, i.e. for the *deviatoric* parts of the stresses. Because the turbulence energy cannot be purely determined from local conditions — Fig. 7.1 being an extreme example — the approximation in Eq. (7.5) implies that the normal stresses will account, at least in some measure, for non-local processes. Indeed, the exact equation for the turbulence energy, Eq. (4.8), demonstrates this non-locality, by virtue of the convection and diffusion terms, in particular.

One important limitation of the proposal in Eq. (7.5) is that it also returns a state of normal-stress isotropy in pure shear strain — a condition representative of the state in a jet, a free shear layer or a boundary layer, for example. However, as demonstrated previously (cf. Figs. 4.1–4.4, 5.3), and verified by many experimental observations, the normal stresses differ greatly in any shear flow, as was discussed in detail in Chapters 4 and 5. In fact, models based on the linear eddy-viscosity concept (Eq. (7.5)) can never be relied upon to return accurate, or even realistic, values for the normal stresses in shear flows.

Yet another (potential) problem with the proposal in Eq. (7.5) is that it can generate unphysical stresses at high strain rates. For example, if a flow accelerates rapidly in the x_1-direction, the streamwise-strain rate may be high enough for $\mu_t \frac{\partial \overline{U}_1}{\partial x_1}$ to exceed $2/3k$, in which case $\overline{u_1^2}$ would become negative. This is an example of the violation of the general principle of 'realisability', and is one of a range of physical and predictive frailties rooted in Eq. (7.5), which are discussed in Chapter 11, as a precursor to more advanced modelling concepts. Suffice it to point out here, by reference to the transport equation for the shear stress, Eq. (4.5), that the generation term in this equation exposes the fact that the shear stress is 'driven' by a range of interactions among stresses and strains, including the (anisotropic) normal stresses, some of which have already been highlighted in Chapter 4. More generally, any one stress is the result of interactions involving all strains and stresses. Hence, proposal Eq. (7.5) is a highly simplified approximation that harbours problems in complex strain fields.

A notional justification for the proposal in Eq. (7.1), additional to that expressed by Eq. (7.2), can be derived from the first equation in the set in Eq. (4.6), which governs the shear stress in a simple shear flow. We note first that the production term of this stress is:

$$P_{\overline{uv}} = -\overline{v^2}\frac{\partial \overline{U}}{\partial y}. \tag{7.7}$$

Next, it is reasonable to assume that the shear stress will, to a first order, increase or decrease in tandem with its production. If we assume that the relationship is linear, and observe that dimensional consistency requires the multiplier to be a time scale, say τ, which we expect to be related to the larger eddies causing the shear stress, then we arrive at the proposal:

$$-\overline{uv} = (C\tau\overline{v^2})\frac{\partial \overline{U}}{\partial y}. \tag{7.8}$$

Given that the bracketed term can be interpreted as an eddy viscosity, Eq. (7.8) is compatible with the proposal in Eq. (7.1). Indeed, it is physically realistic to assume that the cross-flow fluctuations would be the primary cause of cross-flow mixing, in the same way as the eddy viscosity is a surrogate for the mixing effectiveness of turbulence. Eq. (7.8) is, therefore, essentially a primitive 'model' for the shear stress, but its usefulness (even as such a primitive model) hinges on our ability to determine or prescribe the quantities in the bracket. In addition, it is emphatically *not* a tenable model for the normal stresses, if only because Eq. (4.6) shows the generation terms for the normal stresses to be zero for simple shear.

7.2 *Quantification of the eddy viscosity — a first attempt*

Equation (7.8), although not being a closure, shows that the dimensions of the (kinematic) turbulent viscosity are:

$$[\nu_t] = \left[T\frac{L^2}{T^2}\right] = \left[L\frac{L}{T}\right]. \tag{7.9}$$

In other words, the viscosity is linearly related to a length scale and a velocity scale of turbulence via:

$$\nu_t = \ell v, \tag{7.10}$$

most likely associated with the large eddies that dominate the turbulent mixing process.[2] The crucial question is whether we can find relationships that link the length and velocity scales to the mean flow — thus achieving closure. We shall consider this question first by reference to a set of physical arguments that rely on the interpretation of $\rho\overline{uv}$ as cross-flow momentum transport.

The starting point for the considerations to follow is the recognition that the shear stress $-\overline{uv}$ is responsible for the enhanced mixing of momentum in a shear layer due to turbulent motions. In this context, mixing is, essentially, the exchange of streamwise momentum across the shear layer by cross-flow fluctuations (as is, for example, implied by Eq. (7.8)). This process is illustrated in Fig. 7.2.

[2] The absence of a proportionality factor in Eq. (7.10) is of no consequence, as this can be accommodated as a part of either scale, or both, on the right-hand side of the equation.

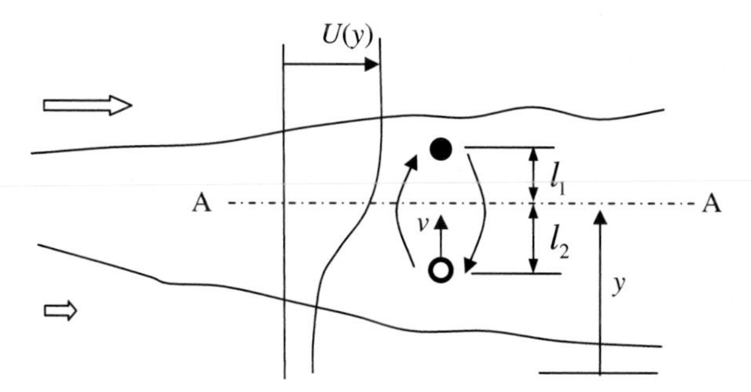

Fig. 7.2: Illustration of mixing across a shear layer by exchange of mass packets across a shear region, relating to Eq. (7.12).

The net cross-flow flux of streamwise momentum across line A–A may be written:

$$F_y = \rho v \left(\overline{U}(y - l_2) - \overline{U}(y + l_1)\right). \tag{7.11}$$

A Taylor-series expansion of $\overline{U}$ and the omission of terms of order 2 and higher gives:

$$F_y \approx -\rho v (l_1 + l_2) \frac{\partial \overline{U}}{\partial y}. \tag{7.12}$$

Assuming approximate isotropy in the turbulent motion, $|v| = |u|$ (a rather poor approximation close to a wall, as conveyed by Fig. 5.3), and noting that:

$$|u| = O\left|\overline{U}(y - l_2) - \overline{U}(y + l_1)\right|, \tag{7.13}$$

followed by Taylor-series expansions of the right-hand-side terms and truncation after the linear terms of the expansions, allows Eq. (7.12) to be written as:

$$F_y \approx \left\{(l_1 + l_2)^2 \frac{\partial \overline{U}}{\partial y}\right\} \frac{\partial \overline{U}}{\partial y}. \tag{7.14}$$

Without loss of generality, we may replace $\ell \leftarrow l_1 + l_2$. Also, the cross-flow flux of momentum is $-\rho\overline{uv}$, and hence:

$$-\rho\overline{uv} = \underbrace{\left\{\rho \ell^2 \frac{\partial \overline{U}}{\partial y}\right\}}_{\mu_t} \frac{\partial \overline{U}}{\partial y}. \tag{7.15}$$

The terms in the curly bracket can, self-evidently, be interpreted as a turbulent viscosity, in which $\ell \frac{\partial \overline{U}}{\partial y}$ represents the turbulent-velocity scale. The equals sign in Eq. (7.15) reflects the fact that ℓ remains to be determined and can be adjusted so as to procure equality. In fact, Eq. (7.15), with an 'appropriately chosen' ℓ, constitutes the 'mixing-length model' of Prandtl (1925) and von Kármán (1930) — the first turbulence model encountered in this text. As to an 'appropriate choice' of ℓ, we note again that this length scale must be expected to represent the energetic, larger-scale eddies in the shear layer, for these are the strongest mixing agents. These eddies can also be expected to be of the order of the thickness of the shear layer, say δ, i.e.:

$$\ell = C_\delta \delta \quad C_\delta = O(0.1 - 1). \tag{7.16}$$

Indeed, experience derived from computational studies shows that, while C_δ varies from one flow type to another, it is in the range of 0.07δ (for a mixing layer) to 0.16δ (for a plane wake), where δ is defined by reference to the locations at which the velocity defect or excess in the shear layer — relative to the outer stream closest to the location in question — reaches 10% of the maximum velocity difference across the shear layer.

A near-wall boundary layer is an interesting special case to which Eq. (7.16) does not apply in the form given, except in the outer part of the layer. A characteristic feature of near-wall turbulence is that the large, energetic eddies, represented by ℓ, progressively decrease in size as the wall is approached (cf. Fig. 5.1). In addition, the length scale associated with the eddy size normal to the wall — the one of relevance here — tends to decrease disproportionately, relative to the overall size, as eddies are deformed into an anisotropic 'pancake'-like shape parallel to the wall, due to the constraining influence of the wall on wall-normal motions. The simplest assumption, corresponding to Eq. (7.16), which reflects this behaviour, is:

$$\ell = C_w y. \tag{7.17}$$

Insertion of Eq. (7.17) into Eq. (7.15) then gives:

$$-\rho\overline{uv} = \left\{ \rho C_w^2 y^2 \frac{\partial \overline{U}}{\partial y} \right\} \frac{\partial \overline{U}}{\partial y}. \tag{7.18}$$

In the case of a zero-pressure-gradient boundary layer, order-of-magnitude considerations readily show that, not too far away from the wall, convection of streamwise momentum is small, relative to wall-normal diffusion,[3] in which case the momentum equation (3.9) (for $i = 1$) reduces to[4]:

$$\frac{\partial}{\partial y}\left(\mu\frac{\partial \overline{U}}{\partial y} - \rho\overline{uv}\right) = 0. \tag{7.19}$$

Integrating this from the wall to any location y yields:

$$\mu\frac{\partial \overline{U}}{\partial y} - \rho\overline{uv} = \tau_w, \tag{7.20}$$

where τ_w is the wall shear stress. Outside the thin viscous sublayer, the turbulent shear stress dominates, so that we get, using Eq. (7.18):

$$-\rho\overline{uv} \approx \tau_w \approx \left\{\rho C_w^2 y^2 \frac{\partial \overline{U}}{\partial y}\right\}\frac{\partial \overline{U}}{\partial y}. \tag{7.21}$$

This is readily integrated between the edge of the viscous sublayer, y_v, to any location above it, y, to give:

$$\overline{U} - \overline{U}_v = \frac{\sqrt{\frac{\tau_w}{\rho}}}{C_w}\ln\frac{y}{y_v}, \tag{7.22}$$

or, with $u_\tau = \sqrt{\tau_w/\rho}$, $y^+ = yu_\tau/\nu$, $\overline{U}^+ = \overline{U}/u_\tau$:

$$\overline{U}^+ = \frac{1}{C_w}\ln y^+ + \left\{\overline{U}_v^+ - \frac{1}{C_w}\ln y_v^+\right\}. \tag{7.23}$$

This is compatible with Eq. (5.23), derived by way of dimensional considerations, and is also confirmed by the log-law plots in Figs. 4.2(a) and 5.5. Hence, in combination with Eq. (7.17), this is also an *a-posteriori* justification for the validity of the mixing-length hypothesis. As already indicated towards the end of Chapter 5, experiments show that C_w should have the value of the von Kármán

[3]This is not the case close to the edge of the boundary layer, where streamwise convection is balanced by turbulent mixing.

[4]As we neglect any variations in x and z, the partial derivative in y is identical to the ordinary or total derivative.

constant $\kappa = 0.4$ (see Eq. (5.23)). Substitution of this value and $\overline{U}_v^+ = y^+ = 11.2$ — which applies to the nominal edge of the viscous sublayer (see Eq. (5.17)) — into Eq. (7.23) gives:

$$\overline{U}^+ = \frac{1}{0.4} \ln y^+ + 5.2, \tag{7.24}$$

which is close to Eq. (5.23).

It needs to be emphasized that we have excluded the viscous sublayer from the integration leading to Eq. (7.22) on two grounds. First, we have neglected the viscous stress in Eq. (7.21). Second, in the viscosity-affected region, the linear assumption of Eq. (7.17) is incorrect. This is easily shown by reference to Eq. (7.15): because the shear stress varies as $\overline{uv} = O(y^3)$ close to the wall (see Eq. (5.3)), and the shear strain varies as $d\overline{U}/dy = O(1)$, the length scale has to vary as $\ell = O(y^{1.5})$ — thus requiring the eddy viscosity to vary as $\mu_t = O(y^3)$, i.e. the length scale has to decline faster than the linear assumption.

Notwithstanding these qualifications and limitations, the above discussion conveys the essential elements of a first, broadly credible, turbulence-modelling approach. While it is restricted to the narrow category of thin shear flows, it does give a flavour of what is meant by 'closure' in the present context. Furthermore, it is historically interesting, and has allowed some important physical and conceptual issues to be explored. Several model variants that are closely related to the original mixing-length model were formulated as late as the 1980s, specifically in order to analyse — quickly and economically — relatively simple boundary layers in external aerodynamic applications (cf. Cebeci & Smith (1974), Baldwin & Lomax (1978) and Granville (1987)). However, these models are now effectively relegated to history, in the wake of rapid advances in computer technology and CFD, and the formulation of more advanced transport models of turbulence.

We end this section by returning to its starting point. The most important fact to take forward is the dimensional relationship — Eqs. (7.9) and (7.10). In this section, we derived, largely by qualitative arguments, a relationship for the velocity scale in terms of a mixing length and the velocity gradient. For very simple shear layers, this might suffice, given a judicious choice of ℓ. However, we wish to put our model on much firmer rational ground, and also make it applicable to more complex flows. A key weakness of algebraic prescriptions of the type in Eq. (7.15) is their failure to account for the fact that turbulence properties are *transported*, as was amply demonstrated in Chapter 4, and that turbulence cannot, in general, be described purely by local relationships. As a case in point, we recall the wind-tunnel flow shown in Fig. 7.1: here, the only important processes are forward convective transport of turbulence energy and the gradual destruction by viscous dissipation. Another example is the centre-line region of the round jet, Fig. 4.4, at which the elevated turbulence energy is due to convection from upstream regions and (mostly) cross-flow diffusion from peripheral high-strain regions towards the centre-line. While in neither case is the shear-stress relevant — in fact, is it zero both in the tunnel and at the centre-line of the jet — the turbulent viscosity must be substantial and cannot possibly be described by the local shear strain (which vanishes). We will, therefore, consider more realistic approaches in the following sections.

7.3 *The turbulent-velocity scale*

The very concept of a turbulent viscosity implies that the velocity and length scales in Eq. (7.10) will represent the intense mixing of the kind that is associated with the large and most energetic eddies. A reasonable velocity scale that meets this association is the root of the turbulence energy,

$$v = k^{0.5}. \tag{7.25}$$

This assumption may be justified by reference to Fig. 2.4, which shows most of the turbulence energy to be associated with the low-wave-number range. This is also an assumption that underpins the large majority of turbulence models that go beyond the algebraic mixing-length concept. The closure problem now translates to determining k and ℓ.

In principle, a route to determining k is offered by the turbulence-energy-transport equation of the type given by Eq. (4.8). The problem with Eq. (4.8) itself is that it is only valid for thin shear flows. To be applicable to more complex flows, we need to derive a more general form. This can be done along a route analogous to that which we followed via Eqs. (4.1)–(4.5).

We return first to Eq. (4.1), which is the 'recipe' for obtaining the set of equations governing the components of the Reynolds-stress tensor. All we need to do is to 'contract' Eq. (4.1), by setting the indices $i = j$, and invoke Einstein's summation convention[5]:

$$\sum_{i=1}^{3} \overline{u_i NS(U_i + u_i) + u_i NS(U_i + u_i)} = 0. \tag{7.26}$$

The result of (7.26) is an equation for the sum $\overline{u^2} + \overline{v^2} + \overline{w^2} = 2k$, so we merely need to divide the result by 2 to obtain the general equation for k. The outcome of this straightforward set of manipulations (with body-force effects ignored at this stage) is[6]:

$$\underbrace{\frac{Dk}{Dt}}_{\text{Convection} — C_k} = -\underbrace{\overline{u_i u_j}\frac{\partial U_i}{\partial x_j}}_{\text{Generation} — P_k} - \underbrace{\frac{\partial}{\partial x_j}\left(\overline{u_j k'} + \frac{2\overline{p u_i}}{\rho}\delta_{ij}\right)}_{\text{Turbulent transport ('diffusion')} — T_k}$$

$$+ \underbrace{\frac{\partial}{\partial x_j}\left(\frac{\mu}{\rho}\frac{\partial k}{\partial x_j}\right)}_{\text{Viscous diffusion} — D_k} \underbrace{- \frac{\mu}{\rho}\left(\overline{\frac{\partial u_i}{\partial x_j}\frac{\partial u_i}{\partial x_j}}\right)}_{\text{Dissipation} — \varepsilon}, \tag{7.27}$$

where k' is the turbulent fluctuation of k. This equation is very similar to Eq. (4.8), and simply contains all contributions associated with all possible strains. Note also that all terms involve repeated i and j indices, so that all expand to scalar sums. Indeed, the last term, representing dissipation, expands exactly to the corresponding sum in Eq. (4.8).

---

[5]As already explained in relation to Eq. (4.1), $NS(U)$ is to be understood as $\frac{DU}{Dt} - F(U) = 0$, where F contains all terms other than advection.

[6]The bars over U are omitted henceforth, as there is no ambiguity in the meaning of U.

It is instructive to identify, at this point, some particular circumstances, or flows, in which only some of the terms in Eq. (7.27) are relevant, and for which the equation then reduces to particular simplified forms. As will emerge in Chapters 8 and 9, some of these specific conditions are useful, in so far as they facilitate the calibration of closure approximations that involve constants determinable from simple laboratory experiments. We consider the following conditions.

- In decaying turbulence, and in the absence of straining, as in the example shown in Fig. 7.1, or in a turbulent strain-free stream bordering a shear layer, $C_k \simeq -\varepsilon$ (with minor streamwise diffusion neglected). Experiments show that the turbulence energy follows $k \propto x^{-n}$ where $n \approx 1.2$ in the initial fully turbulent regime, and $n \approx 2.5$ in the final stage of decay, when the flow approaches a laminar state.

- Close to planes or axes of symmetry of a jet or a wake, $C_k \simeq -\varepsilon - T_k$, with turbulent transport effective predominantly normal to the flow, towards the centre.

- When a flow is suddenly subjected to 'rapid distortion', i.e. in strong straining, $C_k \simeq P_k$, in which case the response of the dissipation is inhibited by the disparity in the time scales associated with the straining (the inverse of the strain) and with the cascade process, which channels turbulence energy towards dissipation.

- In slowly evolving thin shear flows, especially in a boundary layer close to the wall, but above the viscous sublayer, $P_k \simeq \varepsilon$, a condition called 'local turbulence-energy equilibrium'.

- In a fully developed duct or pipe flow, the region at the centre-line or centre-plane is strain-free, and there is no convection. Hence, $D_k \simeq \varepsilon$.

Returning to Eq. (7.27), we note that this equation cannot be solved in the form given, because it contains several unknowns. It is at this stage that we need to embark on what is generally understood to be *turbulence modelling.*

One step we can take ahead of modelling (but only within the linear eddy-viscosity framework) is to insert proposal (7.5) into Eq. (7.27), thus yielding:

$$-\overline{u_i u_j}\frac{\partial U_i}{\partial x_j} = \nu_t\left(\frac{\partial U_i}{\partial x_j} + \frac{\partial U_j}{\partial x_i}\right)\frac{\partial U_i}{\partial x_j} - \frac{2}{3}\frac{\partial U_i}{\partial x_j}\delta_{ij}k. \tag{7.28}$$

However, done so, the last term in Eq. (7.28) vanishes in constant-density flow, because it expands to the sum $\frac{2}{3}k\left(\frac{\partial U_1}{\partial x_1} + \frac{\partial U_2}{\partial x_2} + \frac{\partial U_3}{\partial x_3}\right)$. A further point to note ahead of modelling is that the viscous-diffusion term requires no approximation, because the only unknown in that term is k itself.

Next, we turn our attention to the 'turbulent-transport' term (with the pressure-fluctuation-containing fragment put aside). It is instructive to focus, first, on the physical role of the correlation $\overline{u_j k'}$.[7] This may be interpreted as a 'mixing' process in which k is being redistributed by turbulent-velocity fluctuations in a way that is closely akin to that shown in Fig. 7.2 for the mixing of streamwise momentum. Figure 7.3 illustrates this mixing, with reference to an exchange of fluid packets by velocity fluctuations v across a non-uniform profile $k(y)$. This exchange, with zero net mass flux in the y direction, tends to depress k in the upper part of the profile and elevate k in its lower part, thus increasing the uniformity of the profile. Figure 7.3

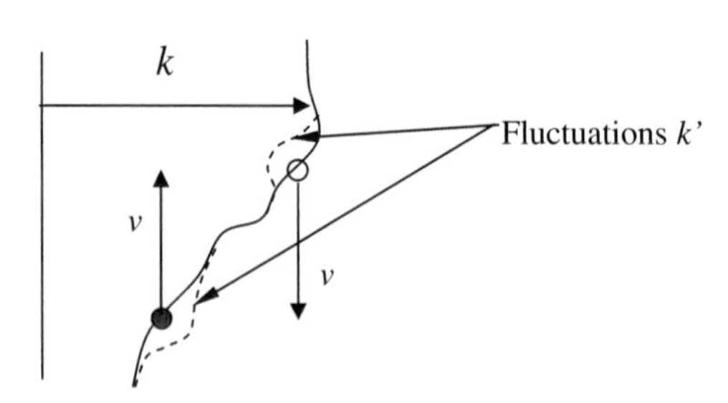

Fig. 7.3: Illustration of the (diffusive) mixing of turbulence energy by the action of velocity fluctuations.

[7]Note again that k' is the instantaneous value of the turbulence energy, different from the time-averaged k.

thus implies that positive v fluctuations are correlated with negative k' fluctuations, and vice versa. Hence, in the present case of positive k-gradient, the correlation $\overline{vk'}$ is predominantly negative. Similarly, the correlation is positive when the gradient of k is negative. By analogy with conductive heat transfer, and also with Eq. (7.1) for the correlation $-\overline{uv}$ (which is identical to $-\overline{vu}$), this suggests the following relationship:

$$-\overline{vk'} = \Gamma_k \frac{\partial k}{\partial y}, \tag{7.29}$$

where Γ_k is an appropriately chosen turbulent diffusivity of the property k, the nature of which will be discussed in Chapter 8, but which we expect to be closely related to ν_t, because all mean properties are mixed by the same large-scale motions as those mixing momentum.

We are now taking the bold step of assuming that the entire turbulent-diffusion vector $\left(\overline{u_j k'} + \frac{2\overline{pu_i}}{\rho}\delta_{ij}\right)$ may be approximated (i.e. 'modelled') by analogy to Eq. (7.29), noting, however, that we must maintain tensor-rank consistency, i.e.:

$$-\left(\overline{u_j k'} + \frac{2\overline{pu_i}}{\rho}\delta_{ij}\right) = \Gamma_k \frac{\partial k}{\partial x_j}. \tag{7.30}$$

In doing so, we hope that the isotropic exchange coefficient Γ_k will cater for the three components of the diffusive flux of k and that the somewhat obscure process represented by $\overline{up}$, $\overline{vp}$ and $\overline{wp}$ is either negligible, or can be adequately represented via the gradient approximation in Eq. (7.30). As is often the case, direct numerical simulation (DNS) data offer insight into questions such as this. Reference to the turbulence-energy budget in Fig. 5.4 demonstrates, at least for a boundary layer, that the diffusion of k by pressure-velocity interaction is very modest, except very close to the wall. However, this is by no means a general conclusion.

As regards Γ_k, we may use, without any loss of generality, the usual definition of the Prandtl–Schmidt number to express Γ_k in terms of the eddy-viscosity:

$$\Gamma_k = \frac{\nu_t}{\sigma_k}, \tag{7.31}$$

with σ_k, the Prandtl–Schmidt number, not necessarily constant. Because the turbulence energy is closely related to the velocity (i.e. momentum) fluctuations, we expect that the diffusivity of k is closely related to the diffusivity of momentum, as already noted. In other words, we expect σ_k to be of order 1; indeed, this is the value assumed in most models.

With the above closure assumptions accepted, and the symbol ε used for the dissipation term, Eq. (7.27) may now be written:

$$\frac{Dk}{Dt} = \nu_t \left(\frac{\partial U_i}{\partial x_j} + \frac{\partial U_j}{\partial x_i} \right) \frac{\partial U_i}{\partial x_j} + \frac{\partial}{\partial x_j} \left(\nu + \frac{\nu_t}{\sigma_k} \frac{\partial k}{\partial x_j} \right) - \varepsilon. \qquad (7.32)$$

To close the system of equations comprising the RANS equations, as well as Eqs. (7.10), (7.25) and (7.32), we need to determine two further quantities: the length scale ℓ and the dissipation rate ε. We shall see in the next section that the two can be linked to each other.

A final point to take forward is that the use of Eq. (7.32), as a key ingredient of the turbulent viscosity, will ensure that the viscosity will not vanish at locations at which the strain vanishes, but will remain appropriately elevated, by virtue of turbulence convection and diffusion, if turbulence in neighbouring regions is high. This reinforces the statement made above that the use of k, as described by Eq. (7.32), establishes a non-local link between the modelled state of the turbulence at any one point and that at neighbouring locations.

7.4 *The turbulent length scale*

We recall that the need to confront the length scale arises from its presence in Eq. (7.10). In contrast to the velocity scale, we do not have an equation that forms the basis for deriving ℓ directly[8] — although we shall see in Chapter 9 that a surrogate equation for a particular length scale can be derived, subject to several assumptions. This being the case, we are forced, at this stage, to make assumptions

[8]There are, however, relevant theoretical derivations based on two-point correlations — see U. Frisch, *Turbulence* (1996), but these are not usable in practical turbulence modelling.

about the length scale, which are very similar to those expressed by Eqs. (7.16) and (7.17). Hence, any model that is based on this approach is likely to suffer constraints almost as serious as those of the mixing-length model in respect of its generality and applicability to any flow beyond the group of thin shear flows. In any event, a prescription of the length scale is only tenable if we can relate it to a key mean-flow scale, as is the case with mixing-length models. Flows that benefit, to a degree, from the use of the turbulence-energy equation include (modestly to moderately) accelerating/decelerating boundary layers, and flows in which a realistic description of the eddy-viscosity is needed to derive realistic turbulent-diffusivity levels (of the type in Eq. (7.31)) for the approximation of heat and mass transfer across weakly strained regions.

If we wish, despite the limitations pointed out above, to construct a model that exploits the fundamental (but limited) advantages of Eq. (7.32), then we face the additional obstacle of not knowing the dissipation rate, ε. We recall, first, that dissipation is associated with the smallest scales of turbulence, in the range C in Fig. 2.4, i.e. those that are annihilated by viscosity — a process that transfers turbulence energy to heat. While we are unable to quantify ε, we may infer from its exact form in Eq. (7.27) that it is dictated by (unknown) velocity and length scales u_η and η, respectively, namely:

$$\varepsilon = O\left(\nu \frac{u_\eta^2}{\eta^2}\right). \tag{7.33}$$

This relationship also follows simply from dimensional reasoning, applied to the assumption $[\varepsilon] = [\nu, u_\eta, \eta]$, where $[\ldots]$ means 'dimensions of the content within of the brackets'. To avoid any doubt, we stress that these scales are drastically different from the corresponding macro-scales $k^{1/2}$ and ℓ, both associated with the high-energy range A in Fig. 2.4. It is now possible to eliminate the velocity scale upon the argument that the strong interaction between the dissipation scales and the viscosity implies that the Reynolds number formed with these scales, $Re_\eta = \frac{u_\eta \eta}{\nu}$, must be of order 1, for at this value, the convective (kinematic) and diffusive processes associated with these scale are of the same order of importance. Hence:

$$u_\eta \approx \frac{\nu}{\eta}. \tag{7.34}$$

Inserting Eq. (7.34) into Eq. (7.33) then yields:

$$\varepsilon \approx \frac{\nu^3}{\eta^4}, \tag{7.35}$$

which is identical to Eq. (2.10). This relationship serves to demonstrate that the dissipation rate is, in effect, a surrogate for a length scale. However, it is not useful for closing Equation (7.32), because we have no knowledge about η.

To achieve closure, we have to return to Fig. 2.4 and the associated discussion. We recall that, within the 'inertial sub-range' B, described by Eq. (2.7), turbulence energy may be thought of as cascading down the eddy-size 'pipeline' at a constant rate, ε, towards the dissipation range C. In this sub-range, turbulence is assumed to be neither produced nor dissipated, the latter clearly implied by the absence of the fluid viscosity in Eq. (2.6). Hence, we may suppose that the amount of turbulence energy dissipated is controlled by the rate which is being fed into sub-range B from sub-range A. If this argument is accepted, it follows that we should be able to *estimate* ε by reference to scales pertaining to the boundary between sub-ranges A and B. But around this boundary, the relevant velocity and length scales are, respectively, of order $k^{1/2}$ and ℓ (the latter being of order Λ, a typical flow dimension), both of which are the macro-scales.

To establish a relationship between ε and the length scale ℓ, we resort again to dimensional analysis. Because viscosity cannot be relevant to the energetic range under consideration, we propose:

$$[\varepsilon] = \left[k^\alpha \ell^\beta\right]. \tag{7.36}$$

It is readily shown that dimensional homogeneity then implies:

$$\varepsilon = \frac{k^{\frac{3}{2}}}{\ell}. \tag{7.37}$$

At this point, we have to exercise some caution in respect of what we mean by ℓ. We expect it to be closely related to the large-eddy scale Λ in Fig. 2.4. However, here, the length scale ℓ plays a dual role: it is a scale that appears in the eddy-viscosity relation (7.10), and it also appears in (7.37). While the two may be presumed to be of the same order, there is no reason to assume that they are identical, so we should strictly use l_μ in Eq. (7.10) and l_ε in Eq. (7.37). We hope, however, that this difference can be accommodated by constant multipliers, or some functional representation, which needs to be determined by computational optimisation — in effect, a trial-and-error process that involves the application of the model to a range of flows. In any event, we now have a simple semi-closed model comprising the following:

$$\nu_t = l_\mu k^{\frac{1}{2}}, \tag{7.38}$$

$$\frac{Dk}{Dt} = \nu_t \left(\frac{\partial U_i}{\partial x_j} + \frac{\partial U_j}{\partial x_i} \right) \frac{\partial U_i}{\partial x_j} + \frac{\partial}{\partial x_j} \left(\nu + \frac{\nu_t}{\sigma_k} \frac{\partial k}{\partial x_j} \right) - \frac{k^{\frac{3}{2}}}{l_\varepsilon}, \tag{7.39}$$

$$l_\mu = f(\delta); \quad l_\varepsilon = g(\delta), \tag{7.40}$$

where the length scales might be prescribed by algebraic functions of the mean-flow scale δ (or wall distance y). This type of closure is generally referred to as a 'one-equation model'. There are several specific variants of this model, and we shall review a few representative forms in Chapter 8. In the large majority of circumstances, these variants have been formulated and applied to boundary-layer flows, in which case δ is the wall distance, y, at least up to a limited value beyond which δ might be a proportion of the outer scale — e.g. the boundary-layer thickness δ_{99}.

The modest predictive potential offered by the framework (7.38)–(7.40) in a general setting is clearly implied by the need to identify appropriate mean-flow length scales, possible only in the case of relatively simple shear layers.

Two supplementary arguments reinforcing the above can be made by comparing the turbulence-transport model to the mixing-length model for limiting conditions. The first argument starts with the mixing-length model, Eq. (7.15):

$$\frac{\partial U}{\partial y} = \frac{(-\overline{uv})^{\frac{1}{2}}}{\ell}. \tag{7.41}$$

For the case of a thin shear flow, the generation of turbulence energy is simply$-\overline{uv}\frac{\partial U}{\partial y}$ (see Eq. (7.27) with $i = 1$, $j = 2$). Equation (7.41), when multiplied by $\overline{uv}$, gives the generation rate as:

$$-\overline{uv}\frac{\partial U}{\partial y} = \frac{(-\overline{uv})^{\frac{3}{2}}}{\ell}. \tag{7.42}$$

Measurements in shear layers for a significant range of fully turbulent conditions (Bradshaw *et al.* (1967)) show:

$$\frac{-\overline{uv}}{k} \approx 0.3\ (\equiv c_\mu). \tag{7.43}$$

The designation of this value as c_μ is emphasised here, because it (or its square value, 0.09, later denoted C_μ) will play a prominent role in considerations to follow, both in this section and later ones. Equation (7.42) may now be written as:

$$-\overline{uv}\frac{\partial U}{\partial y} \approx c_\mu^{\frac{3}{2}}\frac{k^{\frac{3}{2}}}{\ell}. \tag{7.44}$$

In the case of, say, a flat-plate boundary layer, developing subject to a mild pressure gradient, the transport of turbulence is weak (see Fig. 5.4, for $y^+ > 20$), and Eq. (7.39) reduces to:

$$-\overline{uv}\frac{\partial U}{\partial y} = \frac{k^{\frac{3}{2}}}{l_\varepsilon}, \tag{7.45}$$

a condition already referred to as 'turbulence-energy equilibrium'. A comparison of Eqs. (7.44) and (7.45) now shows that the length scale $l_\varepsilon = c_\mu^{-1.5}\ell$, i.e. about 6.1 times the mixing length ℓ. Hence, this demonstrates that the mixing length ℓ is not the correct length scale in Eq. (7.37).

In the second supplementary argument, we examine the length scale l_μ in Eq. (7.38). Insertion of $k^{1/2} \leftarrow c_\mu^{-1/2}(-uv)^{1/2}$ (see Eq. (7.43)), followed by a multiplication of both sides by $\frac{\partial U}{\partial y}$ and the substitution $\nu_t \frac{\partial U}{\partial y} \leftarrow \overline{uv}$, leads to:

$$-\overline{uv} = (c_\mu^{-\frac{1}{2}} l_\mu)^2 \left(\frac{\partial U}{\partial y}\right)^2, \tag{7.46}$$

which shows, by comparison to Eq. (7.15), that $l_\mu = c_\mu^{0.5}\ell$, i.e. 0.54 times the mixing length. Both arguments tell us that, for shear layers, models of the type (7.38)–(7.40) are unlikely to offer material advantages over the mixing-length model, unless conditions are far from turbulence-energy equilibrium — in which case, prescribed length scales are unlikely to be adequate in any event.

In the light of the above manipulations, the simplest prescription that might be valid in a boundary layer — at least for its lower, near-wall portion — is thus:

$$\begin{aligned} l_\mu &= c_\mu^{0.5} \kappa y, \\ l_\varepsilon &= c_\mu^{-1.5} \kappa y, \end{aligned} \tag{7.47}$$

which is consistent with $\ell = C_w y$, Eq. (7.17) and $C_w = \kappa = 0.4$, Eq. (7.24).

One problem with relation (7.47) (already flagged up in Section 7.2) is that additional attention is needed in the model in relation to the semi-viscous sublayer. It is self-evident that the processes in the sublayer must depend on the fluid viscosity. Yet, this does not feature in the set Eqs. (7.38), (7.39) and (7.47), except for the viscous diffusion of the turbulence energy. To appreciate the nature of the modifications required, we need to return our attention to some pertinent physical properties of turbulence close to a wall.

Given:

$$\overline{uv} = -\nu_t \frac{\partial U}{\partial y}, \tag{7.48}$$

the asymptotic variation of the shear-stress (Eq. (5.3)) and the velocity gradient ($O(1)$), we readily deduce:

$$\nu_t\big|_{y\to 0} \propto y^3, \tag{7.49}$$

which is confirmed in Fig. 7.4, showing data extracted from channel-flow DNS data.

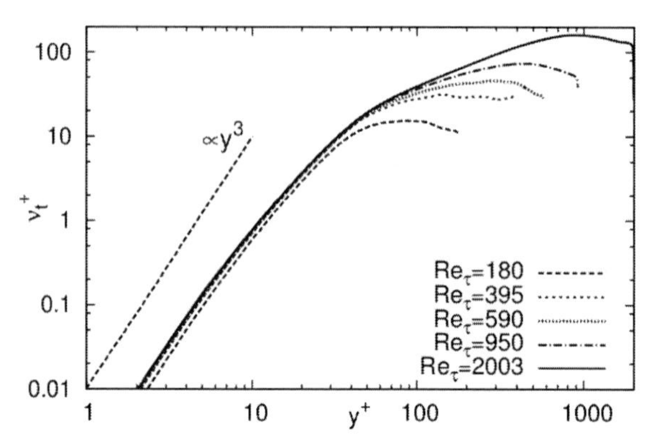

Fig. 7.4: Near-wall variations of the turbulent viscosity ($\nu_t^+ = \nu_t/\nu$) in channel flow at various Reynolds numbers; derived from DNS data by Hoyas and Jimenez (2006) and del Alamo and Jimenez (2003, 2004) (courtesy S. Jakirlić).

As the wall is approached, the turbulence energy k has been shown to decay quadratically with y. Equations (7.38) and (7.49) thus imply:

$$l_\mu\big|_{y\to 0} \propto y^2. \tag{7.50}$$

Clearly, this is incompatible with Eq. (7.47), and the implication is that we need to multiply κy by a damping function, which is at least of order y, to secure the correct asymptotic variation. The qualification 'at least' is meant to convey the suspicion that the decay may need to be even faster than that implied by Eq. (7.50), because Eq. (7.39) is unlikely to secure the correct asymptotic variation of the turbulence energy.

At this stage, we recall that the mixing-length model led us to conclude that the *mixing length* ℓ needed to decline with $O(y^{1.5})$. This incompatibility with Eq. (7.50) might be confusing, but it needs to be appreciated that the required rate of decline in the length scale in the mixing-length model depends upon the particular assumption of how the velocity scale v is related to the mixing length, via Eqs. (7.11)–(7.15).

The usual approach adopted in an effort to reconcile Eq. (7.50) with Eq. (7.47) is to multiply κ by an exponential damping function of the form:

$$\kappa f_w = \kappa \left(1 - e^{-A_w y^+}\right), \tag{7.51}$$

where A_w is a constant. To appreciate the rationale of Eq. (7.51), we can expand the exponential in series form to give:

$$1 - e^{-A_w y^+} = 1 - 1 + A_w y^+ - \frac{1}{2}\left(A_w^+ y^+\right)^2 \ldots . \tag{7.52}$$

We note first that the damping function tends to be linear in y as the wall is approached, which gives the requisite behaviour of Eq. (7.50), to first order. We also recall that, while the nominal edge of the viscous sublayer is at $y^+ \equiv \frac{y u_\tau}{\nu} \approx 11$ (see Chapter 5, Fig. 5.5), in reality, the transition from the viscous to the fully turbulent near-wall region is continuous and extends to around $y^+ \approx 30$. We therefore expect the function to approach a value just lower than 1 at this distance. This suggests a constant $A_w \approx 0.1$, in which case the damping function takes the value $f_w \approx 0.95$. This function thus provides a continuous variation that connects the asymptotic wall behaviour $l_\mu|_{y \to 0} \propto y^2$ with the turbulent-layer variation $l_\mu \propto y$. The precise variation of this transition clearly depends on the choice of A_w, and there is no unambiguous rule for this choice.

An important point to highlight in relation to Eq. (7.51) is the reason for choosing the argument y^+ in the exponential function. This is, in effect, a local, spatially varying Reynolds number formed with the length and velocity scales y and u_τ, respectively. Since the behaviour of l_μ that required the introduction of the damping function is, evidently, not primarily driven by the viscosity, but mostly by the inviscid-blocking effect of the wall on the wall-normal mixing (via the strong suppression of $\overline{v^2}$), it would appear that its introduction is misguided. However, it is important to realise that the use of y^+ (necessarily, a non-dimensional quantity) does not imply that the damping is by viscous action alone. Rather, we use y^+, in conjunction with A_w, to control the dependence of the length scale on y in such a way that l_μ becomes linear in y outside the viscous sublayer. We could have chosen the alternative non-dimensional argument y/y_v, where y_v is the edge of the viscosity-affected near-wall layer. However, this layer is dictated by the constraint $\frac{y_v u_\tau}{\nu} \approx 11 - 30$, depending upon whether we define this layer as the nominal edge or the location at which viscous effects become insignificant. At any rate, substitution of this constraint into y/y_v immediately reintroduces y^+.

The above considerations on the asymptotic behaviour of the length scale pertained specifically to l_μ. Are these compatible with the closely related length scale l_ε? The answer is provided by Eq. (7.37).[9] Since k declines as $O(y^2)$ and ε as $O(1)$ (cf. Eq. (5.7)), it follows that l_ε should decline as $O(y^3)$. So, the answer is 'no'. This is a reason for using a different damping function for l_ε, relative to l_μ. We shall see that this is indeed done in specific forms of one-equation models.

Before we turn our attention to such models we consider the question as to whether we can free ourselves from the shackles of prescribing ε by deriving a transport equation for this quantity.

7.5 *Length-scale transport*

The need to prescribe the length scale in a model is not simply a pragmatic obstacle, but one that rests on a weak physical foundation. If we accept, say by reference to Fig. 7.1, that the turbulence energy is transported, we should be able to recognise that other eddy properties are also transported. Turbulent eddies interact, break up, agglomerate and deform, and do so while being carried along by the mean flow. This then implies that a measure of their size — their length scale — is also transported. The question is whether we can derive a suitable equation for this transport. The answer is suggested by Eq. (7.37), which can in effect be viewed as a definition of the length scale. We have an equation for the turbulence energy. Can we derive an equation for the dissipation rate, ε? The answer is 'yes'.

First, we note that the dissipation rate is purely a function of derivatives of the turbulent-velocity fluctuations (see Eq. (7.27)),

[9]Note that we cannot use Eq. (7.45) to consider the asymptotic state, because this relationship applies only to the turbulent regime above the viscous sublayer.

namely:

$$\varepsilon = \nu \overline{\left(\frac{\partial u_j}{\partial x_j}\right)^2}. \tag{7.53}$$

So, we should be able to manipulate the NS equations to yield an exact equation for the dissipation rate. The formalism is straightforward, in principle, but somewhat laborious in detail. It can be expressed compactly as:

$$2\nu \overline{\frac{\partial u_i}{\partial x_j}\frac{\partial}{\partial x_j} NS(U_i + u_i)} = 0, \tag{7.54}$$

where *NS* stands, as before, for the basic Navier–Stokes equations. Note that both indices i and j are repeating, so the result is a scalar equation. After carrying out the manipulations in Eq. (7.54), the only other step needed is to actually use the definition in Eq. (7.53) to bring out ε as the subject of the equation. The result is (with body-force effects ignored at this stage):

$$\begin{aligned}\frac{D\varepsilon}{Dt} &= -2\nu\left[\overline{\frac{\partial u_i}{\partial x_k}\frac{\partial u_j}{\partial x_k}} + \overline{\frac{\partial u_k}{\partial x_i}\frac{\partial u_k}{\partial x_j}}\right]\frac{\partial U_i}{\partial x_j} - 2\nu\left[\overline{\frac{\partial u_i}{\partial x_k}\frac{\partial u_i}{\partial x_m}\frac{\partial u_k}{\partial x_m}}\right] \\ &\quad -2\nu\overline{u_k\frac{\partial u_i}{\partial x_j}}\frac{\partial^2 U_i}{\partial x_k \partial x_j} - 2\nu^2\overline{\frac{\partial^2 u_i}{\partial x_k \partial x_m}\frac{\partial^2 u_i}{\partial x_k \partial x_m}} \\ &\quad + \frac{\partial}{\partial x_j}\left[-2\nu\overline{\frac{\partial p}{\partial x_m}\frac{\partial u_j}{\partial x_m}} - \nu\overline{u_j\frac{\partial u_i}{\partial x_m}\frac{\partial u_i}{\partial x_m}}\right] + \nu\nabla^2\varepsilon. \end{aligned} \tag{7.55}$$

We shall leave the interpretation of the various terms in Eq. (7.55) to Chapter 9, referring to DNS data that give information on the contribution of the various fragments. Here, we merely note the following basic facts:

- This is a transport equation for ε: it contains convection of ε (left hand side), viscous diffusion of ε (right most term), and various correlations that can be interpreted as generation, turbulent transport and destruction of ε.
- In all terms on the right-hand side, the fluid viscosity is a multiplier. This reinforces the fact that all processes affecting the

dissipation are inherently linked to the smallest scales of turbulence at which the destruction of turbulence occurs by the action of the viscosity.

- This equation is a surrogate for the length scale, for it allows (after it is modelled!) the length scale to be determined directly if k is known.

We return to the length scale equation:

$$l_\varepsilon = \frac{k^{\frac{3}{2}}}{\varepsilon}. \tag{7.56}$$

If we wish to derive a 'transport' equation for this length scale itself, we can do so as follows:

$$\frac{Dl_\varepsilon}{Dt} = 1.5\frac{k^{\frac{1}{2}}}{\varepsilon}\frac{Dk}{Dt} - \frac{k^{\frac{3}{2}}}{\varepsilon^2}\frac{D\varepsilon}{Dt}. \tag{7.57}$$

This is so stated, but not actually executed, merely to highlight the fact that the length scale may, indeed, be thought of as being transported, generated and destroyed by various interactions that can be described exactly, via the terms in the turbulence-energy and dissipation-rate equations. Our preference for the dissipation rate is due to the fact that it is this quantity that appears as the unknown length-scale surrogate in the turbulence-energy equation.

Equation (7.57) serves to introduce the general fact that any quantity of the form:

$$\phi = k^m \varepsilon^n, \tag{7.58}$$

can serve as a length-scale surrogate, and can be derived from:

$$\frac{D\phi}{Dt} = \varepsilon^n m k^{(m-1)}\frac{Dk}{Dt} + k^m n \varepsilon^{(n-1)}\frac{D\varepsilon}{Dt}. \tag{7.59}$$

The most popular alternative to ε itself is $\phi = \varepsilon/k$, the 'specific dissipation', usually denoted by ω.[10] The reason this alternative should

[10]While this is not to be confused with vorticity, it is arguable, on dimensional grounds, that ω is the inverse of the eddy-turnaround time scale, i.e. vaguely representatrive of rotationality.

offer advantages will be discussed in Chapter 9 (Section 9.4, in particular), in which we shall also review other proposals made in the literature.

Another group that is relevant to one of the models covered in Chapter 9 is $\phi = k^2/\varepsilon$. To appreciate the physical significance of this group, we combine the eddy-viscosity relation $\nu_t = k^{0.5} l_\mu$ (cf. Eqs. (7.10) and (7.25)) with the length-scale relation (Eq. (7.56)). Because of Eq. (7.47), $l_\mu = c_\mu^2 l_\varepsilon$, and use of Eq. (7.56) for l_ε leads to:

$$\nu_t = C_\mu \frac{k^2}{\varepsilon}, \tag{7.60}$$

with $C_\mu = c_\mu^2$. Hence, it is also appropriate to regard the turbulent viscosity as a property that is subject to transport, generation and destruction.

7.6 *Summary and lessons*

The eddy-viscosity is an artefact that reflects the notion that the time-averaged effect of turbulence is a diffusion — or mixing — process that is akin to that caused by the fluid viscosity, although the fundamental (microscopic/macroscopic) processes at play are very different. A logical extension of this analogy, by reference to the linear stress-strain relationship applicable to Newtonian fluids, is to adopt the same linear constitutive equations to relate the Reynolds stresses to the strain components. However, this is by no means the only possibility; indeed, as will transpire in Chapters 11 and 12, it is not an especially good one as a foundation for high predictive fidelity and model generality. In Chapter 15, one potentially superior modelling framework we shall consider involves the use of non-linear constitutive equations.

It is reasonable to assume that mixing is associated, primarily, with cross-flow transport by energetic and large-scale turbulent eddies. Dimensional arguments dictate that the eddy-viscosity is proportional to a velocity scale and a length scale, and it is appropriate, therefore, to associate these scales with the turbulence energy k and a representative large-scale eddy-size l_μ, respectively. This association, while conceptually important, is merely the starting point of the turbulence-closure process that involves the formulation of equations that relate the turbulence energy and the length scale to the strain field.

While the turbulence energy is readily derived from rational principles — the turbulent-energy-transport equation — the length scale is much more problematic, because there is no equation for this quality that is equivalent to that which governs k. In simple shear flows, the length scale may be presumed to be related to the thickness of the shear layer, as is done in the mixing-length model, but such a relationship is not a tenable foundation for a general formulation. The problem is further complicated by the presence of the unknown dissipation rate, ε, in the turbulence energy equation. Although we have shown that an exact equation for ε can be derived, this does not provide an obvious route for determining l_μ, because the dissipation process is associated with the smallest scales of the turbulent eddy field. Thus, while the dissipation rate can related, on dimensional grounds, to a length scale l_ε, its value can be expected to be very different from that of the large-eddy scale l_μ. This length-scale conundrum may be 'resolved' by linking the small-scale dissipation process to the large-scale eddy field, via the middle range (the inertial sub-range) of the eddy spectrum (cascade) across which turbulence energy is transferred from the energetic range to the dissipative range; and by modelling the dissipation-rate equation by reference to large-scale processes that give rise to the turbulence-energy, k. This is a crucially important concept, for it justifies the dimensional proposal $l_\varepsilon = \frac{k^{3/2}}{\varepsilon}$, in which case a close (proportional) relationship is established between the length scales l_μ and l_ε.

In the following two chapters, we shall consider specific linear eddy-viscosity models — those based on the linear stress-strain relations — derived either via a determination of the turbulence energy and the length scales, or derived via a more direct route that targets the eddy viscosity itself as the subject of a pertinent closure equation.

Chapter Eight

8. One-Equation Eddy-Viscosity Models

8.1 *Introductory comments*

A substantial amount of preparatory material relating to the present chapter has already been developed in Chapter 7. In particular, the relationship of the eddy-viscosity to the turbulence energy and macro-length scale has been stated, the turbulence-energy equation has been introduced, and various length scales have been discussed. In effect, these considerations amount to an introduction to the category of *one-equation models*, in most of which a single differential transport equation for the turbulence energy is solved, while the length scale is prescribed algebraically. Another model category for discussion (not considered explicitly in Chapter 7) is based on the solution of a transport equation for the eddy-viscosity or (analogously) the turbulent Reynolds number — effectively, the ratio of turbulent-to-fluid viscosities. As will be seen below, models of the latter type also rely on an explicitly prescribed length scale, or analogous assumptions, and they involve the imposition of constraints, or calibrations, that pertain specifically to thin shear flows. Hence, in common with turbulence-energy-based models, eddy-viscosity-transport models do not apply to (or cannot be expected to realistically resolve) separated flows containing regions of recirculation. In such conditions, a model has to include the transport of the length scale, in the sense discussed in Section 7.5.

In recent practice, one-equation models have been used, predominantly, as near-wall models, which are interfaced with more elaborate models for the outer flow — one that might contain complex features, separation in particular. This is a sensible approach — particularly if the outer-flow model is not applicable to the viscosity-affected near-wall region, which then requires an auxiliary near-wall model. Coupling between disparate models poses its own problems, some physical, others computational. One specific physical limitation is that a separated flow almost always involves stagnation regions that extend to the wall. On either side of these regions, the flow is far from a state of equilibrium, being subject to locally strong acceleration and deceleration. Hence, here too, the prescription of a length scale that is rigidly tied to the wall-normal distance, or the imposition of thin-shear-flow constraints, is certain to be questionable, if not unrealistic.

As already explained in Chapter 7, an advantage offered by one-equation models is that the inclusion of convective and diffusive transport allows a realistic description of the turbulent viscosity to be obtained in low-shear portions of thin shear flows, in which the mixing-length formulation fails. This applies, in particular, to turbulent shear-free regions that border shear layers — say, the free-stream above a boundary layer or the central portion of a jet. In the former, a high level of turbulence may be advected from upstream — for example, in the highly turbulent, weakly sheared, central portion of a turbomachine passage, in which boundary layers develop on the blades forming the passage. Simultaneously, turbulence diffuses into the boundary layer, causing it to spread more rapidly and influencing the turbulent structure, possibly right down to the wall. Neither process — turbulence convection or diffusion — is accounted for by mixing-length-type models.

It has to be acknowledged, however, that the 'advantages' discussed above rarely have major consequences to the mean-flow aero/hydro-dynamics, unless the free-stream turbulence above a boundary layer is very high. Thus, the fact that the turbulent viscosity remains finite in shear-free regions is of only marginal importance. Of greater importance is the fact that the turbulent diffusivity of heat, or species concentration, is determined as the viscosity divided by a pertinent Prandtl–Schmidt number. A vanishing diffusivity in a region in which there exists a temperature gradient will, quite obviously, cause serious predictive damage to the thermal field.

8.2 *Turbulence-energy-based models*

If k is chosen as the transported turbulence quantity, then Eq. (7.39) represents the transport of turbulence. Applied to a boundary layer developing along a plane wall, this equation simplifies to:

$$\frac{Dk}{Dt} = \nu_t \left(\frac{\partial U}{\partial y}\right)^2 + \frac{\partial}{\partial y}\left(\left(\nu + \frac{\nu_t}{\sigma_k}\right)\frac{\partial k}{\partial y}\right) - \frac{k^{\frac{3}{2}}}{l_\varepsilon}, \tag{8.1}$$

where y is the wall-normal coordinate. This is supplemented by Eq. (7.38):

$$\nu_t = l_\mu k^{\frac{1}{2}}. \tag{8.2}$$

This closure foundation was first proposed by Prandtl in 1945, but only exploited in real computational predictions for thin shear flows in the 1960s, when digital computers became sufficientyly powerful to solve coupled differential equations for realistic turbulent flows. Several variants based on Eq. (8.2) have been proposed, and these differ mainly in the way in which the length scales are approximated, and the manner in which coefficients and constants are determined and/or calibrated.

It is, unfortunately, an inescapable fact that there is a significant degree of ambiguity and lack of rigour in the determination of model constants, and this applies to *all* classes of models. The standard approach is to apply the model, with constants yet to be determined, to a series of simple (canonical) flows — the constants then being 'calibrated' to give satisfactory agreement with secure experimental or simulation data. Ideally, this process would yield simplified model forms, each containing only one unknown constant, to be associated with as many simple flows as the number of constants to be determined. A combination of analytical solution and 'best fit' with available data for each flow would then be used to determine the constants. More often than not, this process is then followed by an overall 'computational optimisation', which entails an adjustment of the constants to fit as broad a range of experimental and simulation data as possible for flows other than those used for the primary calibration.

We consider two specific models below to illustrate alternative closure processes.

8.2.1 *The Wolfshtein (1969) model*

Although this model is not the earliest of its class actually used to predict the structure of turbulent shear layers, it is one of very few that has maintained, to this day, a degree of popularity in commercial computational fluid dynamics (CFD) codes, as part of a 'two-layer' strategy applied to wall-bounded flows, in which two different, but coupled, models are used for the viscosity-affected near-wall region and the fully turbulent outer flow, respectively.

Based on the considerations in Section 7.4, and Eq. (7.47), in particular, we may write, for a boundary layer, the general length-scale relations:

$$\begin{aligned} l_\mu &= c_{l\mu} y, \\ l_\varepsilon &= c_{l\varepsilon} y. \end{aligned} \tag{8.3}$$

Following the arguments that led to proposal in Eq. (7.51), we may extend this to the viscous sublayer by:

$$\begin{aligned} l_\mu &= c_{l\mu} y\left(1 - e^{-A_\mu y^+}\right), \\ l_\varepsilon &= c_{l\varepsilon} y\left(1 - e^{-A_\varepsilon y^+}\right), \end{aligned} \tag{8.4}$$

in which the distance from the wall is made dimensionless by means of the friction velocity — i.e. $y^+ = yu_\tau/\nu$. This fact is highlighted, because the use of the friction velocity is not the only option. Another obvious velocity scale is $k^{1/2}$, and this is, indeed, the choice made by Wolfshtein. To distinguish this alternative scaling, we replace y^+ by $y^* = yk^{1/2}/\nu$. As an aside, we note that both are, essentially, alternative Reynolds-number definitions, and this is reflected by Wolfshtein's use of the symbol R for y^*.

To clarify the relationship between y^+ and y^*, we take advantage of Eq. (5.18), $-\overline{uv} = u_\tau^2$, and Eq. (7.43), $-\overline{uv}/k \approx 0.3$ ($\equiv c_\mu$).

Hence:

$$y^+ = 0.55y^*. \tag{8.5}$$

Equation (8.4) contains four constants. A fifth is σ_k in Eq. (8.1). Wolfshtein's approach to determining the constants starts from solutions to the one-dimensional momentum equation:

$$\frac{d}{dy}\left((\mu + \mu_t)\frac{dU}{dy}\right) - \frac{dP}{dx} = 0, \tag{8.6}$$

in which the streamwise pressure gradient does not depend on y (i.e. it is treated as constant), and the turbulence-energy equation:

$$\mu_t\left(\frac{dU}{dy}\right)^2 + \frac{d}{dy}\left(\left(\mu + \frac{\mu_t}{\sigma_k}\right)\frac{dk}{dy}\right) - \frac{\rho k^{\frac{3}{2}}}{c_{l\varepsilon} y\left(1 - e^{-A_\varepsilon y^*}\right)} = 0, \tag{8.7}$$

with

$$\mu_t = c_{l\mu}\rho y\left(1 - e^{-A_\mu y^*}\right)k^{\frac{1}{2}}. \tag{8.8}$$

These are solved for four particular flows, all treated as one-dimensional: (i) an equilibrium (non-diffusional) shear layer; (ii) a fully turbulent 'linear-shear layer', subject to a streamwise pressure gradient; (iii) the viscous sublayer, in which the turbulent viscosity is small, relative to the fluid viscosity; and (iv) a 'no-generation' layer, in which the shear stress is small. These solutions, involving combinations of the unknown constants, are juxtaposed with correlations representing experimental data. For example, for case (i) above, the equilibrium layer, the flow is assumed described by the log-law of the wall:

$$U^+ = \frac{1}{\kappa}\ln(Ey^+), \tag{8.9}$$

with $\kappa = 0.4$ and $E = 9.0$,[1] while for case (iii), the viscous sublayer, the empirical correlation:

$$\frac{\mu_t}{\mu} = a(y^+)^\alpha, \tag{8.10}$$

---

[1]The latter value corresponds to an additive constant $\frac{1}{\kappa}\ln(E) = 5.49$, which is close to that in Eq. (7.24).

is exploited, where $a = 8.85 \times 10^{-5}$ and $\alpha = 4$, (which, we note in passing, does not comply with the asymptotic variation given by Eq. (7.49)).

This process yields four relationships interlinking the constants. Computer optimisation by reference to additional sets of data allowed Wolfshtein to converge to the following final set of constants:

$$c_{l\mu} = 0.220; \quad c_{l\varepsilon} = 2.4; \quad \sigma_k = 1.53; \quad A_\mu = 0.016; \quad A_\varepsilon = 0.263. \tag{8.11}$$

As regards the first two constants, insertion of $c_\mu = 0.3$ and $\kappa = 0.4$ into Eq. (7.47) gives $c_{l\mu} = 0.219$, $c_{l\varepsilon} = 2.43$, which are (unsurprisingly) very close to the corresponding values in Eq. (8.11). Because $y^+ = 0.55y^*$, $y^+ = 30$ corresponds to $y^* = 55$. At this distance, the van-Driest damping function in Eq. (8.8) gives a value of 0.59. This is a surprising result, because it suggests that the model applies viscous damping well beyond the viscosity-affected sublayer. In fact, for the damping function $(1 - e^{-A_\mu y^*})$ to increase to 0.95, y^* has to increase to around 180. In other words, the damping function evidently accounts for processes that are not only rooted in the viscosity. The nature of the processes at issue will be discussed in Section 9.3. In contrast to the damping function for l_μ, that for the dissipation length scale l_ε is only effective over a much thinner near-wall layer — around $y^* \approx 11 (y^+ \approx 6)$. This reinforces the conclusion that the damping function associated with l_μ (i.e. that directly linked to ν_t) plays a partially hidden role, unrelated to viscous damping.

8.2.2 *The Norris–Reynolds (1975) model*

In common with Wolfshtein's model, that of Norris and Reynolds is also still used as a part of 'two-layer' strategies. In this model, the derivation starts from Eq. (8.7), but then involves the proposal that the dissipation is:

$$\varepsilon = \frac{k^{\frac{3}{2}}}{c_{l\varepsilon} y} + \frac{2\nu k}{y^2}, \tag{8.12}$$

in which the second term is intended to steer the dissipation rate towards its wall-asymptotic value, as expressed by Eq. (5.7). In effect, this form circumvents the need to introduce the variation of l_ε in the viscous sublayer, via a van Driest damping function. The expression for l_ε follows from the equality:

$$\frac{k^{\frac{3}{2}}}{c_{l\varepsilon}y} + \frac{2\nu k}{y^2} = \frac{k^{\frac{3}{2}}}{l_\varepsilon}, \tag{8.13}$$

resulting in:

$$l_\varepsilon = c_{l\varepsilon}y\frac{yk^{\frac{1}{2}}}{yk^{\frac{1}{2}} + 2\nu c_{l\varepsilon}}. \tag{8.14}$$

The viscosity-related length scale l_μ has, therefore, the same form as that of Wolfshtein, but with a slightly different coefficient in the van Driest damping term:

$$l_\mu = c_{l\mu}y\left(1 - e^{-0.0198y^+}\right). \tag{8.15}$$

The coefficients $c_{l\mu}$ and $c_{l\varepsilon}$ are obtained by reference to conditions pertaining to a fully turbulent boundary layer. First, the log-law, Eq. (8.9), gives, when differentiated:

$$\frac{\partial U}{\partial y} = \frac{u_\tau}{\kappa y}, \tag{8.16}$$

and the mixing-length model, Eq. (7.21), gives:

$$\mu_t = \rho\kappa^2 y^2 \frac{\partial U}{\partial y}. \tag{8.17}$$

The above, when combined with:

$$\mu_t = \rho k^{\frac{1}{2}} c_{l\mu} y, \tag{8.18}$$

gives:

$$c_{l\mu} = \frac{\kappa u_\tau}{k^{\frac{1}{2}}}. \tag{8.19}$$

Next, we may invoke the turbulence-equilibrium concept, which is compatible with the log-law layer, namely:

$$\mu_t \left(\frac{\partial U}{\partial y}\right)^2 = \frac{\rho k^{\frac{3}{2}}}{c_{l\varepsilon}y}. \tag{8.20}$$

Combining this with Eqs. (8.17)–(8.19) gives:

$$c_{l\varepsilon} = \kappa \frac{k^{\frac{3}{2}}}{u_\tau^3}. \tag{8.21}$$

Norris and Reynolds now take advantage of experimental evidence in near-equilibrium boundary layers that suggests:

$$\frac{u_\tau}{k^{\frac{1}{2}}} \approx 0.54, \tag{8.22}$$

which is compatible with Eq. (7.43), except for a slight difference in the numerical value relative to $c_\mu^{1/2}$, because $(\overline{uv})^{1/2} = u_\tau$ in the log-law region. With $\kappa = 0.42$, assumed by Norris and Reynolds, this then results in:

$$\begin{aligned} c_{l\mu} &= c_\mu^{\frac{1}{2}} \kappa = 0.227, \\ c_{l\varepsilon} &= c_\mu^{-\frac{3}{2}} \kappa = 2.67, \end{aligned} \tag{8.23}$$

values which are (unsurprisingly) close to those of Wolfshtein.

A final point to make here relates to the coefficient 0.0198 in the van Driest term in Eq. (8.15). Because of Eq. (8.5), Eq. (8.15) can be written as:

$$l_\mu = c_{l\mu} y \big(1 - e^{-0.011 y^*}\big). \tag{8.24}$$

A comparison of the exponent 0.011 to Wolfshtein's corresponding value 0.016 makes clear that the wall-normal reach of Norris' and Reynolds' viscous damping is even larger than Wolfshtein's. Yet again, the clear implication is, therefore, that this term has a role that goes beyond the physical effects of viscosity.

8.3 *Eddy-viscosity-transport models*

If modelling simplicity is uppermost in a user's mind, that user is entirely reasonable when asking the question as to whether a single equation can be derived that gives directly the turbulent viscosity (as the mixing-length model does) *and* takes transport into account. In other words: is it reasonable and possible to seek a transport equation for the eddy viscosity? One answer is offered by Eqs. (7.58)–(7.60).

Plainly, the answer is 'yes', because it is entirely possible, in principle, to combine the transport equations for the turbulence energy and a length-scale surrogate variable to arrive at a single equation for the turbulent viscosity. Whether it can be adequately closed is another matter. Even if the manipulations in Eq. (7.59) are performed with *modelled* (i.e. closed) forms of the turbulence-energy and length-scale equations, the result is that the parent variables remain, explicitly, in the resulting equation. They thus require elimination by additional closure assumptions and approximations.

Models proposed by Baldwin and Barth (1991), Spalart and Allmaras (1992), Goldberg (1994), Goldberg and Ramakrishnan (1994) and Menter (1997) are all based on the solution of transport equations for the turbulent Reynolds-number $R_t = \frac{k^2}{\nu\varepsilon}$, or on the turbulent viscosity, both quantities being closely related to each other. We consider some reprsentative models in the following sub-sections.

8.3.1 *The Spalart–Allmaras (1992) model*

This is by far the most popular model in the present category — not only because of its extensive use in external aerodynamics as a Reynolds-averaged Navier–Stokes (RANS) model, but also because it has been adopted as the foundation for the 'detached eddy simulation' (DES) method — essentially, a variant of large eddy simulation (LES) that combines LES with RANS, operating in different subdomains. This is referred to, generally, as 'Hybrid LES-RANS modelling'. The principal feature of the model is that it is *not* based on a formal combination of other transport equations of the type derived earlier, but rather on a heuristic approach that proposes terms *analogous* to those found in other transport equations. This route may be criticized as lacking rigour and being 'qualitative', but ultimately it is the performance of the model that is of primary interest, and this can be steered and guided by an elaborate calibration process, by the addition of corrective fragments, and by optimisation (essentially, curve-fitting) with reference to the widest possible set of experimental conditions. The inevitable penalty of this route is, however, a lack of transparency and a somewhat disconcerting decoupling from established principles underpinning 'conventional' models.

The construction of the model progresses along a series of steps, each building upon the previous one, and each targeting a particular generic flow type. The starting point is the conceptual proposition that the requisite equation shall have the form:

$$\frac{D\nu_t}{Dt} = Generation - Destruction + Diffusion. \tag{8.25}$$

With attention focusing, initially, on a free shear layer, the imposition of dimensional homogeneity and the assumption that the turbulent viscosity is driven by the shear strain and the eddy viscosity itself (observe the analogy with Eq. (8.1)) leads to:

$$\frac{D\nu_t}{Dt} = C_1\nu_t\frac{\partial U}{\partial y} - Destruction + Diffusion. \tag{8.26}$$

Next, to make the model co-ordinate invariant, the simple strain has to be replaced by a scalar representative of the general strain tensor:

$$S_{ij} = \frac{1}{2}\left(\frac{\partial U_i}{\partial x_j} + \frac{\partial U_j}{\partial x_i}\right). \tag{8.27}$$

Spalart and Allmaras (henceforth abbreviated S&A) thus choose the replacement:

$$\frac{\partial U}{\partial y} \leftarrow \sqrt{2S_{ij}S_{ij}} \equiv S. \tag{8.28}$$

Next, the basis for choosing the diffusive-transport term is the classical gradient approximation, but S&A argue that provisions have to be made for the possibility that the integral of this term may not be conserved, simply because the eddy viscosity cannot strictly be regarded as a conserved property. Based on this argument, Eq. (8.26) (with Eq. (8.28) incorporated) becomes:

$$\frac{D\nu_t}{Dt} = C_1\nu_t S + \frac{1}{\sigma}\frac{\partial}{\partial x_j}\left(\nu_t\frac{\partial \nu_t}{\partial x_j}\right) + \frac{C_2}{\sigma}\left(\frac{\partial \nu_t}{\partial x_j}\right)^2 - Destruction, \tag{8.29}$$

in which the third term on the right-hand side accounts for the (possibly) non-conservative nature of the diffusive-transport process. Surprisingly, at first sight, the model does *not* require a destruction term for free shear flows — e.g. a mixing layer and a plane wake. This is so, because a judicious choice of the constants C_1, C_2 and σ can be made so as to 'transport', at a sufficiently high rate, the high levels of viscosity generated in high-strain regions towards the edges of the shear layer, thus securing the correct shear-stress levels and spreading rates. The calibration and optimisation process, by reference to experimental observations for a mixing layer and a wake, are not given here, as the details are rather elaborate and do not lend themselves to a transparent exposition. Suffice it to say that S&A give the following 'plausible' set, applicable to free shear flows:

$$\sigma = \frac{2}{3}, \quad C_1 = 0.1355, \quad C_2 = 0.622. \tag{8.30}$$

For any wall-bounded flow, the model requires a destruction term. To this end, S&A use dimensional reasoning:

$$[Destruction] = \left[\frac{L^2}{T^2}\right] = [\nu_t]^\alpha [y]^\beta, \tag{8.31}$$

leading to:

$$\frac{D\nu_t}{Dt} = C_1 \nu_t S + \frac{1}{\sigma}\frac{\partial}{\partial x_j}\left(\nu_t \frac{\partial \nu_t}{\partial x_j}\right) + \frac{C_2}{\sigma}\left(\frac{\partial \nu_t}{\partial x_j}\right)^2 - C_{w1} f_w \left(\frac{\nu_t}{y}\right)^2, \tag{8.32}$$

where f_w is a function that varies between the value '1' in the log-law region and is '0' in free shear layers (the latter reflecting the calibration process outlined earlier). Thus, C_{w1} is chosen so that Eq. (8.32) correctly predicts the log-law layer. With $f_w = 1$, S&A use the log-law, Eq. (8.9), and the one-equation relations, Eqs. (8.17) and (8.18), to show that the imposition of 'turbulent-viscosity equilibrium' on Eq. (8.32) demands:

$$C_{w1} = \frac{C_1}{\kappa^2} + \frac{1 + C_2}{\sigma}. \tag{8.33}$$

The fact that the model requires the introduction of the last (destruction) term, which contains the wall distance, makes clear that this model also depends on the prescription of a length scale (y), analogous to the turbulence-energy models covered earlier.

The f_w function is calibrated so as to extend the applicability of the model to the outermost (wake) part of a boundary layer (which does not follow the log-law!) and to boundary layers subjected to adverse pressure gradient. This calibration then led S&A to propose the following "satisfactory f_w function":

$$f_w = g\left[\frac{1 + C_{w3}^6}{g^6 + C_{w3}^6}\right]^{\frac{1}{6}}, \tag{8.34}$$

where

$$g = r + C_{w2}(r^6 - r); \quad r \equiv \frac{\nu_t}{S\kappa^2 y^2}; \quad C_{w2} = 0.3; \quad C_{w3} = 2. \tag{8.35}$$

The above value of C_{w3} is referred to by S&A as 'reasonable', while the value of C_{w2} was arrived at by matching model predictions to the skin-friction value of a flat-plate boundary layer at the particular momentum-thickness Reynolds number $Re_\theta = 10^4$.

The above form is valid only beyond the viscous sublayer, where the flow is fully turbulent. To extend the model down to the wall, S&A introduce a damping function, defined such that:

$$\nu_t = f_{\nu 1}\widetilde{\nu_t}, \tag{8.36}$$

where $\widetilde{\nu_t}$ is the turbulent viscosity that would prevail *without* the viscous sublayer. The proposed damping function is:

$$f_{\nu 1} = \frac{(\kappa y^+)^3}{(\kappa y^+)^3 + 7.1}. \tag{8.37}$$

S&A argue that a van Driest damping function would be an entirely credible alternative, as the difference to Eq. (8.37) would be small.

Equation (8.32), with the viscous diffusion term added, is now interpreted as governing $\widetilde{\nu_t}$, which is identical to ν_t *outside* the viscous sublayer:

$$\frac{D\widetilde{\nu}_t}{Dt} = C_1\widetilde{\nu}_t S + \frac{1}{\sigma}\frac{\partial}{\partial x_j}\left((\widetilde{\nu}_t + \nu)\frac{\partial \widetilde{\nu}_t}{\partial x_j}\right)$$
$$+\frac{C_2}{\sigma}\left(\frac{\partial \widetilde{\nu}_t}{\partial x_j}\right)^2 - C_{w1} f_w \left(\frac{\widetilde{\nu}_t}{y}\right)^2. \tag{8.38}$$

An inconsistency in Eq. (8.38) arises from the fact that, while $\widetilde{\nu}_t$ is presumed not to account for viscous damping, the occurrence of S in this equation does reflect such damping, because the strain field is ultimately computed with ν_t, i.e. after the application of Eq. (8.36). To counteract this inconsistency, S needs to be modified so as to ensure that the solution of Eq. (8.38) yields a $\widetilde{\nu}_t$ level that satisfies the constraint $\widetilde{\nu}_t \propto \kappa^2 y^2 S$ throughout the near-wall layer. S&A achieve this by replacing S by $\widetilde{S}$:

$$\widetilde{S} = S + \frac{\widetilde{\nu}_t}{\kappa^2 y^2} f_{\nu 2}; \quad f_{\nu 2} = 1 - \frac{\kappa y^+}{1 + \kappa y^+ f_{\nu 1}}. \tag{8.39}$$

It can be argued that this complex sequence of steps — involving 'reasonable' propositions, the imposition of constraints and wide-ranging calibration and optimisation, leading to 'plausible' constants and functions — justifies the statements made at the beginning of this section as to a disconcerting lack of transparency of, and clear formal structure to, the derivation path. This is, unfortunately a significant obstacle to the objective of 'demystifying' the modelling process.

A few final comments are appropriate as regards the range of applicability of the S&A model. We recall a statement made in Section 8.1, to the effect that one-equation models cannot be expected to be valid in separated flows. This has not prevented Spalart and Allmaras and many other practitioners from applying the model to flows that contain separated regions — e.g. shock-induced separation on transonic aerofoils and blunt trailing edges. The point has been made, in relation to Eq. (8.29), that the model has been calibrated for free shear flows, without the need to introduce a flow-specific length scale, and this suggests that it should be applicable to wakes recovering from separation. Yet, it is the view of this author that any one-equation model is too simple and too restricted by its calibration to give a credible description of separated regions, and there are a number of examples in the literature that demonstrate that the model gives far too extensive separation bubbles in massively separated flow. The model works well in two-dimensional boundary layers subjected to adverse pressure gradient, but its application to extensively separated flows is fraught with danger.

8.3.2 *Models combining the k- and ε-equations*

There are several one-equation models that have arisen from a formal combination of the turbulence-energy equation and a length-scale-surrogate equation of the form introduced in Section 7.5. These approaches are very different from that adopted by Spalart and Allmaras. A problem with discussing these models at this stage is that we have not yet covered the closure of any length-scale equation. This closure is part and parcel of 'two-equation models' to be covered in Chapter 9. For this reason, we only outline here the nature of the models in question, deferring a more detailed discussion to Section 9.5.

The desire to construct one-equation models along the present route arose as a response to the complexities that two-equation models were perceived to pose, mainly in the 1980s and the inadequate performance that these models displayed, especially in external aerodynamic flows on streamlined bodies. These inadequacies were assumed to originate mainly from the length-scale-surrogate equation. The hope was that the combination of the two equations, followed by approximations and calibration that tied the resulting model to specific conditions encountered preferentially in external aerodynamics, would result in a framework that has a predictive performance as good as, or better than, that of the parent two-equation model.

The model proposed by Baldwin and Barth is based on a single equation for the turbulence Reynolds number, $R_t \equiv \frac{k^2}{\nu\varepsilon}$, derived by a combination of the k- and ε-transport equations (see Section 9.5), followed by simplifications applicable to turbulence-equilibrium conditions and some calibration by reference to an algebraic model by

Cebeci and Smith (1974). The nature of R_t becomes clear upon noting from Eq. (7.60), $\nu_t = C_\mu \frac{k^2}{\varepsilon}$, that the former is, essentially, the ratio ν_t/ν multiplied by a constant. The equation then derived by Baldwin and Barth is:

$$\frac{DR_t}{Dt} = C_1\sqrt{R_t P_k} + \left(1 + \frac{\nu_t}{\nu\sigma_R}\right)\frac{\partial^2 R_t}{\partial x_j^2} - C_2 \frac{\partial \nu_t}{\partial x_j}\frac{\partial R_t}{\partial x_j}, \tag{8.40}$$

from which the turbulent viscosity is then evaluated as:

$$\nu_t = C_\mu \nu R_t f_{\mu 1} f_{\mu 2}, \tag{8.41}$$

where $f_{\mu 1}$ and $f_{\mu 2}$ are van Driest-type damping functions involving y^+.

Without going into detail, at this stage, we note that the first term represents the convective transport of the turbulent viscosity; the second term represents its generation, by the action of stresses and strains that give rise to the production of turbulence energy, P_k; the third term represents viscous and turbulent diffusion; and the fourth term is analogous to the penultimate term in the S&A equation, Eq. (8.38), which accounts for possible non-conservation of diffusive transport, and which emerged much later. In fact, there are close similarities between all other terms in Eqs. (8.38) and (8.40). The correspondence of convection is obvious. Reference to Eq. (7.32) shows that the generation term in Eq. (8.40) is, essentially:

$$\begin{aligned} C_1\sqrt{R_t P_k} &= C_1\sqrt{\frac{C_\mu^{-1}\nu_t}{\nu\nu_t}\left(\frac{\partial U_i}{\partial x_j} + \frac{\partial U_j}{\partial x_i}\right)\frac{\partial U_i}{\partial x_j}} \\ &= C_1\sqrt{\frac{C_\mu^{-1}}{\nu}\nu_t^2 S^2} = C_1\sqrt{\frac{C_\mu^{-1}}{\nu}}\nu_t S, \end{aligned} \tag{8.42}$$

which has the same form as the S&A model.

To avoid uncertainly in the result of Eq. (8.42), we note the following supplementary derivation. Starting with

$$\left(\frac{\partial U_i}{\partial x_j} + \frac{\partial U_j}{\partial x_i}\right)\frac{\partial U_i}{\partial x_j} = 2S_{ij}\frac{\partial U_i}{\partial x_j},$$

and using the definition of the vorticity tensor:

$$\Omega_{ij} = \frac{1}{2}\left(\frac{\partial U_i}{\partial x_j} - \frac{\partial U_j}{\partial x_i}\right),$$

we can write:

$$\frac{\partial U_i}{\partial x_j} = S_{ij} + \Omega_{ij},$$

and therefore:

$$\left(\frac{\partial U_i}{\partial x_j} + \frac{\partial U_j}{\partial x_i}\right)\frac{\partial U_i}{\partial x_j} = 2S_{ij}(S_{ij} + \Omega_{ij}).$$

However, $S_{ij}\Omega_{ij} = 0$, because S_{ij} is symmetric, while Ω_{ij} is anti-symmetric. Hence:

$$\left(\frac{\partial U_i}{\partial x_j} + \frac{\partial U_j}{\partial x_i}\right)\frac{\partial U_i}{\partial x_j} = 2S_{ij}S_{ij}. \tag{8.43}$$

Equation (8.28) shows this to be S^2.

Finally, the diffusion term in Eq. (8.40) is very close to that in Eq. (8.38).

An advantage of the Baldwin–Barth model is that it avoids an explicit prescription of the length scale, although it does involve the distance from the wall in the damping functions. This last restriction was removed by Goldberg (2001) who proposed a point-wise ('local') version of the model which, for high Reynolds-number flows, is identical to Baldwin and Barth's parent form.

An early one-equation model of Bradshaw, Ferris and Atwell (1967), for the transport of the shear stress, is considered next, mainly to provide a foundation for a discussion of Menter's (1997) one-equation eddy-viscosity-transport model, in Section 9.5, and of his related two-equation model (Menter (1994)), the latter used extensively in computational aerodynamics and discussed in Section 9.4.3. The model is based on Bradshaw *et al.*'s observation that the shear stress in shear flows is more closely connected to other turbulence parameters, especially the turbulence energy, than to the mean-velocity profile. An implication of this observation is that the eddy-viscosity concept is not a good basis for approximating the shear-stress. Instead, Bradshaw *et al.* proposed:

$$-\overline{uv} = ak, \tag{8.44}$$

with a expected to be a function of y/δ, but chosen to be a constant. In fact, two forms of this relationship have been used already, namely Eqs. (7.43) and (8.22), with $a = c_\mu = 0.3$.

Although this fixed ratio applies only to the log-law region, Bradshaw *et al.* suggested that this should be used more generally, because other thin shear flows show a similar behaviour (albeit with a lower constant). They then proceeded to derive a shear-stress-transport equation by inserting Eq. (8.44) into the turbulence-energy equation, approximating the diffusion of the shear stress to be in proportion to the diffusion of k, and replacing the dissipation by u_τ^3/l_ε, essentially a combination of Eqs. (7.56) and (8.22). With $\tau \equiv -\overline{uv}$, the result, for simple shear $U(y)$, is

$$\frac{D}{Dt}\left(\frac{\tau}{a}\right) = \tau\frac{\partial U}{\partial y} - (\tau_{\max})^{\frac{1}{2}}\frac{\partial}{\partial y}(G\tau) + \frac{\left(\frac{\tau}{a}\right)^{\frac{3}{2}}}{l_\varepsilon}, \tag{8.45}$$

in which G is defined as $(\overline{pv} + \rho\overline{kv})/[\tau_{\max}^{1/2}\tau]$, representing the ratio of the diffusive turbulence-energy flux and the diffusive shear-stress flux — a ratio which Bradshaw *et al.* approximate, empirically, in the form of $G = (\tau_{\max}/U_\delta^2)^{0.5} f(y/\delta)$, U_δ being the velocity at the outer edge of the boundary layer.

Following Baldwin and Barth (1991), Menter (1997) derived a single equation for the eddy-viscosity by combining the transport equations for k and ε (the latter discussed in Chapter 9). Insertion of Eq. (7.60), $\nu_t = C_\mu\frac{k^2}{\varepsilon}$, into the resulting equation allowed Menter to replace ε, which left k and ν_t as unknowns. The former was then eliminated using Bradshaw *et al.*'s relation, Eq. (8.44), with $\overline{uv}$ being related to the eddy viscosity via $-\overline{uv} = \nu_t\frac{\partial U}{\partial y}$. The coefficients were determined from those of the parent two-equation $k - \varepsilon$ model of Jones and Launder (1972) (discussed in Chapter 9), with an extension to low-Reynolds-number flows effected by the provision of pragmatic damping functions pre-multiplying the eddy viscosity and the turbulence-production terms (the latter f_ν in Eq. (8.46) below). The use of Eq. (8.44), obviated the need to prescribe a length scale, and the model would appear to be valid for any flow. Indeed, Menter provides the following generalisation of the thin-shear-flow form:

$$\frac{D\nu_t}{Dt} = C_1 f_\nu \nu_t S - C_2 \frac{\nu_t^2}{S^2}\left(\frac{\partial S}{\partial x_j}\right)^2 + \frac{\partial}{\partial x_j}\left(\left(\nu + \frac{\nu_t}{\sigma}\right)\frac{\partial \nu_t}{\partial x_j}\right). \quad (8.46)$$

However, it needs to be emphasized, again, that the key relation in Eq. (8.44), used by Menter to close the eddy-viscosity-transport equation, is rooted in thin-shear-flow considerations, so that the generalisation in Eq. (8.46) has unclear physical implications in non-equilibrium conditions. Nevertheless, Menter demonstrates that the equation performs as well as (indeed, slightly better than) the parent two-equation $k - \varepsilon$ model for two mildly separated flows.

8.4 *Summary and lessons*

The key feature of one-equation models is the use of a single differential transport equation, either for the turbulence energy, to which the eddy viscosity is related, or the eddy viscosity itself. This is a significant improvement on algebraic models, because the inclusion of convective and diffusive transport allows the state of the turbulence at any one location to be sensitised (as is appropriate) to the turbulent state at neighbouring locations. This is especially influential in regions of low or zero strain, at which algebraic models predict a vanishing level of turbulent viscosity, because of the rigid linkage between the viscosity and the strain rate these models entail. However, one-equation models are seriously compromised by the need to prescribe, algebraically and by reference to a global flow dimension, the length scale in models which use the turbulence-energy equation. Although some models based on a transport equation for the eddy viscosity (or the closely related turbulence Reynolds number) do not require an explicit prescription of a length scale, they incorporate assumptions and approximations that are closely tied to thin-shear-flow properties (e.g. turbulence equilibrium and the log-law), which are, implicitly, equivalent to prescribing a length scale.

In complex strain, away from the immediate vicinity of a wall, the turbulent length scale does not, nor is expected to, scale with the wall distance. Similarly, any one-equation model incorporating constraints that rely on the thickness of a free shear layer, or on specific ratios of stress to the turbulence energy, or on the velocity difference (squared) across the shear layer, cannot have general applicability. Rather, the length scale must be expected to be governed by local turbulence mechanisms which evolve in space and time. This implies the need to determine the length scale from its own transport equation — the defining argument forming the foundation for two-equation models covered in the next chapter.

Chapter Nine

9. Two-Equation Models

As pointed out in Chapter 8, the need to prescribe, algebraically, a length scale, or to introduce equivalent constraints applicable to thin shear flow, is a serious limitation of one-equation models. As the length scale is a representative property of the ensemble of eddies at any one spatial location, and as eddies evolve in space by virtue of advection, distortion and breakup (and possibly agglomeration), it is intuitively reasonable to propose models that represent the evolution of the length scale by a differential transport equation of the same type as that governing any other turbulence property — for example, the turbulence energy. The construction of models that follow this rationale is the subject of the present chapter.

9.1 *Options for length-scale surrogates*

The foundation of what follows in this section has already been laid in Section 7.5, where it was shown that the rate of turbulence-energy dissipation, $\varepsilon = \nu \overline{\frac{\partial u_i}{\partial x_j} \frac{\partial u_i}{\partial x_j}}$, is a surrogate of the length scale l_ε, and that an exact transport equation can be derived for this quantity. It was further argued that other surrogate variables can be derived, via the manipulations in Eq. (7.59). Because any two surrogates represent, in principle, precisely the same physical processes, the question as why we might wish to opt for any other variable than ε has no obvious answer at this stage. We shall discuss this subject in Section 9.4.

Table 9.1: Alternative length-scale-surrogate variables and their near-wall behaviour.

ϕ	Dimensions	$\nu_t \propto \ldots$	Wall-asymptotic behaviour	Wall value
ε	$[\mathrm{L}^2/\mathrm{T}^3]$	$\frac{k^2}{\varepsilon}$	$O(1)$	$2\nu\frac{k}{y^2}$
$\omega \equiv \frac{\varepsilon}{k}$	$[\mathrm{T}^{-1}]$	$\frac{k}{\omega}$	$O(y^{-2})$	∞
$\omega^2 \equiv \left(\frac{\varepsilon}{k}\right)^2$	$[\mathrm{T}^{-2}]$	$\frac{k}{\sqrt{\omega^2}}$	$O(y^{-4})$	∞
$\tau \equiv \frac{k}{\varepsilon}$	$[\mathrm{T}]$	$k\tau$	$O(y^2)$	0
$kL \equiv \frac{k^{5/2}}{\varepsilon}$	$[\mathrm{L}^3/\mathrm{T}^2]$	$(kL)/k^{1/2}$	$O(y^5)$	0
$L \equiv \frac{k^{3/2}}{\varepsilon}$	$[\mathrm{L}]$	$k^{1/2}L$	$O(y^3)$	0
$\nu_t \equiv C_\mu \frac{k^2}{\varepsilon}$ or $R_t \equiv \frac{\nu_t}{\nu}$	$[\mathrm{L}^2/\mathrm{T}]$	ν_t	$O(y^4)$	0
$\varsigma \equiv \frac{\varepsilon}{k^{1/2}}$	$[\mathrm{L}/\mathrm{T}^2]$	$\frac{k^{3/2}}{\varsigma}$	$O(y^{-1})$	∞

There are at least eight proposals for the surrogate variables, and these are given in Table 9.1, together with their dimension and their wall-asymptotic behaviour. A general equation for the surrogate ϕ will be given after the closure of the ε-equation. Of the variables given in Table 9.1, the first two are by far the most popular. The reason for ε being popular is obvious, but that for ω is not, at this stage, if only because of its singular wall condition: this variable approaches ∞ at the wall, and thus poses considerable uncertainty in respect of the boundary condition.

An important variant of ε is its so-called 'isotropic' form:

$$\tilde{\varepsilon} \equiv \varepsilon - \frac{\partial}{\partial x_j}\left(\nu \frac{\partial k}{\partial x_j}\right). \tag{9.1}$$

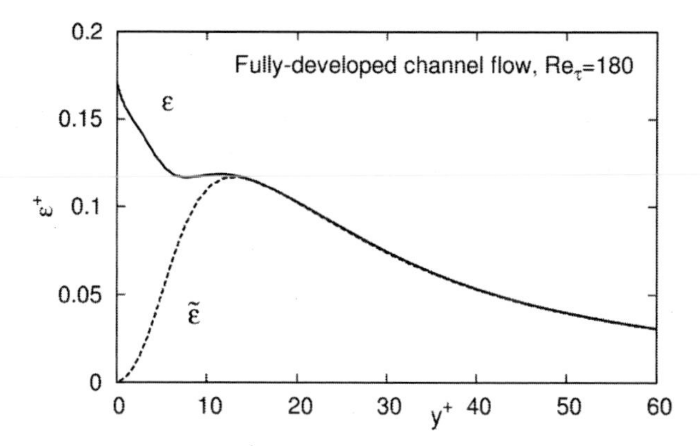

Fig. 9.1: Comparison of dissipation and its 'isotropic' form in a channel flow; derived from DNS data by Kim *et al.* (1987) (courtesy S. Jakirlić).

Its importance lies in the fact that it is used, in some models, in place of ε as the subject of the dissipation-rate equation, as well as in the turbulent-viscosity expression, Eq. (7.60). The motivation for using $\tilde{\varepsilon}$ lies in its asymptotic near-wall behaviour, illustrated in Fig. 9.1 for a channel flow. Because of the quadratic near-wall variation of k at the wall (see Eq. (5.4)), its wall-limiting value is:

$$\tilde{\varepsilon}\big|_{y\to 0} = \varepsilon - 2\nu\left(\frac{\partial k^{\frac{1}{2}}}{\partial y}\right)^2 = \varepsilon - 2\nu\frac{k}{y^2}. \tag{9.2}$$

As shown by Eq. (5.6) and in Fig. 9.1, the wall value $\tilde{\varepsilon}\big|_{y\to 0}$ vanishes, and this decouples the wall-boundary condition for $\tilde{\varepsilon}$ from the solution of k, this coupling being a potential source of numerical instability if ε is the subject of the equation being solved. As seen in Fig. 9.1, $\tilde{\varepsilon}$ is virtually identical to ε outside the viscosity-affected sublayer.

9.2 *The basic $k - \varepsilon$ model*

The route to the formulation of a working model that is based on transport equations for the turbulence energy and its rate of dissipation is marked with a number of influential contributions in the late 1960s and early 1970s, some highlighted in the paper by Jones and Launder (1972). One of the most important milestones is a basic $k - \varepsilon$ model presented by Hanjalić (1970), and applied to a number of high-Reynolds-number flows. However, over the years, the $k - \varepsilon$ model has come to be associated with Jones and Launder whose 1972 paper focuses, primarily, on a low-Reynolds-number form of the model, capable of predicting relaminarisation of boundary layers subjected to strong acceleration.

The 'basic' model consists of Eq. (7.32), governing k, the eddy-viscosity relation, Eq. (7.60),[1] restated below:

$$\nu_t = C_\mu \frac{k^2}{\varepsilon}, \tag{9.3}$$

and a modelled version of the transport equation (7.55) for ε. The adjective 'basic' is meant to convey the fact that this model is only applicable to fully turbulent conditions, and this version is thus usually referred to as the 'high-Reynolds-number' form. This means that any considerations pertaining to the wall-asymptotic behaviour are irrelevant at this stage. Extensions to account for the effects of the fluid viscosity in the near-wall region will be discussed in Section 9.3.

The k-equation requires no further attention, except for the determination of the Prandtl–Schmidt number, σ_k. Moreover, the constant C_μ in Eq. (9.3) remains to be determined.

The exact ε-equation, Eq. (7.55), is reproduced below, in a slightly different form:

$$\frac{D\varepsilon}{Dt} = \underbrace{-2\nu \left[\overline{\frac{\partial u_i}{\partial x_k}\frac{\partial u_j}{\partial x_k}} + \overline{\frac{\partial u_k}{\partial x_i}\frac{\partial u_k}{\partial x_j}} \right] \frac{\partial U_i}{\partial x_j}}_{\text{Generation by mean strain, } P_{\varepsilon 1}+P_{\varepsilon 2}} \underbrace{-2\nu \overline{u_k \frac{\partial u_i}{\partial x_j}} \frac{\partial^2 U_i}{\partial x_k \partial x_j}}_{\text{Generation fragment, } P_{\varepsilon 3}}$$

$$\underbrace{-2\nu \left[\overline{\frac{\partial u_i}{\partial x_k}\frac{\partial u_i}{\partial x_m}\frac{\partial u_k}{\partial x_m}} \right]}_{\text{Generation by turbulent straining, } P_{\varepsilon 4}}$$

[1] C_μ is to be distinguished from $c_\mu = 0.3$, Eq. (7.43).

$$+ \underbrace{\frac{\partial}{\partial x_j}\left[-2\nu\overline{\frac{\partial p}{\partial x_m}\frac{\partial u_j}{\partial x_m}} - \nu\overline{u_j\frac{\partial u_i}{\partial x_m}\frac{\partial u_i}{\partial x_m}}\right] + \nu\nabla^2\varepsilon}_{\text{“Diffusive” Transport, } T_\varepsilon}$$

$$- \underbrace{2\nu^2\overline{\frac{\partial^2 u_i}{\partial x_k \partial x_m}\frac{\partial^2 u_i}{\partial x_k \partial x_m}}}_{\text{Destruction, } D_\varepsilon}. \tag{9.4}$$

A fact that is relevant to both the interpretation of the terms in Eq. (9.4) and their subsequent modelling is that there is a close relationship between the dissipation rate and the 'enstrophy' — the scalar norm of the vorticity fluctuations — defined as $\overline{\omega^2} = \overline{\omega_i\omega_i}$, where ω_i is the curl of the velocity fluctuation $\varepsilon_{ijk}\frac{\partial u_k}{\partial x_j}$ (see Appendix for the definition of ε_{ijk}). Because $\frac{\varepsilon}{\nu} = \overline{\frac{\partial u_i}{\partial x_j}\frac{\partial u_i}{\partial x_j}}$, it is not difficult to recognize, in principle, that the dissipation is closely linked to the vorticity fluctuations. In fact, at sufficiently high Reynolds numbers, $\overline{\omega^2} = \varepsilon/\nu$, and there are several experimental and simulation studies for wall-bounded flows (e.g. Zhu and Antonia (1997), Blackburn, Mansour and Cantewell (1996), and Yeung, Donzis and Sreenivasan (2005)) that show that the statistics of the two quantities are close, except near the wall, where Reynolds-number-related effects are influential. This close relationship makes clear that the dissipation process, which occurs at the smallest length scales, is intimately connected to the way in which vortices (eddies) are stretched by mutual interactions, thus driving the eddies progressively towards smaller scales and hence destruction by viscosity.

Naturally, there is also a close relationship between Eq. (9.4) and the enstrophy-transport equation, which is not given here in full, but can be found in Tennekes and Lumley (1972). For example, the major generation term in the enstrophy equation is $\overline{\omega_i\omega_j\frac{\partial u_i}{\partial x_j}}$, which represents a vortex-stretching process by turbulent-velocity gradients, and this is equivalent to $P_{\varepsilon 4}$ in Eq. (9.4), while the major sink term is $\nu\overline{\frac{\partial \omega_i}{\partial x_j}\frac{\partial \omega_i}{\partial x_j}}$, which is equivalent to the destruction term D_ε in the same equation.[2]

[2]Note that the correspondence is between $\overline{\omega^2}$ and ε/ν.

The following interpretation of the terms in Eq. (9.4) is aided by its juxtaposition with Eq. (4.5) for the shear stress, and with Eq. (7.27) for the turbulence energy. However, in anticipation of what follows, caution is called for, so as not to link the qualitative term-by-term comparisons made below to any quantitative equivalence. In particular, note that all terms in Eq. (9.4) are premultiplied by the fluid viscosity, implying that all processes affecting the dissipation occur, fundamentally, at the high-wave-number range of the eddy-size spectrum (see Fig. 2.4(b)), at which the viscosity is able to interact most effectively with the turbulent strains.

We now discuss the individual groups of terms in Eq. (9.4):

- The first term on the right-hand side, $P_{\varepsilon 1}+P_{\varepsilon 2}$, corresponds to correlations of velocity fluctuations, multiplied by mean strains in the k-equation. Here, the equivalent is a correlation of *derivatives* of velocity fluctuations multiplied by mean-strain terms. Physically, these terms can be interpreted as generating dissipation by stretching of eddies through mean straining, thus making them rotate faster and enhancing the transfer of mechanical energy to heat.

- The physical process which the second term, $P_{\varepsilon 3}$, represents is not obvious, and this term is commonly referred to as 'gradient production'. It is a fragment that is loosely associated with the contribution of the mean velocity to turbulent transport, but it does not satisfy the constraints that allow it to be interpreted as such, and it is thus only interpretable as a source (generation) term that is associated with the mean vorticity (the spatial derivative of the velocity gradient). In any event, this term is a minor contributor, because it is of first order in the turbulent strain, and also because the derivatives of the mean-strain components are generally small.

- The third term, $P_{\varepsilon 4}$, is akin to the first, but it generates dissipation by the action of the more intense turbulent-strain terms (a self-induced vortex-stretching process), as compared to the lower mean-strain components. Therefore, this term can be expected to be the dominant source of generation, because it is cubic in the turbulent strains. As noted earlier, this is also the term equivalent to the major generator of turbulent enstrophy $\overline{\omega_i\omega_j\frac{\partial u_i}{\partial x_j}}$. It is thus expected to rise continuously with Reynolds number, as the smallest scales driving this destruction decline progressively, thus creating ever larger gradients of the velocity fluctuations, which stretch the small-scale vortices.

- The nature of the penultimate term, T_ε, follows from the observation that its integral over any closed domain Ω with boundary Γ vanishes if Neumann (zero-gradient) conditions are applied on Γ. This collection of terms is equivalent to the corresponding turbulent transport and viscous diffusion in Eq. (7.27), except that velocity fluctuations are replaced by *derivatives* of these fluctuations.
- The last term, D_ε, corresponds to the dissipation in Eq. (7.27) — again, with velocity fluctuations replaced by their derivatives. It can be expected to be the major sink term opposing the generation of ε. As is the case with $P_{\varepsilon 4}$, D_ε is also characterized by a dependence on derivatives of velocity fluctuations only. Hence, this term is also expected to rise progressively with increasing Reynolds number.

Prior to any attempt to model this difficult equation, it is instructive to refer to direct numerical simulation (DNS) data that give information on the relative contribution of the terms in Eq. (9.4). Such a budget, given in Fig. 9.2, also helps to confirm — or possibly discount — the notional interpretation presented above, although its limitation, associated with the low Reynolds number of the flow, needs to be noted. The generation of such data, at acceptable accuracy, is a major challenge in itself, because all terms are associated with processes at the small-scale end of the energy spectrum. Hence, such data only exist for relatively low Reynolds numbers.

The budget in Fig. 9.2 suggests the following:

(i) In the log-law region, away from the wall, in which turbulence-energy equilibrium applies, the leading generation term, $P_{\varepsilon 4}$ in Eq. (9.4), is almost balanced by the viscous destruction term,

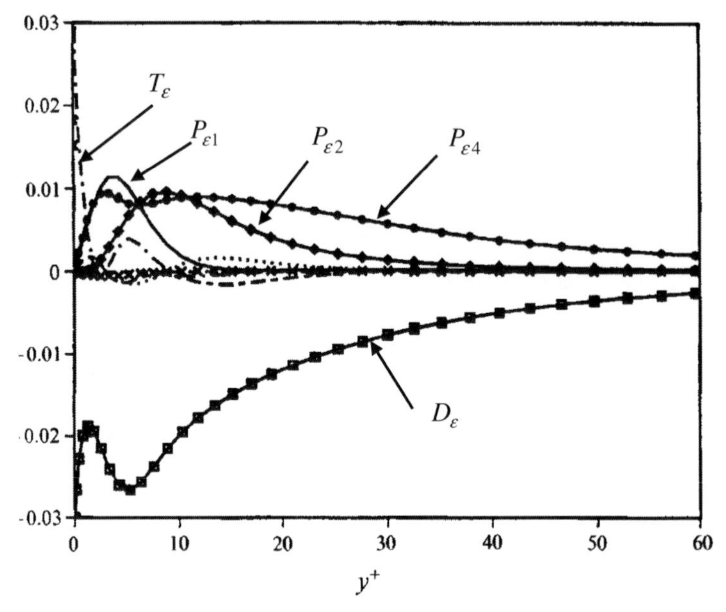

Fig. 9.2: Budget of terms in Eq. (9.4) for channel flow at bulk Reynolds-number of 6670; derived from DNS data (reproduced with permission of Cambridge University Press from Rodi and Mansour (1993)).

D_ε, the difference being made up by an indistinct combination of the remaining production terms and turbulent (diffusive) transport. In essence, this suggests the existence of a k-equivalent 'dissipation-rate equilibrium', wherein generation roughly balances destruction. It is recalled, that a dominance of $P_{\varepsilon 4}$ and D_ε was suggested earlier, purely on the basis of a qualitative examination of these two terms, relative to others.

(ii) In the viscosity-affected layer, away from the immediate vicinity of the wall, viscous destruction is balanced by contributions of similar magnitude from $P_{\varepsilon 1}, P_{\varepsilon 2}$ and $P_{\varepsilon 4}$, and from a relatively small turbulent-transport contribution.

(iii) Very close to the wall, destruction is balanced by viscous diffusion. This is, yet again, similar to the behaviour observed in the energy budget, Fig. 5.4.

While the budget shows complex features near the wall, the simpler behaviour in the outer region can be explained by an order-of-magnitude analysis. Such an analysis was performed by Rodi and Mansour (1993), guided by a corresponding analysis by Tennekes and Lumley (1972) of the enstrophy equation. This shows that, at high values of the Reynolds number (e.g. in the log-law region of a boundary layer), the terms $P_{\varepsilon 4}$ and D_{ε} (see (i) above) both increase with the macro-scale Reynolds number, $Re_t = k^{0.5}l/\nu$ (l being associated with the large, energetic eddies), while their *difference*, as well as T_{ε}, are independent of Re_t, and $P_{\varepsilon 1}, P_{\varepsilon 2}$ decline with Re_t. The above observations suggest that, subject to a sufficiently high Reynolds number, the principal process that needs to be modelled is that represented by $P_{\varepsilon 4} - D_{\varepsilon}$, with the difference balanced by turbulent transport.

Ideally, in the view of the complex budget near the wall in Fig. 9.2, the approach to closing Eq. (9.4) would entail a term-by-term approximation process — a route attempted by Rodi and Mansour (1993) and (to a lesser extent) Jakirlić and Hanjalić (2002), on the basis of DNS data for channel flow. However, this turns out to be a very challenging exercise, not only because of the complexity of the terms involved, but also because their small-scale nature demands extremely high computational fidelity of the DNS data, only available for low-Reynolds-number channel and pipe flow ($Re_h = O(5000)$, h being the channel half-height or pipe radius). The process thus followed by Rodi and Mansour has not led to a working model to replace earlier forms. In contrast, Jakirlić and Hanjalić report modest improvements with their form (albeit within the framework of second-moment closure) in some low-Reynolds-number flows.

As noted above, when attention is restricted to the outer, high-Reynolds-number region of the near-wall layer, our primary task is to approximate $P_{\varepsilon 4} - D_{\varepsilon}$. We know little about the fundamental processes that these terms represent, as they occur at the extreme small-scale end of the eddy-size spectrum. We do know, however, that the difference needs to be finite, at the turbulence-energy-equilibrium state, and that the difference *does not* depend on the Reynolds number. Therefore, the only tenable option we have is to approximate the difference by reference to the Eq. (7.27) for k, based on the notion that $P_{\varepsilon 4} - D_\varepsilon$ is dictated by the cascade of energy across the inertial sub-range (see Fig. 2.4 and related discussion in Chapter 2). This is the point of view advocated by Harlow and Nakayama (1968), Hanjalić (1970), Jones and Launder (1972) and Launder, Reece and Rodi (1975).

The model proposed is, therefore:

$$D_\varepsilon - P_{\varepsilon 4} = \frac{\varepsilon}{k}(C_{\varepsilon 1} P_k - C_{\varepsilon 2}\varepsilon), \tag{9.5}$$

in which the premultiplier ε/k is an inverse of the integral (macro) time scale that secures dimensional homogeneity, and $C_{\varepsilon 1}, C_{\varepsilon 2}$ are constants that remain to be determined. Importantly, the constants have to be different, so as to secure a finite value of the modelled difference at the condition of turbulence-energy equilibrium.

An alternative viewpoint, leading to the same result, is based on the observation that the group of terms within the square brackets in $P_{\varepsilon 1} + P_{\varepsilon 2}$ is a second-rank tensor involving only velocity fluctuations. Approximating this as being proportional to $\overline{u_i u_j}$ ensures tensorial consistency, as well as aiding closure, for if this tensor is multiplied by $\partial U_i/\partial x_j$ the result is clearly P_k. This suggests:

$$P_{\varepsilon 1} + P_{\varepsilon 2} = C_{\varepsilon 1}\frac{\varepsilon}{k} P_k. \tag{9.6}$$

Equation (9.6) appears defensible in view of the behaviour observed in Fig. 9.2 in the buffer layer. However, an objection to this proposal is that $P_{\varepsilon 1} + P_{\varepsilon 2}$ depends, as noted earlier, on the Reynolds number, requiring the former to decrease as the latter increases. Given the assumption of Eq. (9.6), we require a destruction term that opposes the unbounded increase in the production of ε , implied by this equation. We know that the relevant sink is $D_\varepsilon - P_{\varepsilon 4}$. As ε is opposing P_k in the turbulence-energy equation, thus keeping k at the right equilibrium level, it is reasonable to propose, as an extension of Eq. (9.6), that the destruction of ε should be proportional to ε itself. Again, this has to be divided by a time scale, already chosen to be k/ε. Hence:

$$D_\varepsilon - P_{\varepsilon 4} = C_{\varepsilon 2}\frac{\varepsilon}{k}\varepsilon. \tag{9.7}$$

As both sides of Eq. (9.7) are now independent of the Reynolds-number, the objection raised in relation to Eq. (9.6) does not apply here.

Finally, the turbulent-transport terms may be approximated by a gradient-diffusion hypothesis, as done in the case of Eq. (7.32) for k. Hence:

$$T_\varepsilon = \frac{\partial}{\partial x_j}\left(\frac{\nu_t}{\sigma_\varepsilon}\frac{\partial \varepsilon}{\partial x_j}\right) + \nu \nabla^2 \varepsilon, \tag{9.8}$$

in which the second term (although correct) should strictly be nullified, because the modelling is restricted, at this stage, to fully turbulent conditions, in which case viscous effects are assumed negligible.

With the above proposals accepted, the modelled form of Eq. (9.4) is:

$$\frac{D\varepsilon}{Dt} = C_{\varepsilon 1}\frac{\varepsilon}{k}P_k - C_{\varepsilon 2}\frac{\varepsilon^2}{k} + \frac{\partial}{\partial x_j}\left(\frac{\nu_t}{\sigma_\varepsilon}\frac{\partial \varepsilon}{\partial x_j}\right). \tag{9.9}$$

Equations (7.32), (9.3) and (9.9) involve five constants: σ_k, C_μ, $C_{\varepsilon 1}$, $C_{\varepsilon 2}$ and σ_ε. These are determined as follows.

(i) It is recalled that Wolfshtein used $\sigma_k = 1.52$ in his one-equation model — see Eq. (8.11). However, the choice $\sigma_k = 1.0$ has, arguably, stronger rational merit. This is because the turbulence energy is closely tied to the velocity (i.e. linear momentum), in which case, the Prandtl–Schmidt number associated with turbulent transport ought to be 1.0. This is, indeed, the value that is used in the basic model.

(ii) For a thin shear flow, turbulence-energy equilibrium is assumed to prevail outside the viscous sublayer, and this is expressed by:

$$-\overline{uv}\frac{\partial U}{\partial y} = \varepsilon. \tag{9.10}$$

With Eq. (9.3), the shear stress may then be written:

$$-\overline{uv} = C_\mu \frac{k^2}{\varepsilon}\frac{\partial U}{\partial y}. \tag{9.11}$$

Elimination of the velocity gradient from Eqs. (9.10) and (9.11) then leads to:

$$C_\mu = \left(\frac{\overline{uv}}{k}\right)^2. \tag{9.12}$$

But Eq. (7.43) shows this to be simply $c_\mu^2 = 0.09$. It is important to recognise that Eq. (9.12) does not *hard-wire* the model to this condition in the sense of rigidly fixing $\overline{uv}$ to k. Rather, it *anchors* the model to this condition in the particular case when equilibrium prevails.

(iii) The x-wise decay of isotropic turbulence in a uniform stream is governed by:

$$\begin{aligned} U\frac{dk}{dx} &= -\varepsilon \\ U\frac{d\varepsilon}{dx} &= -C_{\varepsilon 2}\frac{\varepsilon^2}{k}. \end{aligned} \tag{9.13}$$

In Section 7.3, it is already stated that experimental evidence shows $k \propto x^{-1.2}$. This allows the variation of ε to be immediately determined from the first relationship in Eq. (9.13) and to be substituted, together with k, in the second, to give $C_{\varepsilon 2} = 1 + \frac{1}{1.2} = 1.83$. However, this value is not rigidly fixed, as the exponent in the decay law varies between 1 and 1.25, depending on the precise experimental data used — i.e. $C_{\varepsilon 2}$ may be as high as 2. In the standard form of the model, the value used is $C_{\varepsilon 2} = 1.92$. As will be seen shortly, it is actually the *difference* between $C_{\varepsilon 2}$ and $C_{\varepsilon 1}$ which is more important than the value of $C_{\varepsilon 2}$ itself.

(iv) In the log-law region of a boundary layer, local turbulence-energy equilibrium $P_k = \varepsilon$ prevails. In this region, Eq. (9.9) reduces to:

$$C_{\varepsilon 1}\frac{\varepsilon}{k}\overline{uv}\frac{dU}{dy} - C_{\varepsilon 2}\frac{\varepsilon^2}{k} + \frac{d}{dy}\left(\frac{\nu_t}{\sigma_\varepsilon}\frac{d\varepsilon}{dy}\right) = 0. \tag{9.14}$$

Note that the transport term cannot vanish, because the production and destruction terms do not balance, unless $C_{\varepsilon 2} = C_{\varepsilon 1}$. This would lead to a pathological form of Eq. (9.9), in which

ε would follow (wrongly) from a balance between convection and diffusion, both of which are small. We can now make the following substitutions:

- The log-law and near-wall constancy of the shear-stress (applicable in zero or mild streamwise pressure gradient) give:

$$\frac{dU}{dy} = \frac{u_\tau}{\kappa y}, \qquad -\overline{uv} = u_\tau^2. \tag{9.15}$$

- Turbulence-energy equillibrium is expressed by:

$$\varepsilon = P_k = u_\tau^2 \frac{u_\tau}{\kappa y}. \tag{9.16}$$

- Combination of Eq. (9.12) with (9.15) gives:

$$k = C_\mu^{-\frac{1}{2}} u_\tau^2. \tag{9.17}$$

Next, a few straightforward algebraic manipulations of Eq. (9.14) yield:

$$C_{\varepsilon 1} = C_{\varepsilon 2} - \frac{\kappa^2}{C_\mu^{\frac{1}{2}} \sigma_\varepsilon}. \tag{9.18}$$

This allows $C_{\varepsilon 1}$ to be determined, provided σ_ε and $C_{\varepsilon 2}$ are known (Note that Eq. (9.18) is, essentially, a constraint on the difference $(C_{\varepsilon 1} - C_{\varepsilon 2})$). The final step is now to perform a range of numerical computations with the model for a boundary layer, a mixing layer, a plane jet and a round jet with difference combinations of σ_ε and $C_{\varepsilon 2}$ to determine an 'optimum' pair of values. This optimisation process turns out to give $C_{\varepsilon 1} = 1.44$, $\sigma_\varepsilon = 1.3$.

We expect the above model to perform well in flows that are not far removed from turbulence-energy equilibrium. Moreover, because of the calibration step (iii) above, we also expect the model to return the correct behaviour when advection is dominant in weak shear. Can we subject the model to a simple test in shear conditions in which the flow is far from equilibrium? The answer is 'yes'. To this end, we consider homogeneous shear $S = \frac{\partial U}{\partial y} =$ constant, a flow that can be created (albeit with some difficulty) in a wind tunnel by

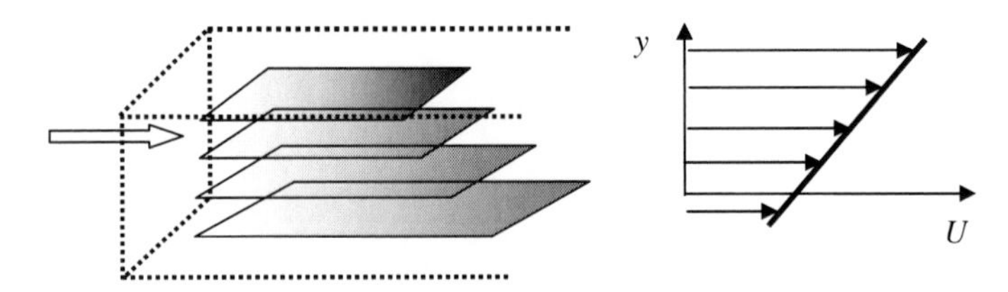

Fig. 9.3: Homogeneous shear flow: $S = \frac{\partial U}{\partial y}$ = constant.

inserting into it a stack of flat plates of different lengths, as shown in Fig. 9.3.

In this case, the $k - \varepsilon$ model reduces to:

$$U\frac{dk}{dx} = C_\mu \frac{k^2}{\varepsilon} S^2 - \varepsilon,$$
$$U\frac{d\varepsilon}{dx} = C_\mu C_{\varepsilon 1} k S^2 - C_{\varepsilon 2}\frac{\varepsilon^2}{k}. \tag{9.19}$$

Rather than following the streamwise evolution of the turbulence, we can choose to move *with* the flow, and follow the behaviour of the turbulence with time, i.e.: $dx = Udt$. Over time, both k and ε increase progressively as production and dissipation are not balanced. However, the time scale $\tau = k/\varepsilon$ turns out to approach a constant value. This emerges upon combining the two equations in Eq. (9.19) to give:

$$\frac{d\tau}{dt} = (C_{\varepsilon 1} - 1)C_\mu(\tau S)^2 + (C_{\varepsilon 2} - 1). \tag{9.20}$$

The solution of this equation is a hyperbolic tangent, asymptoting to $\frac{d\tau}{dt} \to 0$ as $t \to \infty$. Eq. (9.20) can then be written:

$$C_\mu(\tau S)^2 = \frac{P_k}{\varepsilon} = \frac{(C_{\varepsilon 2} - 1)}{(C_{\varepsilon 1} - 1)}. \tag{9.21}$$

With $C_{\varepsilon 1} = 1.44$, $C_{\varepsilon 2} = 1.92$, the result is $\frac{P_k}{\varepsilon} = 2.09$, which is larger than the experimentally observed value of 1.88. However, the choice $C_{\varepsilon 2} = 1.83$, suggested above on the basis of calibration by reference to grid-turbulence decay (Eq. (9.13)), Eq. (9.21) gives the correct experimental value for P_k/ε.

9.3 *Low-Reynolds-number k − ε-model extensions*

The technical literature abounds with model variations that extend the above basic form to the viscous near-wall region. To a significant degree, this reflects lack of insight into the near-wall physics and the absence of reliable data about the near-wall variations at the time most models were formulated, both contributing to a lack of rigour and a hazy modelling path. Some alternative model forms are those of Jones and Launder (1972), Launder and Sharma (1974), Hassid and Poreh (1975), Hoffman (1975), Reynolds (1976), Lam and Bremhorst (1981), Chien (1982), Nagano and Hishida (1987), So *et al.* (1988), Myong and Kasagi (1990), Shih and Mansour (1990), Nagano and Tagawa (1990), Coakley and Huang (1992), Orszag *et al.* (1993), Kawamura and Kawashima (1994), Lien and Leschziner (1994, 1996), and several others. Patel, Rodi and Scheuerer (1985) and Rodi and Mansour (1993) present comparisons of several of the above models, the latter by reference to DNS data for channel flow that had become available around 1990.

Apart from significant variations in the near-wall modifications adopted in the above models, several use non-standard values for the basic constants. An extreme example is the model by Hoffman, which uses the set $\sigma_k = 2.0$, $C_\mu = 0.09$, $C_{\varepsilon 1} = 1.81$, $C_{\varepsilon 2} = 2.0$, and $\sigma_\varepsilon = 3.0$. While this set satisfies Eq. (9.18), and the strain-free turbulence-energy-decay law, it arises from a different optimisation route than that adopted for the basic model and explained in Section 9.2. Another major point of distinction among low-*Re* $k-\varepsilon$ models is that some use the 'isotropic-dissipation' rate $\tilde{\varepsilon} \equiv \varepsilon - 2\nu \left(\frac{\partial k^{1/2}}{\partial x_j}\right)^2$ (see Eq. 9.2) instead of ε , for reasons explained in Section 9.1. Outside the viscous sublayer, this distinction is inconsequential, but close to the wall, there are a number of important implications. One is that the dissipation rate ε in the k-equation requires the determination of the fragment $2\nu\left(\frac{\partial k^{1/2}}{\partial x_j}\right)^2$ that has to be added when ε is replaced by $\tilde{\varepsilon}$.

It is not the purpose of this section to cover the multiplicity of variants and variations that have emerged over a period of around 40 years, some rather unclear in origin. Rather, in harmony with the declared aims of this book, we focus on some basic forms that illustrate the physical principles involved and the rationale underpinning modifications to the basic model.

The need to extend the model given in Section 9.2 arises, principally, from two sources: the use of the constants C_μ, Eq. (9.12), and $C_{\varepsilon 2}$, derived from Eq. (9.13). A third — the omission of the viscous-diffusion terms from the k- and ε-equations — is trivial and readily reversed.

It is recalled, first, that $C_\mu = 0.09$ was derived by reference to the log-law. As the wall is approached, inviscid wall-blocking and viscosity combine to depress wall-normal mixing. Equation (7.8), in Section 7.1, suggests that the shear stress should comply with:

$$-\overline{uv} = C \underbrace{\left(\frac{k}{\varepsilon}\right)}_{\tau} \overline{v^2} \frac{dU}{dy}, \qquad (9.22)$$

in which the bracketed term is assumed to be the appropriate time scale τ. When comparing Eq. (9.3) with Eq. (9.22), we may assume that outside the immediate near-wall layer — say $y^+ > 50$ — the calibration $C_\mu = 0.09$ accounts for the fact that $\overline{v^2}$ is significantly smaller than k, $\overline{v^2}/k$ being, typically, of order 0.3. However, as the wall is approached, the disparity between $\overline{v^2}$ and k increases, with the wall-asymptotic variation following $\overline{v^2}/k = O(y^2)$ in the viscous sublayer. The correct wall-asymptotic variation of $\overline{uv}$ is also important, because it influences the near-wall behaviour of k, via the generation term in Eq. (7.27) governing k.

We have shown earlier, in Chapter 5, that the wall-asymptotic behaviour is described by $\overline{uv} = O(y^3)$, $\overline{v^2} = O(y^4)$, $k = O(y^2)$, $\varepsilon = O(1)$ and $\frac{dU}{dy} = O(1)$. It follows that, provided all quantities on the right hand side of Eq. (9.22) are made to adhere to their respective constraints, an invariant C in Eq. (9.22) implies $\overline{uv} = O(y^6)$. This suggests, in turn, that C must be made to follow $C = O(y^{-3})$. Even with $\nu_t = C_\mu \frac{k^2}{\varepsilon}$, Eq. (9.3), the implication is that we require the variation $C_\mu = O(y^{-1})$. Apart from the fact that Eq. (9.22) is largely qualitative in nature, one uncertainty is that the time scale $\tau = \frac{k}{\varepsilon}[= O(y^2)]$ may be inappropriate at the wall.

If we propose to use $\tau = \frac{k}{\tilde{\varepsilon}}$ as an alternative, we need to examine the asymptotic variation of $\tilde{\varepsilon}$. From Eq. (7.27), we may take the definition of ε as:

$$\varepsilon = \nu \overline{\left(\frac{\partial u_i}{\partial x_j}\right)^2} = \nu \left\{ \overline{\frac{\partial u}{\partial y}\frac{\partial u}{\partial y}} + \overline{\frac{\partial v}{\partial y}\frac{\partial v}{\partial y}} + \overline{\frac{\partial w}{\partial y}\frac{\partial w}{\partial y}} + \cdots \right\}. \tag{9.23}$$

Next, we substitute the Taylor expansions, Eq. (5.1), into the above, noting that u and w vary linearly with y, while v varies quadratically, followed by time averaging. This straightforward process leads to:

$$\varepsilon = \alpha + \beta y + \cdots , \tag{9.24}$$

where $\alpha = 2\nu \frac{k}{y^2} \left(= 2\nu \left(\frac{\partial k^{1/2}}{\partial y}\right)^2\right)$, which we know to be the asymptotic wall value at $y = 0$. Hence, clearly, $\tilde{\varepsilon} = O(y)$ close to the wall, and $\frac{k}{\tilde{\varepsilon}} = O(y)$, but this does not allow an invariant C either, as the right-hand side of Eq. (9.22) would still give the incorrect variation $\overline{uv} = O(y^5)$.

On physical grounds, it is reasonable to argue that scaling very close to the wall should entail the use of the Kolmogorov length scale — see Eqs. (2.8)–(2.10). If this is accepted, an alternative time scale is:

$$\tau = \frac{\eta}{k^{\frac{1}{2}}} = \frac{1}{k^{\frac{1}{2}}} \left(\frac{\nu^3}{\varepsilon}\right)^{\frac{1}{4}}, \tag{9.25}$$

which has the wall-asymptotic variation $O(y^{-1})$, thus yielding the correct asymptotic variation of $\overline{uv}$, through Eq. (9.22), given an invariant C. The question of which time scale is appropriate will be pursued at a later stage. For now, we shall continue to work with Eq. (9.3), if only because it underpins all traditional $k - \varepsilon$ model forms — albeit with ε replaced by $\tilde{\varepsilon}$ in some models, in which case the requisite wall-asymptotic behaviour is secured by $C_\mu = O(y^{-1})$ and $C_\mu = O(1)$, respectively.

The need to modify $C_{\varepsilon 2}$ arises from the fact that the decay rate of the turbulence energy depends on the viscosity at low Reynolds-number values, in which case the energy spectrum is narrow, and the $-5/3$ inertial sub-range, Fig. 2.4, is absent. As noted in Section 7.3, the rate of decay increases from $k \propto x^{-1.2}$ to $k \propto x^{-2.5}$ as the decay rate is in its final stages, thus implying the need for a viscosity-dependent damping function.

A common feature of all low-Reynolds-number modifications is the reliance on viscosity-dependent functions to achieve the desired wall-induced damping. More specifically, this is effected by way of the arguments y^+, y^* or R_t, which represent, in alternative non-dimensional forms, the effects of wall-normal distance and viscosity. It has already been shown, through Eq. (8.5), that $y^* = 1.8y^+$ in the log-law region. Next, we take advantage of two results derived for the log-law region, as part of the determination of $C_{\varepsilon 1}$ in Section 9.2, namely Eqs. (9.16) and Eq. (9.17), which are:

$$k = C_\mu^{-\frac{1}{2}} u_\tau^2, \qquad \varepsilon = P_k = u_\tau^2 \frac{u_\tau}{\kappa y}. \tag{9.26}$$

These can be combined to show,

$$R_t = \frac{k^2}{\nu \varepsilon} = \frac{\kappa}{C_\mu} \frac{y u_\tau}{\nu} = 4.7y^+. \tag{9.27}$$

It is emphasized that these relations only apply to the turbulence-equilibrium region in a canonical near-wall layer.

Whatever parameter is adopted to represent near-wall damping, the use of viscous parameters is physically tenuous, because any compensation for the use of k as a proxy for $\overline{v^2}$ in Eq. (9.3) has to reflect, primarily, inviscid wall blocking. While viscosity undoubtedly plays a role in the damping process, the dominant effect arises from the disproportionate depression of the wall-normal velocity fluctuations. This is also the mechanism which necessitated the use of large values for the constants in the exponential van Driest damping functions in the one-equation models, as discussed in Chapter 8. Unsurprisingly, in view of this non-rigorous use of viscosity-dependent damping models in the present context, there are significant differences among the various proposals put forward.

We now return our attention to Eq. (9.3), multiplied by a damping function f_μ, which may reasonably be assumed to be of the van Driest form:

$$f_\mu = 1 - \alpha \exp(-\beta \, arg), \tag{9.28}$$

where *arg* is either y^+ or y^* or R_t. It is emphasized that there is no physically profound reason for the exponential function; it is used here because it has properties that appeal to our intuition of how a damping process could sensibly be representable mathematically. We wish to secure the correct variation of $\overline{v^2}/k$, the correct y-wise recovery towards 1.0 in the fully turbulent regime, and the correct asymptotic variation $-\overline{uv} = O(y^3)$ or $-\frac{uv}{u_\tau^2} = O(y^{+3})$ — which, we suspect, we shall not be able to achieve along this route. Our hope (perhaps even expectation) is that this violation will not be of major consequence, if only because the turbulent shear stress is dynamically un-influential below $y^+ \approx 5$.[3]

The correct f_μ variation can be obtained, experimentally, or from a DNS solution, using:

$$f_\mu = \frac{-\overline{uv}}{C_\mu k}\left(\frac{\varepsilon}{k}\right)\frac{1}{\frac{dU}{dy}}. \tag{9.29}$$

Data derived upon processing DNS and experimental results for $U, \overline{uv}, k$ and ε are shown in Fig. 9.4. Plot 9.4(b) has been taken from Patel, Rodi and Scheuerer (1985), their experimental data being the only data available at the time when some of the low-Reynolds-number models were formulated. Also included in the figure are the functions used in three models, the large variations being indicative of the non-rigorous nature of the approximation process, in some cases undertaken intuitively, before any experimental data were available.

Both the experimental and DNS data sets show very similar rates of decline of f_μ from $y^+ \approx 60$ down to $y^+ \approx 5$. Although this decline is expected, in view of the foregoing considerations of the relationship between $\overline{v^2}$ and k, it appears to clearly contradict the

[3]Recall that the turbulent and viscous shear-stress components are equal at $y^+ \approx 11$, and the former declines rapidly (in proportion to y^3) as the wall is approached.

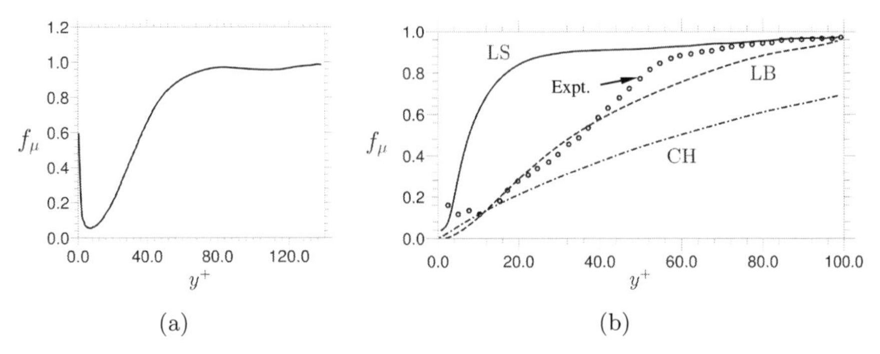

Fig. 9.4: Variations of f_μ, (a) derived from DNS of channel flow at $Re_\tau = 390$ (coutesy S. Jakirlić), and (b) from experiments and the three model variants discussed in Section 9.3.1 (reproduced with permission of The American Institute of Aeronautics and Astronautics (AIAA) from Patel, Rodi and Scheuerer (1985)).

asymptotic requirement $f_\mu C_\mu = O(y^{-1})$. However, below $y^+ \approx 5$, f_μ rises — steeply, in the case of the DNS data — and this does comply, qualitatively, with the $f_\mu C_\mu = O(y^{-1})$ variation. The experimental data also show a rise below $y^+ \approx 5$, but the experimental accuracy is uncertain very close to the wall. The fact that f_μ reverses at $y^+ \approx 5$, and that it rises towards 1 only at $y^+ \approx 60$, suggests (i) that the wall-asymptotic considerations are of very limited utility, since turbulence is very weak below $y^+ \approx 5$, and (ii) that, contrary to the concept underpinning the functional dependence $f_\mu(y^+)$, f_μ does *not* capture the effects of the viscosity on ν_t. Rather, its variation reflects *mainly* the inviscid blocking of wall-normal fluctuations at the wall — a process that extends far beyond the viscous sublayer. The qualification 'mainly' reflects the fact that the time scale $\frac{k}{\varepsilon}$ in Eqs. (9.22) and (9.29) is unlikely to be entirely correct, but should, rather, be some combination of k/ε and the Kolmogorov time scale $\tau_\eta = (\nu/\varepsilon)^{1/2}$ (see Eq. 2.12). In fact, combinations of this type have been proposed in several recent models, more complicated than the two-equation forms considered in this chapter. For example, in a model to be covered in Chapter 14, Durbin (1991) proposes:

$$\nu_t = 0.23\overline{v^2} \max\left(6\left(\frac{\nu}{\varepsilon}\right)^{\frac{1}{2}}; \frac{k}{\varepsilon}\right). \tag{9.30}$$

As seen in Fig. 9.4(b), none of the models attempts to reproduce the asymptotic variation $f_\mu = O(y^{-1})$, and most only capture the variation beyond the wall-asymptotic range qualitatively. Of the seven models examined by Patel *et al.* (1985), we will consider only the three representative variants shown in Fig. 9.4(b): the Lam–Bremhorst model (1981) (LB), the Chien model (1982) (CH) and the Launder–Sharma model (1974) (LS), the first using ε and the second and third $\tilde{\varepsilon}$ as the dissipation-rate variable. These models are presented together, because this allows common features to be highlighted and differences to be contrasted. A fourth model discussed is that by Lien and Leschziner (1994), not for being more recent than others, but because it adopts an approach significantly different from that which leads to the other models.

9.3.1 *The Lam–Bremhorst (1981), Chien (1982) and Launder–Sharma (1974)*[4] *models*

The Lam–Bremhorst model includes:

$$f_\mu = [1 - \exp(-0.0165y^*)]^2 \left(1 + \frac{20.5}{R_t}\right), \tag{9.31}$$

which, of the models presented, adheres most closely to the experimental data. However, the wall-asymptotic variation is clearly wrong, and the model over-estimates the damping around $y^+ \approx 60$.

Chien's model is based on the solution for the 'isotropic dissipation', $\tilde{\varepsilon}$, evaluated from $\tilde{\varepsilon} = \varepsilon - 2\nu\frac{k}{y^2}$. As Eq. (9.1) shows, this is strictly correct *only* if k decays quadratically throughout the viscous sublayer. Also, this model uses coefficients in the dissipation-rate equation which are different from those given in Section 9.2. Thus, $C_{\varepsilon 1} = 1.35$, $C_{\varepsilon 2} = 1.8$, which (with $\sigma_\varepsilon = 1.3$) at least approximately maintain the difference imposed by Eq. (9.18). Chien's damping

[4]The Launder–Sharma model is very similar to the earlier model of Jones and Launder (1972), the only substantive difference being in the form the damping functions f_μ. The "basic" high-Reynolds-number form of the $k - \varepsilon$ model is recovered simply by setting all damping functions to 1.0 and nullifying any terms associated with the fluid viscosity.

function is:

$$f_\mu = 1 - \exp(-0.0115y^+), \tag{9.32}$$

which gives, as is evident from Fig. 9.4(b), seriously excessive damping that extends far beyond $y^+ = 100$. Thus, at this distance, Eq. (9.32) yields $f_\mu = 0.68$.

The Launder–Sharma model is also based on the solution of $\tilde{\varepsilon}$, but evaluated from $\tilde{\varepsilon} = \varepsilon - 2\nu\left(\frac{\partial k^{1/2}}{\partial x_j}\right)^2$. This model is especially popular (i.e. often selected in practice), because it does not require the wall distance to be evaluated (in terms of y^* or y^+). Rather, it uses the turbulent Reynolds number $R_t = k^2/\nu\varepsilon$ in the function:

$$f_\mu = \exp\left[\frac{-3.4}{\left(\frac{1+R_t}{50}\right)^2}\right]. \tag{9.33}$$

Although it clearly gives an inadequate fit to the experimental data, this model has been widely used in practice, due to the ease of its implementation in general codes for complex geometries. A slight ambiguity that arises in the comparison of this and Chien's models with the data in Fig. 9.4(b) is that the modelled f_μ function multiplies an eddy viscosity determined with $\tilde{\varepsilon}$, while f_μ in Eq. (9.29) uses ε. Because $R_t = k^2/\nu\varepsilon$, Eq. (9.33) can readily be plotted in Fig. 9.4, although it needs to be noted that the modelled eddy-viscosity does not comply with Eq. (9.29) in the viscous sublayer, $y^+ < 10$. The fact that the model maintains a finite value of f_μ at the wall is noteworthy. That this behaviour may be physically realistic emerges from Eq. (9.29), due to the fact, derived earlier in this section, that $\tilde{\varepsilon} = O(y)$ (see Eq. 9.24). This shows that the correct asymptotic variation is $f_\mu = O(1)$.

Next, we consider the damping function $f_{\varepsilon 2}$ in $D_\varepsilon = f_{\varepsilon 2} C_{\varepsilon 2} \frac{\varepsilon^2}{k}$. A point to note, first, by reference to Eq. (9.13), is that the decay rate $k \propto x^{-2.4}$, which prevails at low Reynolds numbers, implies $C_{\varepsilon 2} \approx 1.4$. Second, a pertinent condition that needs to be examined is the wall-asymptotic behaviour of the ε-equation:

$$\nu \frac{d^2\varepsilon}{dy^2} = C_{\varepsilon 2} \frac{\varepsilon^2}{k} \quad \text{or} \quad \nu \frac{d^2\tilde{\varepsilon}}{dy^2} = C_{\varepsilon 2} \frac{\tilde{\varepsilon}^2}{k}. \tag{9.34}$$

In both cases, the left-hand side varies as $O(1)$, and Fig. 5.2, for ε^+, also suggests that the asymptotic value of the second derivative is zero (or close to it). With $C_{\varepsilon 2}$ being invariant, the right-hand side, $C_{\varepsilon 2}\frac{\varepsilon^2}{k}$, follows $O(y^{-2})$, while $C_{\varepsilon 2}\frac{\tilde{\varepsilon}^2}{k}$ follows $O(1)$, because $\tilde{\varepsilon} = O(y)$. For the former condition in Eq. (9.34), the damping function should therefore follow as $O(y^2)$, while for the latter condition, $C_{\varepsilon 2}$ is expected to asymptote to some constant wall value.

Yet again, it is reasonable (but by no means rigorous) to propose an exponential-decay function, and several models use:

$$f_{\varepsilon 2} = 1 - \alpha \exp(-\beta R_t^2). \tag{9.35}$$

The Lam–Bremhorst model uses $\alpha = \beta = 1$, while the Launder–Sharma model uses $\alpha = 0.3$ and Chien's model uses $\alpha = 0.22$. For $\alpha = \beta = 1$, the Taylor-series expansion of (9.35) gives:

$$f_{\varepsilon 2} = R_t^2 + \cdots HoTs. \tag{9.36}$$

As $R_t = k^2/\nu\varepsilon$, it follows that $f_{\varepsilon 2} = O(y^8)$, which appears to be far too steep. With $\alpha < 1$, $\beta = 1$, proposed by Launder and Sharma, the wall value is $(1 - \alpha)$, and the Launder–Sharma model thus implies a wall-limiting value $(f_{\varepsilon 2}C_{\varepsilon 2})|_{wall} = 1.92 \times 0.7 = 1.34$, which is close to 1.4 and is also compatible, in principle, with Eq. (9.34), because this model uses $\tilde{\varepsilon}$. Alternatively, in the Chien model, $\alpha = 0.22$, $\beta = 0.028$. Hence, $(f_{\varepsilon 2}C_{\varepsilon 2})|_{wall} = 1.4$, but the damping effect persists to much higher values of R_t, and hence to a much larger distance from the wall.

DNS data, allowing an accurate resolution down to $y^+ < 1$, show that the maximum dissipation rate in a boundary layer occurs at the wall (see Fig. 5.2). However, early measurements, by reference to which models were constructed, indicated that the maximum occurred at around $y^+ \approx 10$, as illustrated by measurements of Laufer (1954), shown in Fig. 9.5. The need to reproduce this 'peak' encouraged modellers to adopt two alternative approaches: either the introduction of an amplification function $f_{\varepsilon 1}$ to the generation term, Eq. (9.6), of the ε equation, i.e.:

$$P_\varepsilon = f_{\varepsilon 1} C_{\varepsilon 1} \frac{\varepsilon}{k} P_k, \tag{9.37}$$

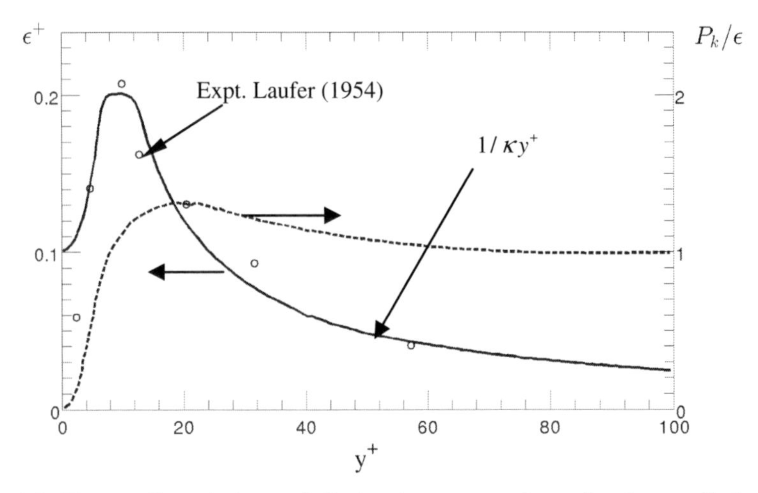

Fig. 9.5: Near-wall variations of dissipation rate and production-to-dissipation ratio in a pipe flow (reproduced with permission of AIAA from Patel, Rodi and Scheuerer (1985)).

or the addition of an extra additive source term, E, to the ε-equation, Eq. (9.9). Launder and Sharma adopted the latter route, introducing:

$$E = 2\nu\nu_t \left(\frac{\partial^2 U}{\partial y^2} \right)^2 . \tag{9.38}$$

The rationale of using this term rests on two facts. First, the term is especially influential in the region in which a high curvature of the velocity profile coincides with a turbulent-viscosity level beginning to rise rapidly from its zero wall value. This region is the buffer layer, in which the dissipation rate features a maximum in Fig. 9.5. In the log-law region, it is easy to show, by differentiating the log-law and using the equilibrium-compatible relations (9.26) for k and ε, that this term falls at the rate $E = O(y^{-3})$ away from the wall. Second, the exact equation for the dissipation rate, Eq. (9.4), features the term $P_{\varepsilon 3}$, which, in the case of a single x-directed wall, simplifies to:

$$P_{\varepsilon 3} = -2\nu \overline{u_2 \frac{\partial u_1}{\partial y}} \frac{\partial^2 U}{\partial y^2} . \tag{9.39}$$

We can now propose the closure assumption:

$$\overline{u_2 \frac{\partial u_1}{\partial y}} \leftarrow \frac{1}{2}\frac{\partial \overline{uv}}{\partial y} = -\frac{1}{2}\frac{\partial}{\partial y}\left(\nu_t \frac{\partial U}{\partial y}\right) = -\frac{1}{2}\nu_t \frac{\partial^2 U}{\partial y^2} + \cdots , \tag{9.40}$$

which, upon substitution into Eq. (9.39), illustrates the rationale of the correction in Eq. (9.40).[5]

The alternative route, of using the function $f_{\varepsilon 1}$, is used, for example, in the Lam–Bremhorst model, their specific form being:

$$f_{\varepsilon 1} = 1 + \left(\frac{0.05}{f_\mu}\right)^3 . \tag{9.41}$$

As the wall is approached, the value of this function rises from 1.3 at $y^+ = 10$ to around 2 at $y^+ = 5$, and then tends to increase much more steeply as the wall is approached, in line with $O(y^{-2})$. However, the remaining fragments in Eq. (9.37) fall more rapidly, so that Eq. (9.41) provides, approximately, the requisite behaviour shown in Fig. 9.5.

9.3.2 *The Lien–Leschziner model (1994, 1996)*

This model is unusual, in so far as it is designed to replicate the behaviour of Wolfshtein's and Norris and Reynolds' one-equation models, covered in Sections 8.2.1 and 8.2.2, respectively, and to return the correct wall-asymptotic value of the dissipation rate. The derivation below relates to the Norris–Reynolds model, with the Wolfshtein-model-based version following along entirely analogous steps.

The first step is to introduce a damping function into the eddy-viscosity expression:

$$\nu_t = C_\mu f_\mu^* \frac{k^2}{\varepsilon} . \tag{9.42}$$

Next, a constraint is introduced so as to force this eddy viscosity to comply with the Norris–Reynolds model (Eqs. (8.12)–(8.24)), near

[5]The factor 2, rather than 1, simply reflects a calibration process designed to procure the behaviour shown in Fig. 9.5.

the wall:

$$\nu_t = k^{\frac{1}{2}} \underbrace{C_\mu f_\mu^* \frac{k^{\frac{3}{2}}}{\varepsilon}}_{l_\mu \text{ 2-eq. model}} (= k^{\frac{1}{2}} l_\mu) = k^{\frac{1}{2}} \underbrace{C_\mu^{\frac{1}{4}} \kappa y (1 - e^{-0.0198 y^+})}_{l_\mu \text{ 1-eq. model}}. \tag{9.43}$$

The group $k^{3/2}/\varepsilon$ may now be replaced by l_ε, with the latter adhering, again, to the Norris–Reynolds expression $l_\varepsilon = c_{l\varepsilon} y \frac{yk^{1/2}}{yk^{1/2} + 2\nu c_{l\varepsilon}}$, Eq. (8.14). This leads to:

$$C_\mu f_\mu^* c_{l\varepsilon} y \frac{yk^{\frac{1}{2}}}{yk^{\frac{1}{2}} + 2\nu c_{l\varepsilon}} = C_\mu^{\frac{1}{4}} \kappa y \left(1 - e^{-0.0198 y^+}\right)$$

or

$$f_\mu^* = \left(1 - e^{-0.0198 y^+}\right) \Big/ \left(C_\mu^{\frac{3}{4}} c_{l\varepsilon} \frac{1}{\kappa} \frac{yk^{\frac{1}{2}}}{yk^{\frac{1}{2}} + 2\nu c_{l\varepsilon}}\right). \tag{9.44}$$

Because of Eq. (8.23) and $C_\mu = c_\mu^2$, it follows that $C_\mu^{3/4} c_{l\varepsilon}/\kappa = 1$, and Eq. (9.44) thus defines one of the two damping functions.

For the damping function $f_{\varepsilon 2}$, Lien and Leschziner adopt Launder and Sharma's proposal in Eq. (9.35):

$$f_{\varepsilon 2} = 1 - 0.3 \exp\left(-R_t^2\right), \tag{9.45}$$

which secures the effective turbulence-decay coefficient $C_{\varepsilon 2}|_{wall} = 1.34$.

The final constraint that the model aims to satisfy is the correct wall-asymptotic variation of ε. The starting point is to introduce an auxiliary (synthetic) production fragment P_k' so that:

$$\frac{\varepsilon}{k}\left(C_{\varepsilon 1}\left[P_k + P_k'\right] - C_{\varepsilon 2}\left[1 - 0.3 \exp\left(-R_t^2\right)\right]\right) = 0, \tag{9.46}$$

with the fragment designed to ensure the requisite wall-dissipation level. Equation (9.46) applied at the wall (at which $P_k \to 0$), together with $\varepsilon = k^{3/2}/l_\varepsilon$, then gives:

$$P_k' = \frac{C_{\varepsilon 2}\left[1 - 0.3 \exp\left(-R_t^2\right)\right] k^{\frac{3}{2}}}{C_{\varepsilon 1} l_\varepsilon}, \tag{9.47}$$

in which l_ε is the Norris–Reynolds length scale, i.e. Eq. (8.14). To ensure that this fragment vanishes outside the viscosity-affected region, the fragment is replaced by:

$$P'_k \leftarrow P'_k \exp(A^*_\mu y^{*2}), \tag{9.48}$$

with $A^*_\mu = 0.00057$ chosen by reference to a calibration for a canonical zero-pressure-gradient flat-plate boundary layer. The above returns the limiting value:

$$\lim_{y\to 0} \varepsilon = \frac{k^{\frac{3}{2}}}{c_{l\varepsilon} y \left\{ \dfrac{y k^{\frac{1}{2}}}{y k^{\frac{1}{2}} + 2\nu c_{l\varepsilon}} \right\}} = 2.12\nu \frac{k}{y^2}, \tag{9.49}$$

in which the numerical multiplier, 2.12, compares with the theoretical value 2.0 (see Eq. (5.7) and following discussion).

Figure 9.6 shows some results for the flow through a complex 'transition' duct, derived from the variant of the Lien–Leschziner model that imposes length-scale constraints from Wolfshtein's rather than the Norris–Reynolds' model. The skin friction plot also contains a solution obtained with a Reynolds-stress-transport model of the type covered in Chapter 14. While this flow is attached, it is nevertheless challenging, in so far as the model is required to resolve correctly the strong distortions in the streamwise flow that arise from the secondary transverse velocity that is induced by the streamwise changes in the duct geometry.

9.4 *Alternative k − φ models*

In Section 7.5, we noted that an equation for any length-scale surrogate combining k and ε can be derived via the manipulation in Eq. (7.59). Applying this to $\phi = k^m \varepsilon^n$, and using Eqs. (7.32) and (9.9), we obtain:

$$\frac{D\phi}{Dt} = \frac{\phi}{k}\left[\underbrace{(m + nC_{\varepsilon 1})}_{C_{\phi 1}} P_k - \underbrace{(m + nC_{\varepsilon 2})}_{C_{\phi 2}} \varepsilon\right] + D_\phi, \tag{9.50}$$

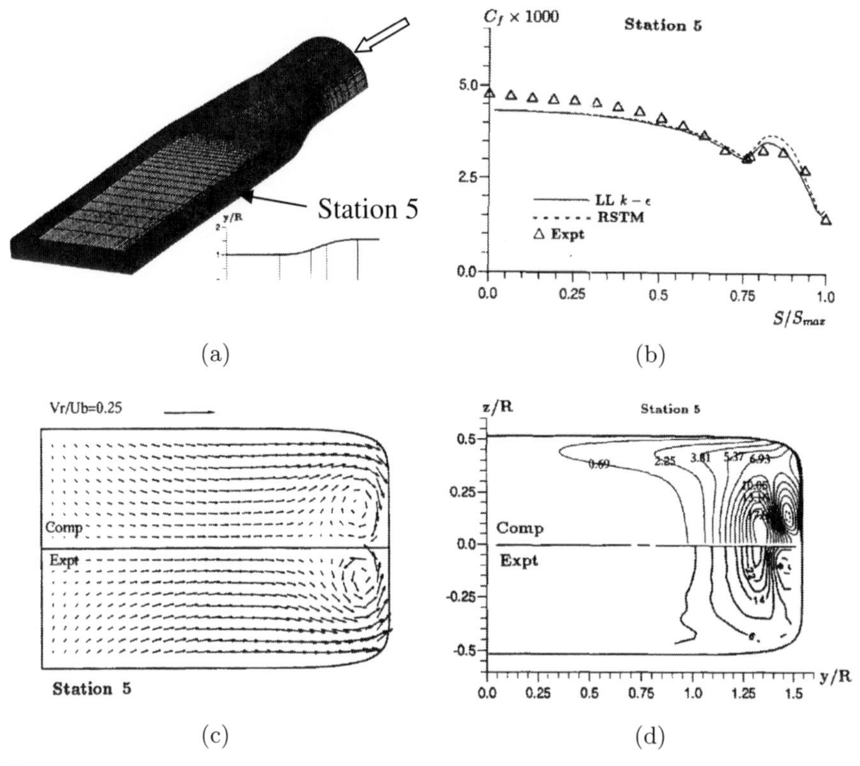

Fig. 9.6: Numerical prediction of flow in a 'transition' duct with the model of Lien and Leschziner (1996), identified as LL, for (a) duct geometry; (b) skin friction around the circumference at station 5 (down stream of transition); (c) secondary velocity at station 5; (d) streamwise-shear-stress contours at station 5 (reproduced with permission of Elsevier from Lien and Leschziner (1996)).

in which D_ϕ is the diffusion term, not expressed in explicit form at this stage.

Table 9.1 lists seven alternative length-scale surrogates to the dissipation rate. If the transformation in Eq. (7.59), leading to Eq. (9.50), is performed exactly, there can be no physical difference between alternative formulations. Hence, at first sight, there seems to be no advantages for opting for any but $\phi = \varepsilon$, except for subjective preferences for the physical meaning of the chosen variables. For example, τ may be interpreted as an eddy-turnover-time scale, ω as the turbulent frequency, and ω^2 as the enstrophy of turbulence. In fact, there are good reasons for suspecting that some of these intuitively reasonable choices may be poor. For example, the choice $\phi = \tau$ ($m = 1, n = -1$) makes the coefficients in Eq. (9.50) negative and renders the destruction term $C_{\phi 2}\varepsilon\frac{\phi}{k}$ a constant, which appears to make this choice pathological in nature. Moreover, reference to Table 9.1 shows that the wall-asymptotic values of three other variables approach infinity, and this is, again, an unattractive feature to be avoided.

Aside from the option of 'recalibrating' the constants $C_{\phi 1}$ and $C_{\phi 2}$ when alternative variables are chosen — a step that modellers almost always take — a very important source of difference between alternative models arises in respect of the diffusion term D_ϕ. To appreciate this, in principle, we return to Eq. (7.59), and note that this imposes:

$$D_\phi = \varepsilon^n m k^{(m-1)} D_k + k^m n \varepsilon^{(n-1)} D_\varepsilon. \tag{9.51}$$

However, almost invariably, D_ϕ is modelled as an independent entity, namely:

$$D_\phi = \frac{\partial}{\partial x_j}\left(\left(\nu + \frac{\nu_t}{\sigma_\phi}\right)\frac{\partial \phi}{\partial x_j}\right). \tag{9.52}$$

This simply reflects the notion that any diffusive process, of whatever variable, is most appropriately — and conveniently — modelled by a gradient-type approximation involving the variable in question. The approximation in Eq. (9.52) can also be defended by the argument that an appropriate choice of σ_ϕ offers the means of securing the correct predictive properties of the model.

If Eq. (9.51) is expanded in terms of k and ε, the result is very different from Eq. (9.52). In particular, fragments containing mixed derivatives arise, and their inclusion or omission can have important predictive consequences, especially at walls and vanishing-shear boundaries — e.g. the edge of a boundary layer or a jet.

Of the variables listed in Table 9.1, the only one (apart from ε) that has gained significant popularity in practice is the turbulent frequency (or specific dissipation) ω. We cover this model in the next section.

9.4.1 *The basic $k - \omega$ model*

Although the proposal to use ω as a length-scale surrogate goes back to the 1940s, (Kolmogorov (1942)), the $k - \omega$ model only attained a significant level of popularity in the 1990s, especially within the aeronautics community, following the publication of the model of Wilcox (1988), leaning on previous work by Saffman (1970), Saffman and Wilcox (1974) and Wilcox and Rubesin (1980). This popularity is mainly due to fact that the model has been observed to return a better near-wall representation than the $k - \varepsilon$ model in boundary layers subjected to adverse pressure gradient — a situation especially pertinent to external aerodynamics, where the accurate prediction of boundary layers developing over wings and other streamlined bodies is a major challenge. While the $k - \varepsilon$ model tends to respond to adverse pressure gradient with excessive values for the length scale $k^{3/2}/\varepsilon$ near walls (and thus turbulent viscosity), the $k - \omega$ model is essentially free from this defect. Surprisingly, especially in view of its singular wall-asymptotic behaviour, it displays favourable near-wall characteristics *without* any viscosity-specific modifications accounting for processes in the low-Reynolds-number near-wall layer, except for the inclusion of the fluid viscosity in the diffusion terms. This aspect will be discussed below.

In the case of ω, $m = -1, n = 1$, and with the constants $C_{\varepsilon 1} = 1.44, C_{\varepsilon 2} = 1.92$ for the basic $k - \varepsilon$ model, Eq. (9.50) becomes:

$$\frac{D\omega}{Dt} = 0.44\frac{\omega}{k}P_k - 0.92\omega^2 + D_\omega. \tag{9.53}$$

As explained by reference to Eq. (9.51), the diffusion term requires careful consideration, for it does not transform compactly to the form:

$$D_\omega = \frac{\partial}{\partial x_j}\left(\frac{\nu_t}{\sigma_\omega}\frac{\partial \omega}{\partial x_j}\right). \tag{9.54}$$

Rather, Eq. (9.51) implies:

$$D_\omega = \frac{1}{k}\frac{\partial}{\partial x_j}\left(\frac{\nu_t}{\sigma_\varepsilon}\frac{\partial \varepsilon}{\partial x_j}\right) - \frac{\varepsilon}{k^2}\left(\frac{\nu_t}{\sigma_k}\frac{\partial k}{\partial x_j}\right). \tag{9.55}$$

Insertion of $\omega = \varepsilon/k$ into Eq. (9.54), followed by some straightforward manipulations, leads to:

$$\frac{\partial}{\partial x_j}\left(\frac{\nu_t}{\sigma_\omega}\frac{\partial \omega}{\partial x_j}\right) = \frac{1}{k}\frac{\partial}{\partial x_j}\left(\frac{\nu_t}{\sigma_\omega}\frac{\partial \varepsilon}{\partial x_j}\right) - \frac{\varepsilon}{k^2}\left(\frac{\nu_t}{\sigma_\omega}\frac{\partial k}{\partial x_j}\right)$$
$$\times\left\{-\frac{2\nu_t}{\sigma_\omega k^2}\frac{\partial k}{\partial x_j}\frac{\partial \varepsilon}{\partial x_j} + \frac{2\nu_t \varepsilon}{\sigma_\omega k^3}\frac{\partial k}{\partial x_j}\frac{\partial k}{\partial x_j}\right\}. \tag{9.56}$$

We note, first, that the leading two terms on the right-hand side agree with Eq. (9.55), but only subject to $\sigma_\omega = \sigma_k = \sigma_\varepsilon$, which is not compatible with the $k-\varepsilon$ model. Even if this inconsistency is ignored, we observe, second, that Eq. (9.56) contains fragments (within the curly braces), which add to the difference between Eqs. (9.54) and (9.55). Hence, an important conclusion is that the form (9.54) for D_ω in Eq. (9.53) inevitably leads to a model that is different from the $k-\varepsilon$ form, even if the $k-\varepsilon$ constants associated with the production and destruction are retained.

To examine the relationship between the constants $C_{\omega 1}, C_{\omega 2}$ and σ_ω in

$$\frac{D\omega}{Dt} = C_{\omega 1}\frac{\omega}{k}P_k - C_{\omega 2}\omega^2 + \frac{\partial}{\partial x_j}\left(\frac{\nu_t}{\sigma_\omega}\frac{\partial \omega}{\partial x_j}\right), \tag{9.57}$$

we can follow exactly the same path as in Section 9.2, which led to Eq. (9.18). Unsurprisingly, this yields:

$$C_{\omega 1} = C_{\omega 2} - \frac{\kappa^2}{C_\mu^{\frac{1}{2}}\sigma_\omega}, \tag{9.58}$$

which is equivalent to Eq. (9.18).

The basic model of Wilcox thus consists of the following equations:

$$\begin{aligned}
\nu_t &= C_\mu \frac{k}{\omega} \\
\frac{Dk}{Dt} &= P_k + \frac{\partial}{\partial x_j}\left[\left(\nu + \frac{\nu_t}{\sigma_k}\right)\frac{\partial k}{\partial x_j}\right] - k\omega, \\
\frac{D\omega}{Dt} &= C_{\omega 1}\frac{\omega}{k}P_k - C_{\omega 2}\omega^2 + \frac{\partial}{\partial x_j}\left[\left(\nu + \frac{\nu_t}{\sigma_\omega}\right)\frac{\partial \omega}{\partial x_j}\right],
\end{aligned} \tag{9.59}$$

where $C_{\omega 1} = 0.555$, $C_{\omega 2} = 0.833$, $\sigma_k = \sigma_\omega = 2$. The first two constants correspond to $C_{\varepsilon 1} = 1.555, C_{\varepsilon 2} = 1.833$, compared to 1.44 and 1.92, respectively, for the basic $k - \varepsilon$ model. The value 1.833 is entirely reasonable, for it corresponds to the decay law of isotropic turbulence $k \propto x^{-1.2}$. Moreover, the constants satisfy the constraint in Eq. (9.58). However, the particular set of constants used in the set in Eq. (9.59), coupled with the diffusion approximation in Eq. (9.54), gives this model different characteristics to those of the $k - \varepsilon$ model. Furthermore, in a computational context, the treatment of the singular wall-boundary condition is also a source of differences.

The choice $\sigma_k = 2$ is curious, if only because the diffusivity of turbulence energy can be expected to be close to that of momentum. As is exemplified by Fig. 5.4, the diffusion term hardly contributes to the turbulence energy in the log-layer, but it does so in the viscosity-affected layer and in the outer layer bordering the free stream or an outer shear layer. Figure 5.4 makes clear that turbulence energy is transported from the buffer layer towards the wall. In the model, this is dictated by the diffusion term, and thus by σ_k. The choice of a high value restricts the transport of energy towards the wall, thus lowering k and ultimately also ε. This is, therefore, one means of reducing the dependence on, or even obviating the need for, the type of damping functions used in $k - \varepsilon$ models.

The differences between Eqs. (9.54) and (9.55) seem to be an especially potent source of differences in model performance in the near-wall region. In the log-law layer, at zero pressure gradient, both the diffusion terms in the ε- and ω-equations display variations that follow $O(y^{-2})$, this commonality reflecting the fact that both ε and $\omega = \varepsilon/k$ vary as $O(y^{-1})$, because $k \simeq C_\mu^{-1/2} u_\tau^2$ (which is nominally constant). However, in the presence of an adverse pressure gradient, $\frac{\partial k}{\partial y} < 0$, while $\frac{\partial \varepsilon}{\partial y} > 0$ remains valid, because ε continues to vary broadly as $O(y^{-1})$. Hence, the additive fragments in Eq. (9.56) become positive at the wall, thus increasing ω, relative to Eq. (9.55), and decreasing the length scale $k^{3/2}/\varepsilon$ below the level predicted by the $k - \varepsilon$ model.

An important and well-known defect of the $k - \omega$ model is its excessive sensitivity to the prescribed value of ω at free-stream

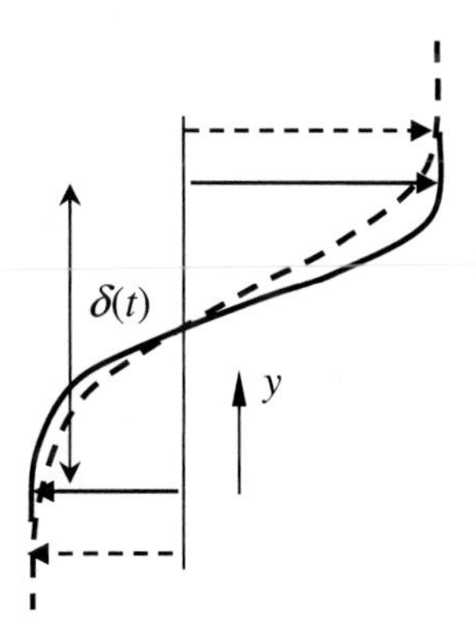

Fig. 9.7: Schematic of a temporal mixing layer (relating to Eqs. (9.60) and (9.61)).

boundaries, if this value is chosen to be too low.[6] This is demonstrated especially well in a numerical study by Menter (1992) for boundary layers and wakes.

Intuition suggests that this must be connected to an ill-conditioned behaviour of the diffusion terms, but the origin is not easily identified by simple physical reasoning. An instructive 'toy-model' flow is the temporal mixing layer: a layer that is formed and thickens in time, by diffusion, as shown in Fig. 9.7. At the edges of this layer, production is negligible, and the $k-\omega$ equations become:

$$\frac{\partial k}{\partial t} = \frac{\partial}{\partial y}\left(\frac{\nu_t}{\sigma_k}\frac{\partial k}{\partial y}\right) - (\omega_o)k, \tag{9.60}$$

$$\frac{\partial \omega}{\partial t} = \frac{\partial}{\partial y}\left(\frac{\nu_t}{\sigma_\omega}\frac{\partial \omega}{\partial y}\right) - (\omega_o)\omega. \tag{9.61}$$

Simplifications introduced into Eqs. (9.60) and (9.61) are the linearisation of the sink terms, via $\omega = \omega_o$, assumed to be locally

[6]In practice, we very rarely have any information on the streamwise variation of k and ε at a free-stream boundary, and we have to estimate or arbitrarily choose these values and specify them as boundary conditions. A possible exception is one in which the free-stream turbulence is generated by an upstream grid of known characteristics, in which case we can solve Eq. (9.13), with assumed or measured initial conditions behind the grid.

'frozen', and $C_{\omega 2} = 1$. Because $\sigma_k = \sigma_\omega$, this simplification brings out a symmetry in the equations, with the solution of one being identical to the solution of the other, within a multiplicative constant. However, this leads to a sensitivity problem.

Consider a situation in which k and ω diffuse from the inner region outwards with low values of both variables prescribed outside the mixing layer. If $\sigma_\omega < \sigma_k$, the 'front' of the elevated ω-field will tend to move more rapidly outwards than the k-front, thus ensuring relatively low values of $\nu_t \propto k/\omega$. In contrast, if $\sigma_\omega > \sigma_k$, the outer edge will be characterised by elevated k-values and low ω-values, causing high ν_t-levels and thus high rates of spread in all properties, including momentum. The condition $\sigma_\omega = \sigma_k$ is a marginally stable state between the two. Kok (2000) presents a formal analysis of the behaviour of Eqs. (9.60) and (9.61) (albeit with the sink terms omitted), which shows that one condition that is necessary to avoid a pathological behaviour of the equations is $(1/\sigma_\omega - 1/\sigma_k) > 0$. Kok then examines the effect of adding to the ω-equation a proportion of the mixed-derivative term $\frac{1}{\omega}\frac{\partial k}{\partial y}\frac{\partial \omega}{\partial y}$, which is one of the terms contributing to the difference between the diffusion of ε and ω, and shows that this addition removes the pathological behaviour that arises from $\sigma_\omega = \sigma_k$. In other words, the ε-equation, when recast into the ω-form, with mixed derivatives retained (through the replacement of $\varepsilon \leftarrow \omega k$ in Eq. (9.55) and rearrangements analogous to those leading to Eq. (9.56)), circumvents the aberrant sensitivity of the $k-\omega$ set. Unfortunatley, this retention robs the model of its advantageous near-wall properties. It is this quandary that has led to the subsequent proposal of a composite model, which combines the $k - \omega$ model in the near-wall region with the $k - \varepsilon$ model in the outer region. Such a model will be discussed in Section 9.4.3.

We close this subsection by pointing out an issue of nomenclature that might cause confusion when consulting supplementary literature. In all above considerations, the variable ω was strictly taken be the specific dissipation ε/k. In contrast, Wilcox has opted to use the definition $\omega = (1/C_\mu)\varepsilon/k$. To distinguish the two definitions, we denote Wilcox's variable by ω^*. The set in Eq. (9.59) may then be written as:

$$\nu_t = \frac{k}{\omega^*},$$

$$\frac{Dk}{Dt} = P_k + \frac{\partial}{\partial x_j}\left[\left(\nu + \frac{\nu_t}{\sigma_k}\right)\frac{\partial k}{\partial x_j}\right] - \underbrace{C_\mu}_{\beta^*} k\omega^*, \tag{9.62}$$

$$\frac{D\omega^*}{Dt} = \underbrace{C_{\omega 1}}_{\alpha}\frac{\omega^*}{k}P_k - \underbrace{(C_{\omega 2}C_\mu)}_{\beta}\omega^{*2} + \frac{\partial}{\partial x_j}\left[\left(\nu + \frac{\nu_t}{\sigma_\omega}\right)\frac{\partial \omega^*}{\partial x_j}\right].$$

The symbols below the under-braces are those used by Wilcox to identify the model constants. One further trivial difference of nomenclature that has *not* been changed in set Eq. (9.62) relates to the Prandtl–Schmidt numbers σ_k and σ_ω. Wilcox does not use σ_k and σ_ω in the physical sense implied by the designation 'Prandtl–Schmidt numbers', but opts to use these same symbols as multiplicative constants of ν_t, representing $1/\sigma_k$ and $1/\sigma_\omega$, respectively, in set (9.62).

9.4.2 *Low-Reynolds-number $k - \omega$-model extensions*

The eddy-viscosity relation in the set in Eq. (9.59), on its own, is no different to that used in the $k - \varepsilon$ model. Hence, the considerations in Section 9.3 relating to f_μ, by reference to Fig. 9.4 and Eq. (9.29), apply here as well. In other words, we cannot expect that the quantities on the right-hand side of Eq. (9.29) would be correctly returned by the model in the viscosity-affected layer without the use of f_μ, as identified by the experimental and DNS data in Fig. 9.4. Yet, surprisingly, the $k - \omega$ model involves no such function. Indeed, it involves no viscosity-dependent terms at all, except for the fluid-viscosity in the diffusion terms in set (9.62). For this to make sense at all, the diffusion terms and constants chosen (especially σ_k and σ_ω) must combine to give this model some peculiar characteristics that allow it to return acceptably realistic near-wall conditions without any of the measures needed to adapt the basic $k - \varepsilon$ model — although the likely role of $\sigma_k = 2$ in achieving this has already been noted in Section 9.4.1. The absence of damping functions is yet another reason for the popularity of the $k - \omega$ model. It is important to emphasise, however, that this must mean that one or more variables in Eq. (9.29) is/are not correctly reproduced by the model — for example, the near-wall variation of k.

As regards the asymptotic variation of the model, we have noted already in Table 9.1 that ω varies as $O(y^{-2})$, thus implying. $\omega|_{y\to 0} \to \infty$. This requires the wall condition to be prescribed at some distance away from the wall. A prescription via the wall-asymptotic condition $\omega = \frac{2\nu}{y^2}$ is possible, but this presumes that $\varepsilon = \frac{2\nu k}{y^2}$ applies from the wall to the y^+-position at which the boundary condition is prescribed. As seen in Fig. 5.2, however, this is only secured within $y^+ < 2$.

An alternative is to examine the ω-equation in the set in Eq. (9.59), reduced to its wall-asymptotic form:

$$\nu \frac{d^2\omega}{dy^2} = C_{\omega 2}\omega^2. \tag{9.63}$$

Use of $\omega = \frac{2\nu}{y^2}$ shows that this ε-consistent asymptotic variation does not satisfy Eq. (9.63), which is, yet again, a consequence of some fragments being omitted from a consistent transformation of the ε- to the ω-diffusion term. Both sides are at least consistent to the extent that both vary as $O(y^{-4})$. This is in contrast to the corresponding condition (9.34), pertaining to the $k - \varepsilon$ model, which provided an argument for a damping function that varies as $O(y^2)$. A solution that does satisfy Eq. (9.63) is:

$$\omega = \frac{6\nu}{C_{\omega 2} y^2} = \frac{7.2\nu}{y^2}, \tag{9.64}$$

the numerical validity of which is contingent on the correctness of $C_{\omega 2}$. What we conclude is that there is a significant amount of uncertainty about the wall condition with this model. This is not a minor issue, because ω is large and changing rapidly with y, and this can have a significant influence on the solution.

We finally discuss, briefly, a particular low-*Re* variant of the $k-\omega$ model (Wilcox, 1994) that is specifically meant to allow the prediction of transition. As a matter of principle, it has to be emphasized, first, that the prediction of transition over a broad range of conditions by the sole use of a conventional turbulence model is not a credible proposition. Transition, whether of a natural or bypass type (the latter provoked by free-stream turbulence) is an inherently unsteady phenomenon, involving instabilities and their amplification. A turbulence model can be calibrated (say, by use of viscosity-related damping functions) to mimic a form of (usually jump-like) transition from a laminar state to a turbulent state in a boundary layer, but this behaviour reflects a mathematical bifurcation of the turbulence-model equations, and will not give the correct description in any but a few carefully selected sets of conditions. In particular, no Reynolds-averaged Navier–Stokes (RANS) model will give a credible description of natural transition, a consequence of the onset of Tollmien–Schlichting waves, followed by chaotic instability.

In the particular case of bypass transition, provoked by free-stream turbulence, an appropriately calibrated model will 'predict' transition due a bifurcation, as a consequence of the diffusion of turbulence energy from the free stream into the near-wall region. In the initial thin boundary layer, turbulence is suppressed by destruction of turbulence energy exceeding its production (aided by a judicious use of damping functions), thus maintaining the boundary layer in a laminar state. As the boundary layer grows, damping diminishes, and at some critical state, dictated by the mathematical structure of the model, production exceeds destruction, and turbulence is amplified. The dissipation rate ε, or its specific form ω, follows suit (by virtue of its increasing production), eventually leading to an equilibrium that represents the fully turbulent state. The large majority of transition models, almost invariably target bypass transition, and involve additional model elements over and above the default equations for turbulent flow — for example, an equation for the transport of intermittency and/or empirical correlations (see Langtry and Menter (2009) and Lardeau and Leschziner (2006) for two alternative approaches). However, this subject is outside the scope of this book.

Wilcox (1992) shows that the difference between the production and destruction terms in Eq. (9.59) (or Eq. (9.62)), subject to laminar upstream conditions in a simple boundary layer, switches from negative (damping) to positive (amplification) at

$Re_x = U_\infty x/\nu = 8100$ and 12150, for the turbulence energy and specific dissipation, respectively. Either value is about two orders of magnitude too low, demonstrating the disparity between the mathematical properties of the equations and the physics of transition. He then proceeds to introduce three damping functions, with the principal argument being $Re_T = k/\nu\omega^*$, one multiplying α, the second β^* and the third ν_t in Eq. (9.62). The augmented model is then shown to replicate the transition location in a flat-plate boundary layer for a range of free-stream turbulence levels, subject to assumptions about the length scale $k^{1/2}/\omega$ in the free stream. However, for reasons noted earlier, this framework cannot be expected to offer broader generality in complex multi-dimensional strain.

9.4.3 *Hybrid $k-\omega/k-\varepsilon$ modelling — the SST model*

As explained in Sections 9.4.1 and 9.4.2, the $k-\omega$ model displays some advantageous properties in the near-wall region, without the use of damping functions. On the other hand, it is afflicted with defects in the outer, low-shear region, associated especially with its sensitivity to the boundary condition for ω, whereas the $k-\varepsilon$ model fares much better in this respect. This provides an obvious motivation for blending the two models — a route taken by Menter (1994).

Formally, this blending can be (and, indeed, *is*) effected by way of a weighted-average formula operating on corresponding $k-\varepsilon$- and $k-\omega$-model coefficients:

$$C_{\mathit{eff}} = FC_{k-\omega} + (1-F)C_{k-\varepsilon}, \tag{9.65}$$

with F being a prescribed blending function, which ensures a dominance of the $k-\omega$ model in the region below $y^+ \simeq 70$ and a dominance of the $k-\varepsilon$ model above this value. There are some extra terms, however, that require specific action beyond the blending in Eq. (9.65). To realize this equation, the two models need to be cast into a common framework, one of two options being based on the $k-\omega$ form. The $k-\omega$ model itself, in the form given by Eqs. (9.59), can be used directly in Eq. (9.65). However, the $k-\varepsilon$ model needs to be cast into its $k-\omega$ equivalent. Eq. (9.53), reproduced below with a slight variation in the last term, already offers a partial ω-based representation of the ε-equation:

$$\frac{D\omega}{Dt} = (1 - C_{\varepsilon 1})\frac{\omega}{k}P_k - (1 - C_{\varepsilon 2})\omega^2 + D_{\varepsilon\to\omega}. \tag{9.66}$$

Next, we need to recast $D_{\varepsilon\to\omega}$, the ω-equivalent of the ε-diffusion term. This can be done readily by using transformation in Eq. (7.59):

$$D_{\varepsilon\to\omega} = \varepsilon^n m k^{(m-1)} D_k + k^m n \varepsilon^{(n-1)} D_\varepsilon, \tag{9.67}$$

which, with $m = -1, n = 1$, gives:

$$D_{\varepsilon\to\omega} = \frac{1}{k}(D_\varepsilon - \omega D_k). \tag{9.68}$$

Insertion of the $k - \varepsilon$ diffusion terms into Eq. (9.68), and some straightforward algebraic manipulations, to express $\frac{\partial\varepsilon}{\partial x_j}$ in terms of $\frac{\partial\omega}{\partial x_j}$, leads (after the omission of some minor fragments) to:

$$D_{\varepsilon\to\omega} = \frac{\partial}{\partial x_j}\left[\left(\nu + \frac{\nu_t}{\sigma_\varepsilon}\right)\frac{\partial\omega}{\partial x_j}\right] + 2\frac{C_\mu}{\sigma_\varepsilon}\frac{1}{\omega}\frac{\partial\omega}{\partial x_j}\frac{\partial k}{\partial x_j}. \tag{9.69}$$

Hence, the $k - \varepsilon$ model can be written:

$$\frac{Dk}{Dt} = P_k + \frac{\partial}{\partial x_j}\left[\left(\nu + \frac{\nu_t}{\sigma_k}\right)\frac{\partial k}{\partial x_j}\right] - k\omega, \tag{9.70}$$

$$\begin{aligned}\frac{D\omega}{Dt} &= (1 - C_{\varepsilon 1})\frac{\omega}{k}P_k - (1 - C_{\varepsilon 2})\omega^2 \\ &\quad + \frac{\partial}{\partial x_j}\left[\left(\nu + \frac{\nu_t}{\sigma_\varepsilon}\right)\frac{\partial\omega}{\partial x_j}\right] + 2\frac{C_\mu}{\sigma_\varepsilon}\frac{1}{\omega}\frac{\partial\omega}{\partial x_j}\frac{\partial k}{\partial x_j}.\end{aligned} \tag{9.71}$$

Thus, blending consists of combining the coefficients in Eqs. (9.70) and (9.71) with the corresponding equations in the set in Eq. (9.59), and adding the mixed-derivative (or 'cross-diffusion') term $2(1 - F)\frac{C_\mu}{\sigma_\varepsilon}\frac{1}{\omega}\frac{\partial\omega}{\partial x_j}\frac{\partial k}{\partial x_j}$ as the outer layer is entered. The blending factor F will be discussed below.

An influential addition to the hybrid model is a correction which limits the shear stress in accordance with Bradshaw's relation, Eq. (7.43) or Eq. (8.44), which has also been used to construct the one-equation models expressed by Eqs. (8.45) and (8.46). Here again, the basic idea is rooted in the difference between Eq. (7.43), reflecting experimental observations for shear layers, and Eq. (9.3), which, when multiplied by the shear strain, gives a shear stress consistent with the $k - \varepsilon$ model. Menter thus introduced a 'shear-stress limiter' via:

$$\nu_t = \frac{C_\mu^{\frac{1}{2}} k}{\max(C_\mu^{-\frac{1}{2}}\omega; \alpha \partial U/\partial y)}, \tag{9.72}$$

where α is a function with extreme limits of 1 for boundary-layer flow and 0 for free shear flow.[7] In the boundary layer, the switch occurs when:

$$C_\mu^{-\frac{1}{2}}\omega = \frac{\partial U}{\partial y} \quad \rightarrow \quad C_\mu^{\frac{1}{2}} = C_\mu \frac{k}{\varepsilon}\frac{\partial U}{\partial y} = -\frac{\overline{uv}}{k}, \tag{9.73}$$

which corresponds to the equilibrium condition $P_k/\varepsilon = 1$ (see Eqs. (9.10)–(9.12)). Hence, in high strain, beyond the equilibrium state, Eq. (9.72) yields:

$$\nu_t = \frac{C_\mu^{\frac{1}{2}} k}{\partial U/\partial y} \quad \rightarrow \quad -\overline{uv} = C_\mu^{\frac{1}{2}} k, \tag{9.74}$$

as is intended with the limiter. If the limiter is not imposed, the shear stress would, for all conditions, be:

$$-\overline{uv} = C_\mu \frac{k^2}{\varepsilon}\left(\frac{\partial U}{\partial y}\right). \tag{9.75}$$

If the strain is removed with $P_k = -\overline{uv}\frac{\partial U}{\partial y}$, the shear stress arises as:

$$-\overline{uv} = C_\mu^{\frac{1}{2}} \sqrt{\frac{P_k}{\varepsilon}} k, \tag{9.76}$$

which, when compared to Eq. (9.74), demonstrates that the standard $k - \varepsilon$ model returns excessive levels of shear stress in highly strained

[7] Menter actually uses the vorticity Ω in place of the velocity gradient; these are identical in thin shear flow. Moreover, with this limiter, Menter prefers $\sigma_k = 1.18$ instead of 2.

flow, relative to the expression linking the shear stress linearly to the turbulence energy.

This composite formulation — termed 'shear-stress-transport' (SST) model[8] — has enjoyed significant popularity in aerodynamic computations, because of its favourable response to adverse pressure gradient. In a decelerating boundary layer, Eq. (9.72) comes into action and reduces the shear-stress level relative to that of the non-limited $k-\omega$ model, and this is extremely effective in promoting separation, including shock-induced separation (Batten *et al.* (1999), Leschziner *et al.* (2001)). In fact, it has been observed to be excessively sensitive in some separated flows, both subsonic (Apsley & Leschziner (2000, 2001)) and trans/supersonic flows (Liou *et al.* (2000)), in which it gave too early separation and excessively long recirculation regions.

We finally return to the blending factor F. There is, inevitably, a substantial degree of arbitrariness in choosing this function. A reasonable proposal is to use an error function or a hyperbolic-tangent function to transit smoothly between the two models. The argument must be chosen such that F is '1' close to the wall, is '0' in the outer region, and is appropriately non-dimensionalised. Menter uses the following function:

$$\begin{aligned} F &= \tanh(arg^4), \\ arg &= \min\left\{\max\left(\frac{\sqrt{k}}{0.09\omega^* y}; \frac{500\nu}{y^2\omega^*}\right); \frac{4k}{\sigma_\varepsilon y^2 CD_{k\omega}}\right\}, \\ CD_{k\omega} &= \max\left(\frac{2}{\sigma_\varepsilon \omega}\frac{\partial k}{\partial x_j}\frac{\partial \omega}{\partial x_j}; 10^{-20}\right), \end{aligned} \tag{9.77}$$

with ω^* being defined in Eq. (9.62). To appreciate the basic rationale of the function in Eq. (9.77), it is instructive to replace ω^* in terms of k and ε. In the expectation that the last argument in the

[8]This is, arguably, a misnomer. Ultimately, the model implements Bradshaw's relation (7.43) into a standard linear eddy-viscosity model, and there is no relationship to the stress-transport equations.

min{...} function is very large and never selected, it may be assumed that $arg = \max(\ldots;\ldots)$. This last term in arg is a 'safety switch' (see Menter, 1994) guarding against the occurrence of a degenerate solution at the edge of shear layers, associated with the wrong sensitivity of the $k-\omega$ model to the boundary value of ω. With attention focused on the max (...;...) function, the two major arguments can be written as follows:

$$\max\left(\frac{\sqrt{k}}{0.09\omega^* y};\frac{500\nu}{y^2\omega^*}\right)=\frac{\left(\frac{k^{\frac{3}{2}}}{\varepsilon}\right)}{y}\max\left(1;\frac{45}{\left(\frac{yk^{\frac{1}{2}}}{\nu}\right)}\right). \tag{9.78}$$

The pre-multiplier is the macro length scale divided by the distance from the wall. In the log-law region, this ratio is (see second equation in the set (7.47), noting that $l_\varepsilon = k^{3/2}/\varepsilon$):

$$\frac{k^{\frac{3}{2}}}{\varepsilon y}=C_\mu^{-\frac{3}{4}}\kappa=2.5. \tag{9.79}$$

This ratio declines beyond the log-law region, tending to '0' as the edge of the boundary layer is approached, hence yielding $\max(\ldots;\ldots)\to 0$, $arg \to 0$ and $F \to 0$. Beyond the viscosity-affected layer, $yk^{1/2}/\nu > 45$, the second argument in the bracket assumes values below '1', so that the 'max' function chooses the argument '1', giving $arg \approx 2.5$ in Eq. (9.77) and hence $F \approx 1$. Finally, in the viscous sublayer, the second argument in the bracket becomes larger than 1. With Eq. (9.64) governing the variation of the specific dissipation in the viscous sublayer, the right-hand side of Eq. (9.78) becomes much larger than '1', so that $F \approx 1$.

9.5 *Reductions of two-equation models to one-equation forms*

In Chapter 8, we considered, among a number of one-equation models, formulations that involve the solution of an eddy-viscosity-transport equation. Two particular models, those of Baldwin and Barth (1991) and Menter (1997), were outlined in Section 8.3.2, and were said to derive from simplified forms of two-equation models, but no detailed derivations were given at that stage. Now, that the subject of two-equation models has been covered, we can return to these models and provide a more detailed explanation of how they arise. We only consider Menter's model in detail, highlighting a single substantive difference between this and the Baldwin–Barth model.

The first step of the process in both models is to derive a basic transport equation for the eddy-viscosity by differentiating the eddy-viscosity relation $\nu_t = C_\mu k^2/\varepsilon$:

$$\frac{D\nu_t}{Dt} = C_\mu \left(2\frac{k}{\varepsilon}\frac{Dk}{Dt} - \frac{k^2}{\varepsilon^2}\frac{D\varepsilon}{Dt} \right). \tag{9.80}$$

This yields an equation of the form:

$$\frac{D\nu_t}{Dt} = f(\nu_t, k, \varepsilon), \tag{9.81}$$

from which k and ε need to be eliminated in order to close the equation. The dissipation rate can be removed upon using the definition of the eddy viscosity, i.e.:

$$\varepsilon = C_\mu \frac{k^2}{\nu_t}. \tag{9.82}$$

Next, Menter proposes to remove the turbulence energy by using Bradshaw's relation (8.44):

$$|-\overline{uv}| = \nu_t \left| \frac{\partial U}{\partial y} \right| = ak. \tag{9.83}$$

At this stage, it is important to point out that the use of Eq. (9.83) limits the model to relatively simple near-wall shear flows, because this equation arises from observations of boundary layers. In fact, Eq. (9.76) shows that this constraint is equivalent to the imposition $P_k/\varepsilon = 1$, i.e. turbulence-energy equilibrium, onto the $k - \varepsilon$ model. In contrast to the substitution (9.83), Baldwin and Barth impose, explicitly, the constraint $P_k = \varepsilon$.

The outcome of the steps in Eqs. (9.80)–(9.83), subject to the assumption of a single strain $\frac{\partial U}{\partial y}$, is the equation:

$$\frac{D\nu_t}{Dt} = C_1\nu_t\left|\frac{\partial U}{\partial y}\right| - C_2\nu_t^2\left(\frac{\partial}{\partial y}\left|\frac{\partial U}{\partial y}\right| \Big/ \left|\frac{\partial U}{\partial y}\right|\right)^2 + \frac{\partial}{\partial y}\left(\frac{\nu_t}{\sigma_\varepsilon}\frac{\partial \nu_t}{\partial y}\right)$$
$$+ 2\frac{(\sigma_\varepsilon - \sigma_k)}{\sigma_k\sigma_\varepsilon}\left\{\nu_t\frac{\partial^2\nu_t}{\partial y^2} + \left(\frac{\partial \nu_t}{\partial y}\right)^2 + \nu_t^2\frac{\partial^2}{\partial y^2}\left|\frac{\partial U}{\partial y}\right| \Big/ \left|\frac{\partial U}{\partial y}\right|\right.$$
$$\left. + 3\nu_t\frac{\partial \nu_t}{\partial y}\left(\frac{\partial}{\partial y}\left|\frac{\partial U}{\partial y}\right| \Big/ \left|\frac{\partial U}{\partial y}\right|\right)\right\}, \tag{9.84}$$

where

$$C_1 = (C_{\varepsilon 2} - C_{\varepsilon 1})\sqrt{C_\mu}, \quad C_2 = \frac{C_1}{\kappa^2} + \frac{1}{\sigma},$$

which entails the use of Eq. (9.18).

Next, Menter introduces the approximation $\sigma_\varepsilon = \sigma_k \equiv \sigma = 1$, purely in order to procure greater simplicity through the removal of the entire group of terms within the curly braces in Eq. (9.84), thus yielding:

$$\frac{D\nu_t}{Dt} = C_1\nu_t\left|\frac{\partial U}{\partial y}\right| - C_2\frac{\nu_t^2}{L_K^2} + \frac{\partial}{\partial y}\left(\frac{\nu_t}{\sigma}\frac{\partial \nu_t}{\partial y}\right), \tag{9.85}$$

where $L_K^{-1} = \left(\frac{\partial}{\partial y}\left|\frac{\partial U}{\partial y}\right| \Big/ \left|\frac{\partial U}{\partial y}\right|\right)$ is clearly the inverse of a length scale, known as the von Kármán length scale.

Menter finally proposes the extension of the above high-Reynolds-number form, to become applicable to the viscous sublayer, by introducing two damping functions:

$$\nu_t \leftarrow f_{\nu 1}\nu_t, \tag{9.86}$$

$$C_1\nu_t\left|\frac{\partial U}{\partial y}\right| \leftarrow f_{\nu 2}C_1\nu_t\left|\frac{\partial U}{\partial y}\right|, \tag{9.87}$$

where $f_{\nu 1} = 1 - \exp\left(-A_w\frac{\nu_t}{\nu}\right)$, $f_{\nu 2} = \frac{f_{\nu 1}\nu_t + \nu}{\nu_t + \nu}$ and $A_w = 0.19$.

A generalisation of the model, to apply to an arbitrary strain field, can be effected by replacing $\left|\frac{\partial U}{\partial y}\right|$ by S, while L_K^{-1} is replaced by:

$$L_K^{-1} = \frac{\dfrac{\partial}{\partial y}\left|\dfrac{\partial U}{\partial y}\right|}{\left|\dfrac{\partial U}{\partial y}\right|} \leftarrow \frac{\left(\dfrac{\partial S}{\partial x_j}\dfrac{\partial S}{\partial x_j}\right)^{\frac{1}{2}}}{S}, \tag{9.88}$$

which, with the addition of the fluid viscosity in the diffusion term, yields Eq. (8.46).

9.6 *Summary and lessons*

Two-equation linear eddy-viscosity models represent the lowest level of closure that can be claimed to be applicable to general two- and three-dimensional flows, i.e. those more complex than thin shear flows. This generality is due to the fact that two-equation models do not require a prescription, by reference to any global flow quantity, of either the velocity scale or the length scale that are present in the eddy-viscosity relation. Instead, both scales are determined from respective differential transport equations that link the scales to the local strain field, and account for their generation, destruction and transport by diffusion and convection.

While the velocity scale is invariably represented by the turbulence energy, a number of alternative surrogate variables for the length scale feature in different models proposed in the literature. The most prominent surrogates are the dissipation rate, ε, and its specific form, $\omega = \varepsilon/k$. Although the respective models display different predictive properties, these differences are relatively modest, and reflect, primarily, particular decisions taken in the course of the approximation of various terms in the equations governing the surrogate variables (diffusion, in particular), and additional variations in the calibration process that leads to the respective sets of model constants.

The faithful representation of near-wall turbulence properties, especially across the viscous sublayer, has been shown to be an especially challenging task, requiring much care if the asymptotic variations of the properties are to be modelled to an acceptable accuracy. This is especially so in the case of $k-\varepsilon$-type models. While the $k-\omega$ model is observed to be more 'robust', displaying an acceptable near-wall behaviour without wall-specific correction terms, the reasons for this are not entirely clear: they are likely to be rooted in the diffusion approximation adopted, in combination with the particular constants used, and the rather artificial boundary condition that needs to be applied close to the wall, because ω tends to infinity as the wall is approached. The different characteristics and properties of the $k-\varepsilon$ and $k-\omega$ models have been exploited in the construction of the hybrid SST model that blends the former model in wall-remote regions with the latter model operating close to walls.

Despite the advantages that arise from the use of differential transport equations for the velocity and length scales, any model of the type considered in this chapter is, as we have seen, subject to serious limitations. These arise from two major sources: (i) the use of scalar properties to derive a single scalar (isotropic) quantity — the eddy viscosity — to represent the influence of turbulence on the mean-flow field, and (ii) the assumption of linear (Boussinesq) stress-strain relationships that link the stresses to the respective strains. In reality, turbulence properties are anisotropic, and the relationship between stresses and strains is much more complex than is assumed in linear eddy-viscosity models. Chapter 11 considers some of the above limitations, by reference to a variety of strain types that are encountered in complex flows, as a precursor to the presentation, in Chapter 12, of more elaborate and physically robust models that do not rely on the eddy viscosity, and take the anisotropy of turbulence into account.

Chapter Ten

10. Wall Functions for Linear Eddy-Viscosity Models

10.1 *The purpose of 'wall functions'*

The models presented in Chapters 8 and 9 involved no restrictions that exclude their use from any part of a turbulent flow. However, we had to devote much attention to the semi-viscous near-wall region, in an effort to make the models applicable to this region. We recall, in particular, that the buffer layer, around $y^+ = 10 - 15$, is characterised by steep variations in all properties, as viscous and turbulent transport compete for dominance, and that the turbulent stresses decay as the wall is approached at rates that vary within the range $O(y^2) - O(y^4)$. All these characteristics demand elaborate modelling adjustments to ensure that the models return an adequate representation in this thin — but extremely influential — layer. These efforts are never entirely successful, if only because near-wall turbulence is highly anisotropic and exceptionally complex in structure, characterised by quasi-organised streaks and quasi-streamwise vortices. Also, even if a model performs well in the near-wall layer, the steep gradients therein demand a high density of computational nodes — typically 10 within $y^+ = 10 - 15$, and a further 10 below this layer, with the wall-closest node at $y^+ = O(1)$. In complex industrial applications, this can be extremely expensive, can pose major difficulties with meshing around complicated three-dimensional geometries, and can provoke numerical stability and convergence problems. There is, therefore, a strong motivation for bypassing the need of a fine resolution of the near-wall region, by the use of a semi-analytical bridge between the wall and the fully turbulent region. Such a bridge is, in fact, indispensable in the case of models that do *not* include viscosity-related elements, and are, therefore, not applicable to the near-wall layer. Approximations that bridge the near-wall layer are referred to as 'wall functions', and their formulation is the subject of this chapter.

10.2 *Log-law-based wall functions*

The simplest wall-function formulation is based on the assumption that the flow near the wall complies with a channel-flow boundary layer, flow in which case:

$$\underbrace{\mu\frac{dU}{dy} - \rho\overline{uv}}_{\tau} = \tau_w\left(\frac{1-2y}{h}\right), \tag{10.1}$$

where τ is the (total) shear stress, τ_w is the wall shear stress, and h is the channel half-height. Outside the viscosity-affected sublayer, the viscous shear stress is negligible, and $\overline{uv}$ thus varies linearly with y. Close to the wall, the total shear stress, τ, is almost constant and thus, $\tau \simeq \tau_w$. In fact, in the absence of a pressure gradient, this is an identity, if convection is absent or assumed negligible, as is the case in a zero-pressure-gradient boundary layer close to the wall. In the above circumstances, the flow near the wall is well described by the universal velocity profile and associated universal turbulence properties, Fig. 4.2. This fact is used to derive the requisite semi-analytic bridge across the near-wall layer.

Consider the situation shown in Fig. 10.1, applicable to a numerical scheme that is based on the finite-volume approach. Suppose P is the wall-nearest computational node at which we solve the Reynolds-averaged Navier–Stokes (RANS) equations, as well as the equations governing k and ε. This node is assumed to lie in the fully turbulent region, above the viscous sublayer y_v. At the wall, the boundary

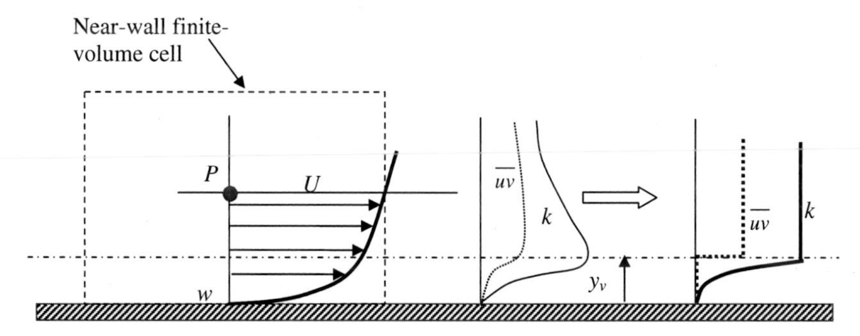

Fig. 10.1: Schematic of a boundary layer, with the wall-nearest computational node P, its associated finite-volume cell (left-hand-side sketch), realistic distributions of k and $\overline{uv}$ (middle sketch), and the assumed simplified profiles of k and $\overline{uv}$ (right-hand-side sketch).

conditions are:

$$U = 0, \quad V = 0, \quad k = 0, \quad \varepsilon = \frac{\mu}{\rho}\frac{d^2 k}{dy^2}\bigg|_{y\to 0} = 2\,\frac{\mu}{\rho}\frac{k}{y^2}\bigg|_{y\to 0}, \tag{10.2}$$

the last condition having been explained in Section 5.1 (Eq. (5.6)). In the context of a finite-volume-based numerical scheme — the one used most frequently in practice — the task of the wall law is two-fold:

(i) to provide the correct wall-shear stress, which is the diffusive flux of momentum into the near-wall cell through its lowest face;

(ii) to provide a realistic estimates for the computational values at the node P, i.e. k_P, ε_P, which represent the cell-averaged values.

The simplest wall laws are based on the assumption that the turbulent shear stress is constant outside the viscous sublayer, and zero within this layer, while the turbulence energy is constant outside the viscous layer and declines quadratically within it (see Eq. (10.2)).

As the node P is in the log-law region, the wall-shear stress can be determined from:

$$\frac{U_P}{u_\tau} = \frac{1}{\kappa}\ln\frac{y_P u_\tau}{\nu} + B. \tag{10.3}$$

A seemingly subtle, but important, modification is based on the relationship (see Eq. (9.12)):

$$k = C_\mu^{-\frac{1}{2}} u_\tau^2, \tag{10.4}$$

which arises upon imposing the assumption of turbulence-energy equilibrium. Eq. (10.4) can also be written:

$$u_\tau \equiv \sqrt{\frac{\tau_w}{\rho}} = C_\mu^{\frac{1}{4}} k^{\frac{1}{2}} \equiv u_k. \tag{10.5}$$

Clearly, in equilibrium conditions, there is no difference between $u_\tau \equiv \sqrt{\tau_w/\rho}$ and $u_k \equiv C_\mu^{1/4} k^{1/2}$. However, in a near-wall layer, which is perturbed by external turbulence or by deceleration or acceleration, there *will* be a difference, and this can be important.

Equation (10.3) can now be written as[1]:

$$\frac{U_P u_k}{\frac{\tau_w}{\rho}} = \frac{1}{\kappa} \ln\left(\frac{y_P u_k}{\nu}\right) + B = \frac{1}{\kappa} \ln\left(\frac{E y_P u_k}{\nu}\right), \tag{10.6}$$

or as:

$$U_P^* = \frac{1}{\kappa} \ln C_\mu^{\frac{1}{4}} E y^*, \tag{10.7}$$

where the use of the superscript '*' indicates scaling with k rather than u_τ.

The substitution of u_k into the right-hand side as well as the left-hand side of Eq. (10.6) is based on three arguments:

(i) It allows an explicit calculation of the wall-shear stress, given U_P and k_P.

(ii) In cases in which the near-wall state is perturbed, as stated above, scaling y_P with k_P gives a more realistic wall-shear stress than using u_τ. This is especially so when the near-wall layer is close to separation, in which case $u_\tau \to 0$ and Eq. (10.3) breaks down.[2]

---

[1]Note that, since y^* has previously been defined as $k^{1/2}y/\nu$, it follows that the argument in Eq. (10.6), $u_k y/\nu = C_\mu^{1/4} y^* = 0.547 y^*$.

[2]It must be acknowledged, however, that the validity of the bridging process is very questionable in such circumstances.

(iii) With $-\rho\overline{uv} = \mu_t \frac{dU}{dy}$, Eq. (10.1) is easily shown to yield $\frac{(U_P - U_w)}{\tau_w} = \int_o^{y_P} \frac{1}{\mu + \mu_t} dy$, which implies that the ratio of the velocity difference $(U_P - U_w)$ to the wall-shear stress is purely dependent on the turbulence properties in the layer. This is compatible with Eq. (10.6), but not with Eq. (10.3).

Next, we need to determine representative values for k_P and ε_P. The simplest approach appears to be to specify k_P from Eq. (10.5), using u_τ, but this would imply, inappropriately, that Eq. (10.6) is identical to Eq. (10.3). Moreover, it would force k_P to always assume a value that is rigidly constrained to the equilibrium state. Thus, if the near-wall flow is close to being separated, or it is redeveloping after reattachment, k_P would be inappropriately low. Finally, as shown in Fig. 10.2(a), k is far from constant in the log-law region, even at relatively high Reynolds numbers.

A more appropriate approach is to solve the transport equation for k over the near-wall volume, as if it were an internal cell, but with modifications to the production rate $P_{k,P}$. To this end, the cell is divided into a lower part, $P-w$, with the remainder above it. In the former, the production can be determined from $-\overline{uv} = C_\mu^{1/2} k = u_k^2$

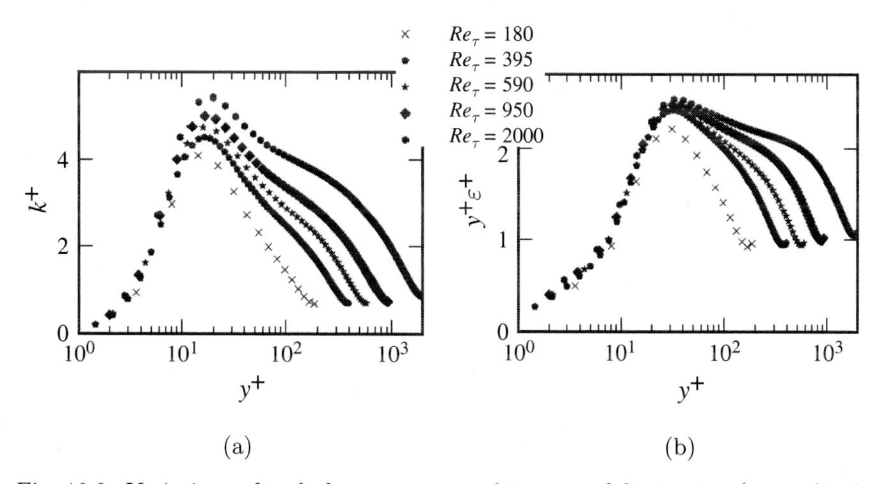

Fig. 10.2: Variations of turbulence energy and its rate of dissipation (normalised by u_τ) in channel flow for a range of Reynolds-number values (reproduced with permission from Billard (2012)).

and the velocity gradient derived by differentiating Eq. (10.6) (with subscript 'P' omitted):

$$\begin{aligned} P_k &= u_\tau^2 u_k \frac{1}{\kappa y}; \quad y \geq y_v, \\ P_k &= 0; \quad\quad\quad\quad y < y_v, \end{aligned} \tag{10.8}$$

where $y_v^+ = 11.2$.

The dissipation rate is required in order to solve the turbulence-energy equation in the near-wall cell, as well as to provide the nodal value necessary for the numerical solution at the cell above. In the log-law region, the assumption $P_k = \varepsilon$ provides the requisite variation, while in the viscous sublayer, Eq. (10.2) may be used. Thus:

$$\begin{aligned} \varepsilon &= u_\tau^2 u_k \frac{1}{\kappa y}; \quad\quad\quad\quad y \geq y_v, \\ \varepsilon &= \frac{2\mu}{\rho}\frac{k_v}{y_v^2} = \frac{2\mu}{\rho}\frac{k_P}{y_P^2}; \quad y < y_v. \end{aligned} \tag{10.9}$$

Figure 2(b) shows that $y^+\varepsilon^+$ does not vary greatly in the log layer when the Reynolds-number is high, so prescription Eq. (10.9) is defensible.

With Eqs. (10.8) and (10.9), the semi-cell average production and dissipation rates then become:

$$\begin{aligned} \overline{P}_k^{w-P} &= \frac{u_\tau^2 u_k}{y_P}\left(\frac{1}{\kappa}\ln\frac{y_P}{y_v}\right), \\ \overline{\varepsilon}^{w-P} &= \frac{u_\tau^2 u_k}{y_P}\left(\frac{1}{\kappa}\ln\frac{y_P}{y_v} + \frac{2\mu}{\rho k_P^{\frac{1}{2}} y_v}\right). \end{aligned} \tag{10.10}$$

It is emphasized that these averages only cover half the near-wall cell. Contributions pertaining to the half-cell above it need to be derived by numerical interpolations between values at P and neighbouring nodes, as is done with any other internal cell. An alternative to the above two-part-cell treatment is to assume that the distributions shown in Fig. 10.1 are applicable across the entire near-wall cell, in which case the averages (10.10) can easily be computed across this cell.

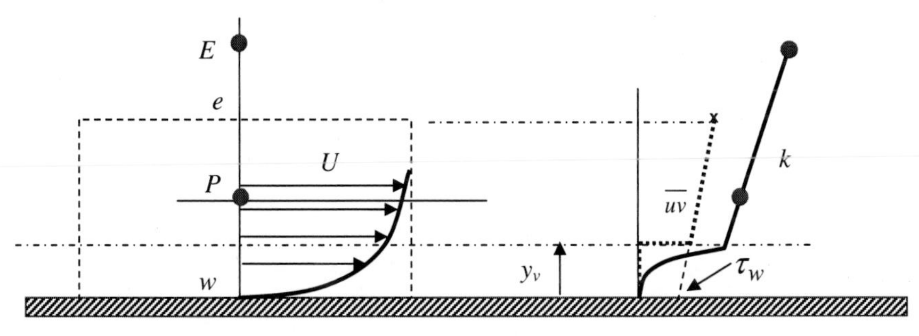

Fig. 10.3: Shear-stress and turbulence-energy variations assumed by Chieng and Launder (1980).

As regards the dissipation-rate equation, this is not normally solved over the near-wall volume — although this can be done, in principle, with Eqs. (10.8) and (10.9) used to evaluate volume-averaged source and sink terms respectively. Instead, the nodal value ε_P is prescribed explicitly by way of the first equation in the set (10.9).

A slightly more elaborate route, proposed by Chieng and Launder (1980), is based on linear variations of the shear stress and turbulence energy in the region beyond y_v, across the remainder of the near-wall cell, as shown in Fig. 10.3. The velocity is assumed to be governed, as before, by Eq. (10.6), from which the wall-shear stress can be computed. To determine the linear variation of k, Chieng and Launder propose the use of the computational node in the cell lying above the near-wall cell P, and to extrapolate the linear variation to y_v, thus giving k_v. As k varies quadratically in the viscous sublayer, it follows:

$$k = k_v \left(\frac{y}{y_v} \right)^2 ; \quad y \leq y_v. \tag{10.11}$$

Since the shear-stress in the viscous sublayer is assumed to be zero, as in Fig. 10.1, $P_k = 0$ in this layer. The linear variation of the turbulent shear stress is determined from τ_w and the value of $\overline{uv}$ computed from the numerical solution at the cell face 'e' (note that this is simply the turbulent-diffusion flux $\frac{\mu_t}{\rho}(\frac{\partial U}{\partial y})$ at the cell face, determined by numerical interpolation, as with any other internal cell). The cell-averaged generation rate can now be determined by integration from:

$$\overline{P}_k^{w-e} = \frac{1}{y_e}\int_{y_v}^{y_e}\left[\tau_w + (\tau_e - \tau_w)\frac{y}{y_e}\right]\frac{\partial U}{\partial y}dy$$
$$= \frac{\tau_w(U_e - U_v)}{y_e} + \frac{\tau_w(\tau_e - \tau_w)}{\rho\kappa u_{k,v}y_e}\left(1 - \frac{y_v}{y_e}\right), \quad (10.12)$$

where $u_{k,v}$ is determined from k_v via Eq. (10.5).

To determine the average dissipation rate, the variation of ε over the near-wall cell is assumed to be:

$$\varepsilon = \frac{2\nu k_v}{y_v^2}; \quad y \leq y_v,$$
$$\varepsilon = \frac{k^{\frac{3}{2}}}{c_{l\varepsilon}y}; \quad y > y_v, \quad (10.13)$$

the former following directly for Eq. (10.2), and the latter following from the length scale definition $l_\varepsilon = \frac{k^{3/2}}{\varepsilon}$ (Eq. (7.56)), together with the mixing-length-type assumption $l_\varepsilon = c_{l\varepsilon}y$ (Eq. (8.3)), with $c_{l\varepsilon} = 2.55$. The integration of Eq. (10.13), in two parts, and with k in the second equation replaced by its linear variation, is somewhat elaborate, and results in:

$$\overline{\varepsilon}^{w-e} = \frac{2}{y_e}\frac{\nu k_v}{y_v} + \frac{1}{y_e c_{l\varepsilon}}\left[\frac{2}{3}(k_e^{\frac{3}{2}} - k_v^{\frac{3}{2}}) + 2a(k_e^{\frac{1}{2}} - k_v^{\frac{1}{2}}) + b\right], \quad (10.14)$$

where

$$a = k_P - \frac{(k_P - k_E)}{(y_P - y_E)}y_P,$$

$$b = a^{\frac{3}{2}}\log\left[\frac{(k_e^{\frac{1}{2}} - a^{\frac{1}{2}})/(k_e^{\frac{1}{2}} + a^{\frac{1}{2}})}{(k_v^{\frac{1}{2}} - a^{\frac{1}{2}})/(k_v^{\frac{1}{2}} + a^{\frac{1}{2}})}\right]; \quad a > 0,$$

$$b = 2(-a)^{\frac{3}{2}}\left[\tan^{-1}\frac{k_e^{\frac{1}{2}}}{(-a)^{\frac{1}{2}}} - \tan^{-1}\frac{k_v^{\frac{1}{2}}}{(-a)^{\frac{1}{2}}}\right]; \quad a < 0.$$

Apart from the constraints associated with turbulence-energy equilibrium and the omission of any wall-parallel gradients due to pressure gradient and convective transport, a problem arises with the above log-law-based wall functions in circumstances in which the wall-nearest node falls within the viscosity-affected layer. This can easily occur in complex geometries in which the near-wall-cell structure is very rarely adapted to changing near-wall-flow conditions. If the wall-nearest cell is within the viscous sublayer — say, $y^+ < 8$ — then the conditions at the node P may be assumed to comply with:

$$\begin{aligned} U_P^+ &= y_P^+, \\ k_P &= k_v \left(\frac{y_P}{y_v}\right)^2, \\ \varepsilon_P &= \frac{2\mu}{\rho}\frac{k_P}{y_P^2}, \end{aligned} \tag{10.15}$$

with k_v determined from the numerical solution above the cell P, or as indicated by the extrapolation in Fig. 10.3. However, even if these values are assumed valid for node P, the region above this node is affected by the viscosity, and includes the complex buffer layer, in which case the turbulence model needs to be of a low-Reynolds-number variety, applied in such a way that this region is numerically resolved adequately, with a sufficiently fine grid. This is difficult, if not impossible, to achieve in practice. Even more serious difficulties arise when the wall-nearest node falls into the buffer region, in which case we have to rely on correlations derived from DNS data, such as shown in Fig. 10.2, or from a high-fidelity, low-Reynolds-number turbulence model, applied in conjunction with a fine wall-layer-resolving grid. These limitations clearly provide strong motivation for more refined formulations. Some options are discussed below.

10.3 *Eddy-viscosity-based wall functions*

Any wall-function formulation requires assumptions about, or a prescription of, the variation of some key property (or properties) across the near-wall layer, allowing an analytical or semi-analytical derivation of relationships that bridge the semi-viscous near-wall layer.

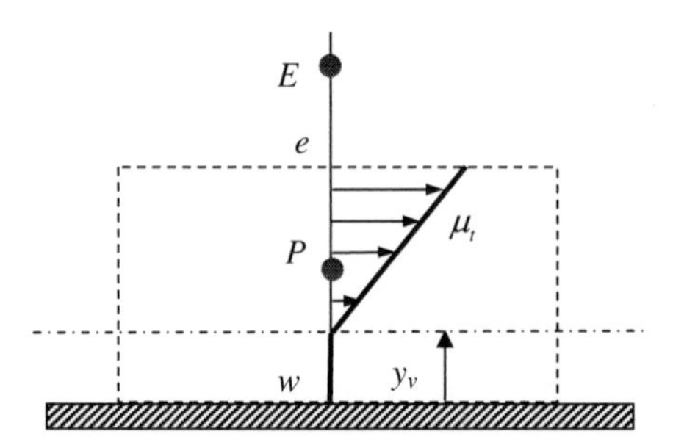

Fig. 10.4: Assumed variation of the eddy-viscosity across the near-wall cell, as a basis of the wall function of Craft *et al.* (2002).

One alternative option to prescribing, explicitly, a velocity profile — as done in Section 10.2 — is to start from a prescribed variation of the eddy-viscosity, as shown in Fig. 10.4, and given by:

$$\nu_t^+ = \begin{cases} 0 & 0 < y < y_v \\ \kappa(y^+ - y_v^+) \quad \text{or} \quad C_\mu^{\frac{1}{4}}\kappa(y^* - y_v^*), & y_v < y < y_e \end{cases}, \tag{10.16}$$

This is the proposal made by Craft *et al.* (2002) as a basis for deriving their 'analytical wall functions'.

We know the above assumption to be a major simplification of reality. First, the eddy viscosity rises cubically from the wall, i.e. is finite in the viscous sublayer. Second, the eddy-viscosity relation, Eq. (7.38), in conjunction with the length-scale expression, Eq. (7.47), shows that a linear variation of the eddy viscosity in the fully turbulent region is only justified if the turbulence energy is assumed to be constant, as shown in Fig. 10.1 — i.e. the turbulence energy used in y^* is assumed to be k_P^*. However, the advantage gained from making these approximations is that we can now set out to derive an expression for the velocity variation across the entire near-wall cell, and that we can take wall-parallel gradients into account, at least approximately. To this end, Craft *et al.* propose that the velocity profile across the cell be derived by solving the simplified momentum equation:

$$\frac{\partial}{\partial y}\left[(\mu + \mu_t)\frac{\partial U}{\partial y}\right] = \left\{\rho U\frac{\partial U}{\partial x} + \rho V\frac{\partial U}{\partial y} + \frac{\partial P}{\partial x}\right\}_P = Const_P, \tag{10.17}$$

where the right-hand side terms are estimated from the prevailing numerical solution (i.e. determined 'on the fly', as the solution evolves), and are assumed to be constant in y across the cell. This allows Eq. (10.7) to be integrated analytically to give the requisite relationship $U^+ = f(y^+)$ or $U^* = f(y^*)$. The details of the method and the (algebraically complex) wall functions, including the manner in which the turbulence equations are modified in the near-wall cell, may be found in Craft *et al.* (2002).

10.4 *Numerical wall functions*

If no assumptions are made about any wall-normal property variations, a numerical solution in the wall-nearest cell cannot be avoided. Arguably, this disqualifies the method in question as being included under the heading 'wall functions' in the conventionally understood sense. However, such a method, proposed by Craft *et al.* (2004), is described here, because it can be regarded as an extension of the ideas underpinning the method in Section 10.3. It is designed to overcome the same limitations as those motivating wall functions, and hence may be regarded as closely related to wall functions. The method offers, albeit at a price, both higher fidelity in conditions such as those shown in Figs. 10.1, 10.3 and 10.4, and in more difficult situations, in which the flow is subject to acceleration or deceleration.

The key feature of the method is conveyed schematically in Fig. 10.5.

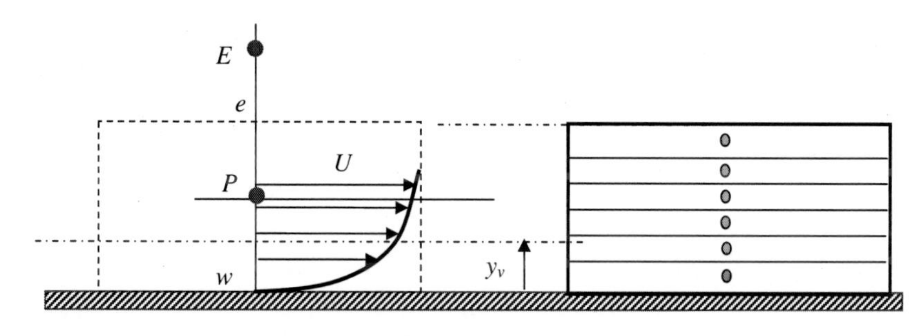

Fig. 10.5: Division of the near-wall main cell into sub-cells in the method of Craft *et al.* (2004).

As noted already, the method requires no assumptions about the variation of near-wall properties. Instead, any near-wall cell is divided, in the wall-normal direction only, into a number of sub-cells. Simplified versions of the equations governing the wall-parallel momentum and turbulence quantities are then solved, numerically, over the set of sub-cells, with the wall-parallel pressure gradient taken from the wall-nearest nodes of the outer grid (i.e. from node P and its neighbours) and then assumed to be identical for all sub-cells — corresponding to the assumption in Eq. (10.17), but only for the pressure. The two numerical solutions — that pertaining to the flow as a whole and that applied to the sub-cells — then progress in parallel, but are pressure-decoupled. Importantly, as the sub-cells cover the viscous near-wall layer, the turbulence-model equations need to be of the low-Reynolds-number variety. The outer model can be a high-Reynolds-number formulation, however, provided the outer node E lies well outside the viscous layer.

To enhance clarity, we consider the sub-cell momentum equation governing U:

$$\rho U \frac{\partial U}{\partial x} + \rho V \frac{\partial U}{\partial y} = -\frac{dP}{dx} + \frac{\partial}{\partial y}\left(\mu \frac{\partial U}{\partial y} - \rho \overline{uv}\right), \tag{10.18}$$

where the shear-stress is determined from any low-Reynolds-number model. As noted earlier, the pressure gradient is constant across all sub-cells and is prescribed on the basis of the outer-grid solution. Craft *et al.* take into account convection in all sub-cells, and this requires the properties at all sub-grid nodes, along the entire wall, to be stored simultaneously. The streamwise term is simply evaluated by interpolation of velocities at neighbouring nodes, while the wall-normal velocity in the wall-normal convection term is evaluated by an explicit cell-by-cell application of the mass-conservation principle, starting at the sub-cell closest to the wall, at the lower face of which $V = 0$. A significant simplification could be achieved by neglecting convection, but this would obviously diminish the accuracy of the method in situations where the near-wall flow is subjected to significant acceleration or deceleration — say, close to a stagnation point. The sub-cell solution allows any cell-averaged property to be determined for the main cell surrounding node P, in the sense explained in Sections 10.2 and 10.3.

10.5 *More general wall functions*

With the numerical method described in Section 10.4 set aside, more general formulations than those described in earlier sections require the prescription of flow conditions that are continuous, and physically realistic, across the near-wall cell. These allow the node P to fall into any portion of the near-wall layer. Formulations based on continuous variations are referred to as 'adaptive (or 'blended') wall functions'.

One key element of such wall functions is a general prescription of the near-wall velocity, such as that of Reichardt (1951):

$$U^+ = \frac{1}{\kappa}\ln(1+\kappa y^+) + 7.8\left[1 - \exp\left(-\frac{y^+}{11}\right) - \frac{y^+}{11}\exp\left(-\frac{y^+}{3}\right)\right]. \tag{10.19}$$

However, as noted in relation to the path leading from Eq. (10.3) to Eq. (10.6), this is a rather restrictive expression, only applicable to canonical and benign conditions.

Next, we need general expressions for the turbulence and dissipation rates. As noted earlier, these could be taken from curve fits to direct numerical simulation (DNS) data. However, Fig. 10.2 shows that this is made problematic by the Reynolds-number dependence of the distributions. An alternative is to derive the distributions in the viscous and buffer layers by solving simplified channel-flow forms of an appropriate (selected) low-Reynolds-number model, and then fit curves to the resulting distributions. If, for example, we wish to derive distributions applicable to the viscous sublayer, we might choose a low-Reynolds-number $k-\varepsilon$ model, in which we nullify the turbulent viscosity and production terms, i.e.:

$$\begin{aligned} &\mu\frac{d^2k}{dy^2} - \rho\varepsilon = 0, \\ &\mu\frac{d^2\varepsilon}{dy^2} - \rho f_{\varepsilon 2}C_{\varepsilon 2}\frac{\varepsilon^2}{k} = 0, \end{aligned} \tag{10.20}$$

in which $f_{\varepsilon 2}$ is a viscous damping function (see Eq. (9.35), for example). For $f_{\varepsilon 2} = 1$, this set can be solved analytically, but not otherwise. In any event, such a solution will not suffice, as we also require solutions for the buffer layer. This route is unlikely to yield accurate results, if only because the large majority of models do not return the correct near-wall variation of the dissipation rate. Nevertheless, an approach following the above lines was adopted, most recently, by Billard, Osman and Laurence (2012) who used a particular version of the $v^2 - f$ model — a three-equation eddy-viscosity model described in Chapter 14 — to derive general wall functions, in which values at the nodal position of the near-wall cell, rather than cell-averaged values, were derived by consolidating log-law-compatible conditions, numerical buffer-layer data and analytical viscous-sublayer solutions. However, the detailed derivation of this formulation is complex and outside the scope of this text.

A broadly similar route to that of Billard *et al.* above, but one that requires no explicit assumptions for the universal velocity profile, was proposed by Kalitzin *et al.* (2005). To understand the key point of difference, we return to Eq. (10.1), introduce the eddy-viscosity hypothesis for $\overline{uv}$ and nullify the streamwise pressure gradient. Eq. (10.1) then becomes:

$$(\mu + \mu_t)\frac{dU}{dy} = \tau_w = \rho u_\tau^2, \tag{10.21}$$

which, when scaled with u_τ, takes the form:

$$(1 + \nu_t^+)\frac{dU^+}{dy^+} = 1, \tag{10.22}$$

where $\nu_t^+ = \nu_t/\nu$. In principle, we wish to obtain $U^+ = f(y^+)$ from this equation, but this can only be achieved if we have the variation $\nu_t^+ = g(y^+)$. A focus on the turbulent viscosity, rather than the velocity, was earlier proposed by Craft *et al.* (2002) as a foundation of the 'analytical wall functions', described in Section 10.3. In contrast to Craft *et al.*'s algebraic prescription, Kalitzin *et al.* proposed that the functional dependence $\nu_t^+ = g(y^+)$ be obtained by performing a one-dimensional (y-wise) numerical solution for the near-wall layer, at zero pressure gradient and zero advection, with any model applicable to the full near-wall layer, including the viscous sublayer (e.g. a low-Reynolds-number $k - \varepsilon$ or the $k - \omega$ model). In fact, given a particular choice of model, the functional variation $\nu_t^+ = g(y^+)$ would be specific to that model, the inevitable consequence being that each model gives rise to a different function. With the function determined, $U^+ = f(y^+)$ can be obtained, and thus u_τ from an iterative inversion of $U_P^+ = f(y_P^+)$.

As regards turbulence quantities — k and ε, in particular — Kalitzin *et al.* again take advantage of numerical solutions for these quantities, derived with the specific model selected and represented as functions $k^+ = f(y^+)$ and $\varepsilon^+ = g(y^+)$. We have already encountered variations of this type, namely those shown in Fig. 10.2, but these are DNS-derived variations, rather than model solutions. The latter show a much weaker sensitivity to the Reynolds number, this insensitivity being one of several predictive model defects. Hence, as already noted in relation to Billard *et al.*'s approach, the use of model-derived functions is certain to be subject to error, albeit no worse than other errors that arise from the application of the related model to the flow outside the near-wall layer.

10.6 *Wall functions for heat transfer*

Finally, we consider how to deal with heat transfer, at least in principle. Essentially, all that we require, additional to previous considerations, is a means of relating the temperature at P to the wall flux — in the same way as Eq. (10.3) or (10.6) relates U_P to τ_w.

The simplest expression is one that is applicable to the log-law region only, and is equivalent to Eq. (10.6):

$$T^* \equiv \rho C_p u_k \frac{(T_w - T_P)}{q_w} = \sigma_{T,t} \left(\frac{1}{\kappa} \ln \frac{E y_P u_k}{\nu} + P_j \right), \qquad (10.23)$$

where q_w is the wall heat flux (into the flow); $\sigma_{T,t}$ is the turbulent Prandtl number; σ_T is the fluid Prandtl number; C_p is the specific heat at constant pressure; and P_j is the so-called 'Jayatilleke function' (see Jayatilleke (1969)[3]), of which there are several forms, one being:

$$P_j = \frac{u_{k,v} y_v}{\nu} \left(\frac{\sigma_T}{\sigma_{T,t}} - 1 \right), \tag{10.24}$$

in which the pre-multiplier identifies the virtual thickness of the viscous sublayer — i.e. the intersection point of the linear and logarithmic velocity profiles.

For $\frac{u_{k,v} y_v}{\nu} = 11.2$, $\sigma_T = 0.7$ (for air) and $\sigma_{T,t} = 0.9$, $P_j = -2.4$. From a physical perspective, this function accounts for the difference between the fluid-dynamic and thermal viscous sublayers, and hence for the different levels of 'resistance' to momentum and heat transfer of the two sublayers. Self-evidently, $P_j = 0$ when the fluid and turbulent Prandtl-number values are identical, in which case the intersection points of the linear and logarithmic profiles are identical for both velocity and temperature.

An equation that applies across the whole near-wall layer, rather than only to the log-law layer, was proposed by Howard (2012), based on an adaptation of Eq. (10.19) and its combination with correlations derived by Kader (1981) from experiments (subject to $\sigma_{t,T} = 0.85$). This is as follows:

$$T^* = \frac{1}{\kappa} \ln(1 + \sigma_T \kappa y^+) + \left[(\sqrt{7.8} + 1.3)\sigma_T^{0.3} - 1.3 \right]^2 \times \left[1 - \exp\left(-\sigma_T^{0.5} \frac{y^+}{11} \right) - \frac{\sigma_T^{0.3} y^+}{11} \exp\left(-\sigma_T^{0.125} \frac{y^+}{3} \right) \right], \tag{10.25}$$

With T^* given in Eq. (10.23), Eq. (10.25) allows the heat flux to be determined from the temperature difference $(T_w - T_P)$, the wall distance y_P, the wall-shear stress and the turbulence energy k_P.

10.7 *Summary and lessons*

The subject of wall functions is not, strictly, one that fits neatly into a text on turbulence modelling. The reason is simple: wall

[3] Jayatilleke (1969) and Spalding (1967) give more complex forms, but the differences in the respective numerical values are modest.

functions are computational approximations designed to circumvent resource constraints posed by the fine grids needed to resolve the viscosity-affected near-wall layer. On the other hand, wall functions are important elements in some practical implementations of turbulence models, and can have a major impact on the predictive capabilities of any model applied to the flow domain outside the near-wall layers; in other words, a model that performs in some known manner when applied over a high-quality, wall resolving mesh, can perform quite differently (not necessarily worse) when combined with wall functions. This is the rationale for covering the subject in this book.

Wall functions are *always* approximations of the near-wall physics — not simply of the *real* physics, but also of the *modelled* physics, as derived from wall-resolving low-Reynolds-number models. They require a great deal of care and a clear appreciation of the limits of their validity. A most important issue to appreciate is that numerical resolution is almost always compromised, to some degree, when wall functions are used. Even if a refined treatment is implemented in the cell closest to the wall, the cell above it — often also lying in a region of substantial gradients — has to be of similar coarseness as the near-wall cell, to avoid excessive expansion ratios in the intermodal distances. Therefore, the numerical resolution above the wall-function region is also an important issue to keep in mind.

When the wall-closest node is outside the viscosity-affected layer, and the near-wall flow adheres broadly to the condition of a slowly evolving boundary layer, log-law based formulations can provide a perfectly adequate representation. However, the validity of this approach can decline quickly when the flow is more complex, involving strong acceleration and deceleration near the wall, as occurs near stagnation points. Fortunately, the influence of the frictional wall properties on the aero/hydrodynamic-flow conditions around stagnation regions is often small — except in flows separating from gently curving surfaces. However, the thermal field can be far more sensitive to the validity of the modelled frictional characteristics.

In complex conditions, the numerical grid is very rarely generated against the background of any *a priori* knowledge of the flow. In such conditions, the wall-nearest node — even several nodes — may fall into the viscosity-affected layer. In this case, quite complex wall functions are needed to achieve an acceptable representation of the flow. Moreover, this requires the adoption of a low-Reynolds-number model, not withstanding the use of wall functions, simply because the flow outside the wall-nearest cell is (or may be) affected by the viscosity, in such circumstances.

With computing speed and memory size rising rapidly, and computing costs dropping equally fast, the importance of wall functions is diminishing steadily. It is only in very complex industrial flows, and situations in which low-Reynolds-number models suffer from numerical stability problems, that wall functions will continue to be of significant interest.

Chapter Eleven

11. Defects of Linear Eddy-Viscosity Models, Their Sources and (Imperfect) Corrections

11.1 *The need for corrections*

When introducing the eddy-viscosity concept, through Eq. (7.5) in Section 7.1, we emphasised the fact that this was a gross simplification of reality, which could be expected to harbour major sources of predictive defects, regardless of the closure approximations made in respect of the turbulent-velocity and length scales — which are, in themselves, the cause of additional errors. At that stage, we pointed to two particular problems, (i) the (wrong) implication of normal-stress isotropy in simple shear, and (ii) the possibility of predicting negative normal stresses in strong acceleration (lack of 'realisability'). A third problem was identified in Section 9.4.3, in which the shear-stress transport (SST) model of Menter was summarised — namely, the need for the shear-stress limiter, Eq. (9.72), introduced to avoid excessive levels of shear stress in strong shear. Finally, we observed, in Chapter 6, Eq. (6.8), by reference to Fig. 6.1, that the eddy-diffusivity approximation does not capture correctly the scalar fluxes, even in the simplest case of a single scalar gradient across a simple shear layer.

In this chapter, we expand on this theme, elaborating on the subject of realisability, prior to highlighting a range of other defects and identifying their origin, as a preliminary to introducing 'fixes', intended to partially address, mostly in a pragmatic manner, the consequences of these defects. The many corrections and fixes proposed, none of which is general, bear witness to the inescapable fact that a more universal framework than linear eddy-viscosity modelling is highly desirable for complex flow conditions. Indeed, as will emerge below, some of the corrections are derived from simplified forms of more complex closure models.

A helpful concept, which we have already exploited in order to arrive at Eq. (7.8), is that the primary drivers of a stress is its production rate. We have also argued that a first, if rather crude, approximation for the stresses would be a quasi-linear relationship linking them to their respective production rates and a turbulent time scale, e.g. the macro-scale k/ε. Generalising Eq. (7.8), we conjecture:

$$\overline{u_i u_j} - \frac{2}{3}\delta_{ij}k = C\frac{k}{\varepsilon}\left(P_{ij} - \frac{2}{3}\delta_{ij}P_k\right). \tag{11.1}$$

A question arising immediately from this equation is why we propose to model the stress-anisotropy tensor, rather than the stress tensor itself. The answer is simple: we have a model for determining k and P_k, namely the turbulence-energy equation (7.32). If we were to propose $\overline{u_i u_j} = C\frac{k}{\varepsilon}P_{ij}$, its contraction would yield $k = C\frac{k}{\varepsilon}P_k$, which is incompatible with the k-transport equation.

In simple shear, Eq. (11.1) is clearly compatible with Eq. (7.8) for the shear stress. For the normal stresses, Eq. (11.1) gives:

$$\begin{aligned}
\overline{u_1^2} - \frac{2}{3}k &= C\frac{k}{\varepsilon}\left(P_{11} - \frac{2}{3}P_k\right) = C\frac{k}{\varepsilon}\frac{4}{3}P_k, \\
\overline{u_2^2} - \frac{2}{3}k &= -C\frac{k}{\varepsilon}\frac{2}{3}P_k, \\
\overline{u_3^2} - \frac{2}{3}k &= -C\frac{k}{\varepsilon}\frac{2}{3}P_k,
\end{aligned} \tag{11.2}$$

i.e. a finite level of anisotropy — specifically the ratio 2:1:1 — which is quite close to that shown in Fig. 4.4 for a jet, and also for a simple boundary layer in a region well away from the wall.

While we shall not, in fact, make use of the 'model' in Eq. (11.1), we can derive from it the important message that there is a good case for arguing that the stress productions drive the stresses. This will serve to highlight relationships that are, as we shall find, incompatible with the eddy-viscosity concept.

To proceed with the arguments relating to Eq. (11.1), we need to know how P_{ij} is related to the strains. At this stage, we only know this for the shear stress in the case of statistically two-dimensional turbulence, Eq. (4.5), and for all stresses in the case of simple shear, Eq. (4.6), both without the influence of body forces. The latter are particular forms of the general Reynolds-stress-transport equations, given symbolically by Eq. (4.1), and treated in detail in Chapter 12. It will be shown, in particular, that the general production of the stresses is given by:

$$P_{ij} + G_{ij} + R_{ij} = \underbrace{-\left(\overline{u_i u_k}\frac{\partial U_j}{\partial x_k} + \overline{u_j u_k}\frac{\partial U_i}{\partial x_k}\right)}_{P_{ij}} + \left\{\underbrace{(\overline{f_i u_j} + \overline{f_j u_i})}_{G_{ij}} + \underbrace{2\Omega_k\left(\overline{u_j u_m}\varepsilon_{ikm} + \overline{u_i u_m}\varepsilon_{jkm}\right)}_{R_{ij}}\right\}, \quad (11.3)$$

in which P_{ij} is the production due to straining; G_{ij} is the generation due to body forces (with f_i representing fluctuations is the body force arising from buoyancy or electromagnetic forces, for example); and R_{ij} is the generation resulting from rotating the flow (and thus the coordinate system) at the rotational speed and direction given by Ω_k. The tensor ε_{ikm} is a third-rank equivalent of the Kronecker delta δ_{ij} (see Appendix). It has a value '1' when the indices jkm are cyclic (e.g. '123', '231'), '-1' when the indices are non-cyclic (e.g. '132', '213') and '0' when any two of the three indices are repeated. Hence, a generalisation of Eq. (11.1) would include G_{ij} and R_{ij} on the right-hand side.

We thus make the general and crucially important observation that the relationship between stresses and strains is much more complicated, even in the absence of body forces and rotation, than is suggested by the eddy-viscosity hypothesis $-\rho\overline{u_i u_j} = \mu_t\left(\frac{\partial U_i}{\partial x_j} + \frac{\partial U_j}{\partial x_i}\right) - \frac{2}{3}\delta_{ij}\rho k$. This linear relationship implies, among a number of unphysical properties, that the eigenvectors of the stress and strain tensors are mutually aligned, which is incorrect. In general, any one stress component is driven by *all* strains, and the weight of any one strain is governed by the associated pre-multiplying stress. Because of anisotropy, any one strain has a different impact on the stress than other strains, hence the importance of resolving the anisotropy correctly.

We shall now examine a number of strain conditions, or simple flows, to illustrate the limitations of the eddy-viscosity framework. This is followed a consideration of proposed corrections that aim to partially address these limitations. Importantly, several of these are 'fixes' for specific flow types (e.g. shear combined with curvature and swirl), and are not invariant, as they do not apply to general flow conditions.

11.2 *Realisability*

The term 'realisability' has established itself in the area of statistical turbulence modelling to characterize any state of the stress field that is physically tenable. This subject will be discussed later in more rigorous terms in the context of Reynolds-stress modelling, because realisability requires considerations that involve the relationship between invariants (scalar contractions) of the stress tensor.

In simplified terms, appropriate to the present level of discussion and modelling, a realisable state is one in which none of the normal stresses becomes negative, i.e.:

$$\left\{\overline{u^2}, \overline{v^2}, \overline{w^2}\right\} \geq 0, \tag{11.4}$$

and in which the cross-correlations (shear-stress components), when normalised by the appropriate auto-correlations (normal stresses), are less than '1', i.e.:

$$\left\{\frac{\overline{uv}}{\sqrt{\overline{u^2}}\sqrt{\overline{v^2}}}, \frac{\overline{uw}}{\sqrt{\overline{u^2}}\sqrt{\overline{w^2}}}, \frac{\overline{vw}}{\sqrt{\overline{v^2}}\sqrt{\overline{w^2}}}\right\} < 1, \tag{11.5}$$

referred to as the Schwartz inequalities. These relations essentially express the fact that the cross-correlation cannot be perfect, reflecting the stochastic nature of turbulence.

In the context of $k - \varepsilon$ eddy-viscosity modelling, the focus of realisability is on the properties of

$$\overline{u_i u_j} = -C_\mu \frac{k^2}{\varepsilon} \underbrace{\left(\frac{\partial U_i}{\partial x_j} + \frac{\partial U_j}{\partial x_i} \right)}_{2S_{ij}} + \frac{2}{3}\delta_{ij} k. \tag{11.6}$$

As stated already, this permits, in principle, negative normal stresses in high normal strain rates.

Consider, first, the particular case of homogeneous streamwise straining:

$$\overline{u_1^2} = -0.09 \times k \left(\frac{2k}{\varepsilon} \frac{\partial U}{\partial x} \right) + \frac{2}{3} k. \tag{11.7}$$

The bracketed term is the non-dimensional strain rate, generally denoted by (see Eq. (8.43)):

$$S^* \equiv \frac{k}{\varepsilon} S, \quad S = \sqrt{2 S_{ij} S_{ij}}. \tag{11.8}$$

It is readily seen from Eq. (11.7) that $\overline{u^2} < 0$ when $S^* > 7.4$. To avoid a negative value, we have to allow C_μ to vary with S^*. Thus, Eq. (11.7) leads to the realisability constraint ($\overline{u^2} \geq 0$):

$$\begin{aligned} S^* C_\mu &\leq \frac{2}{3}, \\ C_\mu &\leq \frac{0.66}{S^*} = \frac{0.33}{\frac{k}{\varepsilon}\left(\frac{\partial U}{\partial x}\right)}. \end{aligned} \tag{11.9}$$

Second, consider the case of simple shear in a boundary layer. Equation (9.76), repeated below,

$$-\frac{\overline{uv}}{k} = C_\mu^{\frac{1}{2}} \sqrt{\frac{P_k}{\varepsilon}}, \tag{11.10}$$

which shows that the shear stress rises steadily with increasing departure from equilibrium due to strong shear straining. If this ratio is to be maintained at 0.3 — as done by Menter in his SST model (see Section 9.4.3) on the basis of experimental observations — Eq. (11.10) yields[1]:

$$C_\mu = \frac{0.3}{S^*}, \tag{11.11}$$

where in this case, $S^* = \frac{k}{\varepsilon}\frac{\partial U}{\partial y}$. At turbulence-energy equilibrium, $C_\mu = 0.09$, so $S^*_{eq} = 3.33$.

Both examples illustrate that lack of realisability requires C_μ to be made indirectly proportional to S^*, with C_μ fixed at 0.09 at $S^* = S^*_{eq} = 3.33$, in the case of shear. In fact, this type of inverse relationship was first observed by Rodi (1972) who correlated experimental data and proposed an empirical function of the form $C_\mu = f(P_k/\varepsilon)$. As $P_k/\varepsilon = C_\mu S^{*2}$, the variation $C_\mu \propto 0.3S^{*-1}$ suggests $C_\mu = f(P_k/\varepsilon) = f(0.3S^*)$, which explains why Rodi's relationship shows a very similar variation to that given by Eq. (11.11).

A model that is specifically designed to secure realisability in the sense above is that of Shih, Zhu and Lumley (1995). This combines a number of non-standard elements, which are not discussed in detail here, including a dissipation-rate equation that is derived from a dynamic equation for the mean-square vorticity fluctuations. Most pertinent to the present focus is the fact that the model contains the following functional relationship for C_μ, adopted from earlier work by the authors on a non-linear eddy-viscosity formulation (see Chapter 15):

$$C_\mu = \frac{1}{4 + \alpha\sqrt{S_{ij}S_{ij} + \Omega_{ij}\Omega_{ij}}\,\frac{k}{\varepsilon}}, \tag{11.12}$$

in which $\Omega_{ij} = \frac{1}{2}\left(\frac{\partial U_i}{\partial x_j} - \frac{\partial U_j}{\partial x_i}\right)$ is the vorticity tensor,[2] and α is a fairly complicated function of the third invariant of the strain tensor $(S_{ij}S_{jk}S_{ki})$. In the case of a simple shear flow, $(\Omega_{ij})^2 = (S_{ij})^2$, so that:

$$C_\mu = \frac{1}{4 + \alpha S^*}, \tag{11.13}$$

---

[1]Durbin & Peterson-Reif (2001) give a more general constraint that arises from the formal requirement that none of the eigenvalues of $\overline{u_i u_j} = -2\nu_t S_{ij} + 2/3\delta_{ij}k$ must fall below zero. This leads to a constraint on the maximum eigenvalue of S_{ij}being less than $(2/3)^{0.5}S^*$ and hence $C_\mu < 1/(\sqrt{6}S^*) \approx 0.4/S^*$.

[2]Not to be confused with Ω, which is used as a rotation rate of a channel or coordinate system attached to it.

which is of a similar form to Eq. (11.11), the difference reflecting a particular calibration route adopted by Shih, Zhu and Lumley. In irrotational straining, $\Omega_{ij} = 0$, as is the case in a free-vortex flow or pure normal straining. This merely reduces the pre-multiplier of S^* by a factor $\sqrt{2}$. The use of Ω_{ij} as well as S_{ij} in Eq. (11.12) is intended to render the expression more general across a variety of conditions that combine strain and vorticity.

The essential point to take forward is that the eddy-viscosity hypothesis is unrealistic, even for very simple flows, if a constant $C_\mu = 0.09$ is used. To make it (more) realistic, we need to sensitize C_μ to the strain, reducing its value beyond the equilibrium strain level in proportion to S^{*-1}.

Next, we return to the subject of normal straining, in order to examine the relationship between the failure of eddy-viscosity models to resolve normal-stress anisotropy and lack of realisability. The type of straining under consideration here is shown schematically in Fig. 11.1. It arises, for example, as the flow approaches an obstacle, thus being subjected to an adverse pressure gradient, or if it is accelerating — say, over an aerofoil, or in a constriction.

For these cases, Eq. (11.3) simplifies to:

$$\begin{aligned} P_{11} &= -2\overline{u^2}\frac{\partial U}{\partial x}, \\ P_{22} &= \overline{v^2}\frac{\partial U}{\partial x}, \\ P_{33} &= \overline{w^2}\frac{\partial U}{\partial x}, \\ P_k &= -\frac{1}{2}\left(2\overline{u^2} - \overline{v^2} - \overline{w^2}\right)\frac{\partial U}{\partial x}, \end{aligned} \tag{11.14}$$

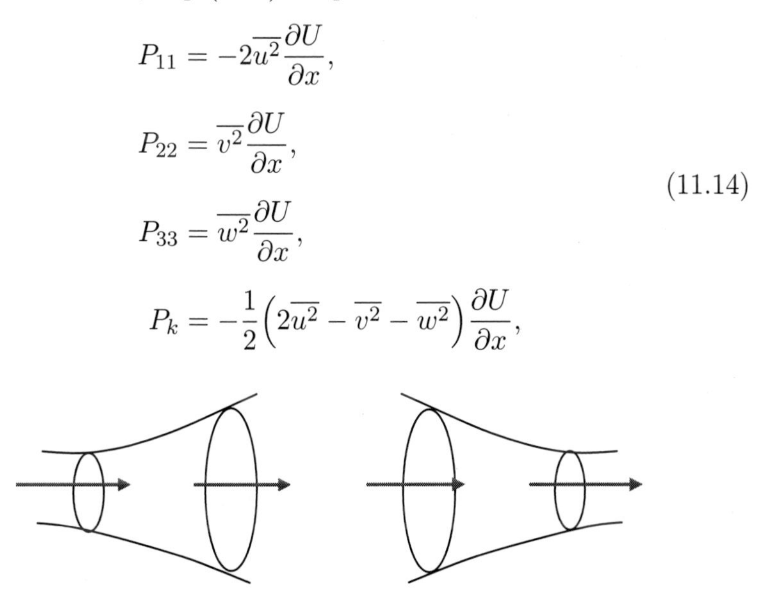

Fig. 11.1: Normal straining due to deceleration or acceleration.

in which the mass-conservation principle, in combination with axial symmetry $\frac{\partial V}{\partial y} = \frac{\partial W}{\partial z} = -0.5\frac{\partial U}{\partial x}$, has been used.

For the case of deceleration, $\frac{\partial U}{\partial x} < 0$, the following may be inferred from Eq. (11.14):

(i) The production P_{11} is positive, thus driving $\overline{u^2}$ upwards. Conversely, $\overline{v^2}$ and $\overline{w^2}$ are driven downwards by negative productions P_{22} and P_{33}, respectively. Because of axial symmetry, $\overline{v^2} = \overline{w^2}$ and $P_{22} = P_{33}$ given isotropic and homogeneous conditions upstream.

(ii) Realisability dictates that all three normal stresses must remain positive. Hence, the production of turbulence energy is expected to be modestly positive, as $P_{22} = P_{33}$ oppose P_{11}. In fact, when turbulence is close to being isotropic, there is hardly any generation of turbulence energy.

For the case of acceleration, $\frac{\partial U}{\partial x} > 0$, the reverse occurs. As $\overline{v^2}$ and $\overline{w^2}$ increase, while $\overline{u^2}$ decreases, P_k is expected to remain positive.

To give the above arguments some quantitative substance, direct numerical simulation (DNS) data are shown in Fig. 11.2 for the temporal evolution of the anisotropy $a_{ij} \equiv \frac{\overline{u_i u_j}}{k} - \frac{2}{3}\delta_{ij}$ for both

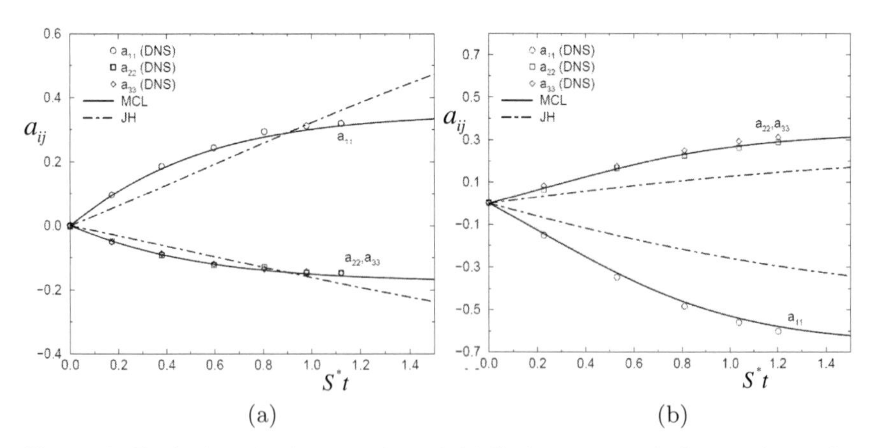

Fig. 11.2: Evolution of anisotropy in an initially isotropic turbulence subjected to (a) compressive and (b) extensive straining. Lines indicate predictions from two Reynolds-stress-transport models (reproduced with permission from the Royal Society from Leschziner (2000)).

deceleration (compression) and acceleration (extension) when an initially isotropic turbulence ($a_{ij} = 0$) is subjected to constant straining at the rate $\frac{k}{\varepsilon}\frac{\partial U}{\partial x}$. In terms of the anisotropy tensor a_{ij}, the production of turbulence energy can be written:

$$P_k = -\frac{1}{2}(2a_{11} - a_{22} - a_{23})k\frac{\partial U}{\partial x}, \tag{11.15}$$

and inspection of Fig. 11.2 shows that P_k indeed remains positive in both cases, with the asymptotic states being $P_k \approx -0.33k\frac{\partial U}{\partial x}$ and $P_k \approx 0.66k\frac{\partial U}{\partial x}$ in cases (a) and (b), respectively. Importantly, the production is (quasi-)linearly dependent on the strain, which clearly contradicts the eddy-viscosity concept (with constant C_μ). Thus, the application of the eddy-viscosity hypothesis to the conditions in Fig. 11.2 yields:

$$a_{11} = -2\frac{\nu_t}{k}\frac{\partial U}{\partial x}; \quad a_{22} = a_{33} = \frac{\nu_t}{k}\frac{\partial U}{\partial x},$$
$$P_k = 3\nu_t\left(\frac{\partial U}{\partial x}\right)^2. \tag{11.16}$$

While there is a degree of correspondence, at least in terms of the sign, between the anisotropy components in Eq. (11.16) and the variations in Fig. 11.2, a major defect is that the production P_k is a quadratic function of the strain and is the sum of unconditionally positive contributions $P_{11} = 4\frac{\nu_t}{k}\left(\frac{\partial U}{\partial x}\right)^2$ and $P_{22} = P_{33} = \frac{\nu_t}{k}\left(\frac{\partial U}{\partial x}\right)^2$. This relationship causes very high levels of turbulence-energy amplification in both acceleration and deceleration. Consistently, it is observed in practical applications that eddy-viscosity models return very high levels of turbulence energy close to stagnation points, exceeding the measured level by a factor of order 10.

A 'harsh' correction, proposed by Launder and Kato (1993), counteracts the excessive production of turbulence energy through the replacement:

$$P_k = \nu_t S^2 \leftarrow \nu_t S\Omega, \tag{11.17}$$

where $\Omega = \sqrt{\Omega_{ij}\Omega_{ij}}$. This correction aims, specifically, at stagnation flows in which S is large while Ω is low, in which case turbulence production is suppressed. However, the correction is not valid for general shear flows (except in thin layers, where $S \approx \Omega$), and is also formally inconsistent with the eddy-viscosity relation, which unconditionally leads to $P_k = \nu_t S^2$.

A better approach is that covered earlier, involving limits on C_μ. Ultimately, the present defect arises because of the excessive response of the turbulence-energy production, and hence the eddy viscosity, to normal straining. This can be — and, indeed, is — alleviated by reducing the implied quadratic dependence of the eddy viscosity on the strain rate to a linear dependence, through the inverse relationship between C_μ and S^*.

11.3 *Curvature*

Experiments (e.g. by Bradshaw (1973)) show that even modest curvature, when imposed on simple shear, has a strong effect on turbulence. Curvature can either stabilise or destabilise turbulence, depending on the sign of the radius of curvature relative to the flow — more specifically, relative to its shear-strain orientation. One aspect of the mechanism by which this interaction occurs has been indicated, in principle, in Chapter 2 (see Eqs. (2.24) and (2.25)). Here, we expand our discussion, focusing specifically on the implications of the eddy-viscosity hypothesis.

We start by considering, in the most transparent manner possible, the effects of curvature within a Cartesian framework, as indicated in Fig. 11.3. We impose a (weak) curvature strain, $\frac{\partial V}{\partial x}$, onto a

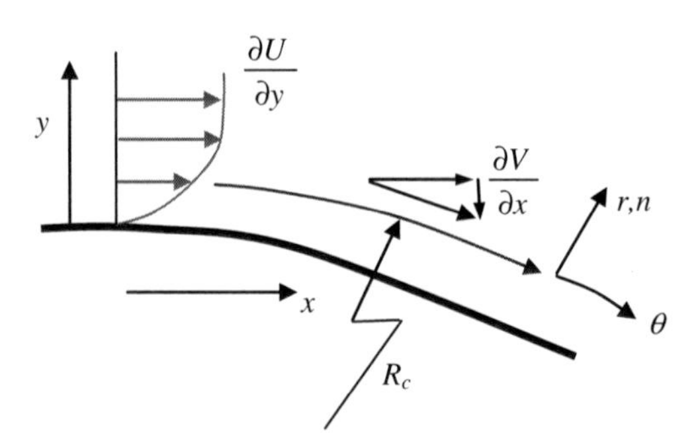

Fig. 11.3: Curvature strain imposed on a simple boundary layer.

simple shear layer, $\frac{\partial U}{\partial y}$. For this case, the eddy-viscosity hypothesis (Eq. (7.5)) gives:

$$-\overline{uv} = \nu_t \left(\frac{\partial U}{\partial y} + \frac{\partial V}{\partial x} \right). \tag{11.18}$$

Hence, the eddy-viscosity relation attaches the same *weight* to the main and to the curvature strains, merely raising the shear stress by a marginal amount, commensurate with the weak curvature strain. Moreover, any indirect effect, via the eddy viscosity, is also minor, because the turbulence-energy production makes use of Eq. (11.18), and is thus hardly modified by the curvature strain.

To gain a view of the real effects of curvature, we need to consider the Reynolds-stress-production rates, Eq. (11.3). Simplified for the conditions in Fig. 11.3,[3] these give:

$$\begin{aligned} P_{11} &= -2\overline{uv}\frac{\partial U}{\partial y}, \\ P_{12} &= -\overline{v^2}\frac{\partial U}{\partial y} - \overline{u^2}\frac{\partial V}{\partial x}, \\ P_{22} &= -2\overline{uv}\frac{\partial V}{\partial x}, \end{aligned} \tag{11.19}$$

which highlights three key interactions:

(i) Since $\overline{u^2} >> \overline{v^2}$, especially close to the wall (see Fig. 4.3), the impact of the curvature strain is much greater than its magnitude suggests.

(ii) In convex curvature, shown in Fig. 11.3, $\frac{\partial V}{\partial x}$ is negative, so that its effect is to materially reduce the magnitude of the shear stress, i.e. the curvature damps the turbulence intensity. In concave curvature, the effect is reversed, and turbulence is enhanced.

(iii) The wall-normal stress $\overline{v^2}$ is reduced, due to the fact that its production is negative, as both $\overline{uv}$ and $\frac{\partial V}{\partial x}$ are negative. In turn,

[3]Subject to the omission of the minor contributions from $\frac{\partial U}{\partial x}$ and $\frac{\partial V}{\partial y}$.

this reduces the primary shear-induced fragment $-\overline{v^2}\frac{\partial U}{\partial y}$ in P_{12}. Hence, the disparity in the weighting of $\frac{\partial U}{\partial y}$ and $\frac{\partial V}{\partial x}$ increases further, over and above that noted under (i).

A more rigorous discussion requires the Reynolds-stress-transport equations to be transformed from the Cartesian to the polar-cylindrical — $r-\theta$ or $n-\theta$ — coordinate system, as indicated in Fig. 11.3. Upon the assumption $\frac{\partial U_\theta}{\partial \theta} = U_n \approx 0$ (locally circular motion), the relevant production rates[4] are:

$$\begin{aligned} P_{\theta\theta} &= -2\overline{u_n u_\theta}\frac{\partial U_\theta}{\partial n} - \underline{2\overline{u_n u_\theta}\frac{U_\theta}{R_c}}, \\ P_{nn} &= \underline{4\overline{u_n u_\theta}\frac{U_\theta}{R_c}}, \\ P_{n\theta} &= -\overline{u_n^2}\frac{\partial U_\theta}{\partial n} + \underline{\left(2\overline{u_\theta^2} - \overline{u_n^2}\right)\frac{U_\theta}{R_c}}, \end{aligned} \tag{11.20}$$

in which the underlined terms are extra curvature-related contributions. It is instructive to consider the implications of Eq. (11.20) in the presence of solid-body rotation, $\frac{\partial U_\theta}{\partial n} = \frac{U_\theta}{R_c}$. In this case, the curvature fragments make no contribution to the turbulence energy, while the curvature fragment in the shear-stress production is identical and opposite to the primary shear production, thus nullifying $P_{n\theta}$. This suggests that any model based on the turbulence-energy equation offers little scope for returning the correct sensitivity of the turbulent stresses to curvature, unless corrections are introduced, either directly to the eddy viscosity — say by sensitising C_μ to the curvature strain — or indirectly, to the transport equations contributing to the eddy viscosity.

[4] A subtle point, merely mentioned here in passing, is that some of the curvature-related 'production' fragments actually arise from the transformed *stress-convection* terms. These are indistinguishable from the true production fragments, however, and system-invariance considerations suggest that they should be treated as production fragments.

The interactions identified above provide some of the most powerful arguments for using closure forms other than those based on the eddy-viscosity framework. They demonstrate that even weak departures from simple shear strain can have profound effects on the turbulent stresses, not captured by the type of eddy-viscosity models discussed so far. As noted earlier, while the curvature strain is included in Eq. (11.18), and also in the turbulence-energy production P_k (via $0.5P_{22}$), its impact, as expressed through these terms, is very weak, and the turbulence field thus predicted is hardly influenced by the curvature strain.

Unfortunately, it is rather difficult to derive a fix for this defect, within the eddy-viscosity framework, that is rational, is applicable to general three-dimensional conditions, is system-invariant and is acceptably economical — as is illustrated in studies by Hellsten (1998), Pettersson-Reif, Durbin and Ooi (1999) and Spalart and Shur (1997). To adhere to the spirit of this book, we restrict ourselves here to relatively simple corrections, derived by reference to idealised conditions and phenomenological arguments.

One approach, derived from a simple Reynolds-stress model, is that of Leschziner and Rodi (1981), applicable to two-dimensional (plane or axisymmetric) flows. From a fundamental perspective, this correction can be claimed to be the most rational, insofar as it takes into account the influence of curvature on the shear-stress, resulting in a correction to the eddy-viscosity, via a modification of C_μ. This is also the option adopted by Pettersson-Reif *et al.* (1999), albeit pursued along a more rigorous and general route. The method of Leschziner and Rodi consists of five key steps:

(i) the calculation of the local radius of curvature, R_c;

(ii) the use of a simplified second-moment closure (see Chapter 12), subject to the assumption of zero stress transport (i.e. local stress equilibrium), leading to an algebraic relationship that links the stresses to their production rates P_{ij} and to k and ε;

(iii) the transformation of the production terms from their Cartesian form to the streamline-oriented system (s, n) — i.e. in the direction of the velocity vector and normal to it, resulting in strains such as $\frac{\partial U_\theta}{\partial n}$ (the principal shear strain), $\frac{U_\theta}{R_c}$ (the curvature strain), $\frac{\partial U_\theta}{\partial \theta}$ and $\frac{\partial U_n}{\partial n}$, the last two eventually ignored;

(iv) the derivation of an explicit expression for the shear stress:

$$-\overline{u_n u_s} = \underbrace{\left\{ \frac{K_1 K_2}{\left[1 + 8K_1^2 \frac{k^2}{\varepsilon^2}\left(\frac{\partial U_\theta}{\partial n} + \frac{U_\theta}{R_c}\right)\frac{U_\theta}{R_c}\right]} \right\}}_{C_\mu} \frac{k^2}{\varepsilon} \underbrace{\left(\frac{\partial U_\theta}{\partial n} - \frac{U_\theta}{R_c}\right)}_{\text{Strain normal to n}}, \tag{11.21}$$

where K_1 and K_2 are constants associated with, and derived from, the underlying Reynolds-stress model;

(v) the interpretation of the leading group between the curly braces in Eq. (11.21) as a curvature-dependent C_μ, which is then used within the standard eddy-viscosity expression (11.6).

Leschziner and Rodi provide a range of results for recirculating annular and plane jets to demonstrate the effectiveness of this correction. Its extension to three-dimensional conditions has not been attempted, however, and is probably too complicated, unless pursued via general tensor algebra, which respects system-invariance (see Durbin (2011)). One other problem is that it is entirely unclear whether and how it could be combined with a realisability fix of the type (11.12). No such combination appears to have been investigated.

A much simpler option, albeit fundamentally inferior, is to introduce non-dimensional curvature indicators into the dissipation-rate equation. The basic idea is to reduce or increase the eddy viscosity by increasing or reducing the dissipation rate, respectively, via additional curvature-driven extra sink terms in the dissipation-rate equation. Alternative parameters proposed are the 'gradient' and 'flux' Richardson numbers (Launder, Priddin and Sharma (1977),

Rodi (1979), Rodi and Scheuerer (1983)), defined respectively, as[5]:

$$Ri_g = \frac{k^2}{\varepsilon^2} \frac{U_\theta}{R_c^2} \frac{\partial (R_c U_\theta)}{\partial n}, \tag{11.22}$$

$$Ri_f = \frac{-2\overline{u_n u_\theta} \dfrac{U_\theta}{R_c}}{P_k} = \frac{2\nu_t \dfrac{U_\theta}{R_c} \left(\dfrac{\partial U_\theta}{\partial n} - \dfrac{U_\theta}{R_c} \right)}{P_k} = \frac{2\nu_t U_\theta \left(\dfrac{\partial U_\theta / R_c}{\partial n} \right)}{P_k}. \tag{11.23}$$

Both may be viewed as expressing, albeit in different ways, the relative magnitude of the curvature-related strain, $\left(\frac{U_\theta}{R_c}\right)/\left(\frac{\partial U_\theta}{\partial n}\right)$, either in terms of the strain itself (Eq. 11.22)) or the extra turbulence production arising from that strain (Eq. 11.23)) — i.e., the sum of the underlined terms in the first two equations of the set in Eq. (11.20).

A basic fault in the logic of using the curvature-related production of the turbulence energy in arriving at the flux Richardson number (Eq. (11.23)) is rooted in the fact that our primary interest is *not* in the effect of curvature on the turbulence energy, but rather its effects on the shear stress. However, as is seen from the third equation in the set (11.20), the shear stress is affected by curvature fragments which are entirely different from those associated with the turbulence energy, and which are directly linked to the anisotropy of the normal stresses. This gives support to the argument that curvature corrections should, rationally, be effected via modifications of the form of Eq. (11.21).

The gradient Richardson number (11.22) may readily be expanded to:

$$Ri_g = \frac{k^2}{\varepsilon^2} \left(\frac{\partial U_\theta}{\partial n} + \frac{U_\theta}{R_c} \right) \frac{U_\theta}{R_c}, \tag{11.24}$$

[5]Several variations of these definitions appear in the literature, but this is of little importance in the present context. In any event, the use of these parameters involves calibration and the introduction of empirical coefficients which are specific to variations used.

which appears in the fix in Eq. (11.21), and tends to depress C_μ for the stabilizing curvature shown in Fig. 11.3. With the explicit form of the turbulence-energy production, P_k, inserted into the flux Richardson number, Eq. (11.23), this becomes:

$$Ri_f = 2\frac{\partial U_\theta}{\partial n}\frac{U_\theta}{R_c}\bigg/\left(\frac{\partial U_\theta}{\partial n} - \frac{U_\theta}{R_c}\right)^2. \tag{11.25}$$

Either parameter can now be used, pragmatically, to augment or diminish the dissipation rate, ε, via a modification of the source or the sink term of the ε-equation. As Ri_f is associated with the generation process, expressing the extra curvature-related production, it is sensible to use it to modify the source term of the ε-equation, while if Ri_g is used, it is conventional to modify the sink term. In both cases, this can be effected via the following replacements of the constants $C_{\varepsilon 1}$ or $C_{\varepsilon 2}$:

$$\begin{aligned} C_{\varepsilon 1} &\leftarrow C_{\varepsilon 1}(1 + C_{\varepsilon 3} Ri_f), \\ C_{\varepsilon 2} &\leftarrow C_{\varepsilon 2}(1 - C'_{\varepsilon 3} Ri_g), \end{aligned} \tag{11.26}$$

where $C_{\varepsilon 3}$ and $C'_{\varepsilon 3}$ are constants that need to be determined by computational calibration for some generic conditions. Analogous corrections to the $k - \omega$ model have been proposed by Hellsten (1998).

The above corrections, based on Richardson numbers, are too coarse for any but simple curved flow, and have, indeed, only been used in curved boundary layers, in which the local radius of curvature can be easily determined by adding to the local wall radius the wall-normal distance of the location at which either of the Richardson numbers has to be computed.

11.4 *Swirl*

This is a more complicated form of the curved flow considered in the previous section. It is discussed separately, because it is frequently encountered in engineering applications — for example, swirl combustors, such as the one shown in Fig. 11.4, in which swirl causes a toroidal recirculation zone that stabilizes the combustion process.

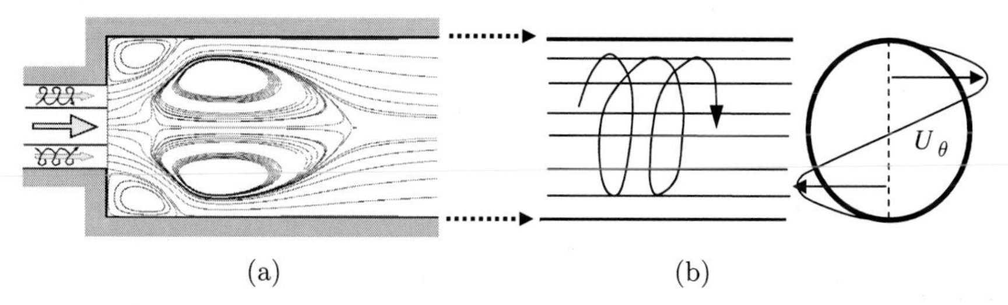

(a) (b)

Fig. 11.4: (a) Swirl combustor (courtesy S. Jakirlić) and (b) related generic swirl flow.

Figure 11.4(a) makes clear that any curved streakline has a highly complex spiralling shape, due to the combination of swirl and the distortions associated with the non-uniform flow pattern in cross-sectional plane. In dealing with the effect of swirl in this section, we restrict ourselves to the simpler configuration 11.4(b) in which the curvature is due to the swirling motion alone, the flow being fully developed ($\partial\{\ldots\}/\partial x = 0$) and axially symmetric. In effect, this is the state that would approximately arise far downstream of the recirculation zone in Fig. 11.4(a).

As done in Section 11.3, we consider, here too, the effects of swirl by reference to the stress-production terms, Eq. (11.3), transformed to cylindrical-polar co-ordinates. The relevant production terms for the case shown in Fig. 11.4(b) are as follows[6]:

$$\begin{aligned} P_{\theta\theta} &= -2\overline{u_r u_\theta}\frac{\partial U_\theta}{\partial r} - \underline{2\overline{u_r u_\theta}\frac{U_\theta}{r}}, \\ P_{rr} &= \underline{4\overline{u_r u_\theta}\frac{U_\theta}{r}}, \\ P_{r\theta} &= -\overline{u_r^2}\frac{\partial U_\theta}{\partial r} + \underline{\left(2\overline{u_\theta^2} - \overline{u_r^2}\right)\frac{U_\theta}{r}}, \\ P_{xr} &= -\overline{u_r^2}\frac{\partial U_x}{\partial r} + \underline{2\overline{u_x u_\theta}\frac{U_\theta}{r}}, \end{aligned} \tag{11.27}$$

[6]See footnote 4.

in which U_θ/r is the curvature strain analogous to U_θ/R_c in Section 11.3. We note the following features:

(i) The first three production terms — and, in particular, the curvature corrections therein — are identical those in Eq. (11.20).

(ii) $P_{r\theta}$, driving $\overline{u_n u_\theta}$, is enhanced or damped by the curvature strain $\frac{U_\theta}{r}$, depending upon the sign of the principal swirl strain $\frac{\partial U_\theta}{\partial r}$. Close to the wall, $\frac{\partial U_\theta}{\partial r} < 0$, $\overline{u_r u_\theta}$ is positive, and curvature increases this shear stress, i.e. it has a destabilizing effect. In the inner core, $\frac{\partial U_\theta}{\partial r} > 0$, $\overline{u_r u_\theta}$ is negative, and the *magnitude* of this negative stress is decreased by the curvature strain, so curvature has a damping influence. As the curvature strain is pre-multiplied by $\left(2\overline{u_\theta^2} - \overline{u_r^2}\right)$, the influence of this strain is much larger that its relative magnitude implies. All this is entirely consistent with the considerations in Section 11.3.

(iii) P_{rr} is positive in the near-wall layer and negative in the core region, thus increasing $\overline{u_r^2}$ near the wall and decreasing it in the core. This further increases the relative importance of the curvature strain, via the primary shear productions in $P_{r\theta}$ and P_{xr}. Again, this is consistent with the interactions discussed in Section 11.3.

(iv) The curvature fragment in P_{xr} is positive in the near-wall layer and negative in the core region. Thus, the production of the shear stress $\overline{u_x u_r}$ is correspondingly increased or decreased. However, the effect on the shear stress itself depends on the sign of this stress, which is clearly linked to the sign of the shear strain $\frac{\partial U_x}{\partial r}$.

As in the case considered in Section 11.3, the influence of the curvature strain on the shear stresses, when expressed via the eddy-viscosity model, is very modest. Here too, the weight attached to the curvature strain is identical to that of the principal shear strains, and its modest contribution to the production of turbulence energy is restricted to $0.25P_{rr}$.

The set in Eq. (11.27) shows that the extra production of turbulence energy due to swirl is[7] $2\overline{u_r u_\theta}\frac{U_\theta}{r}$, which, with the eddy viscosity used to approximate $\overline{u_r u_\theta}$, becomes $2\nu_t \frac{U_\theta}{r}\left(\frac{\partial U_\theta}{\partial r} - \frac{U_\theta}{r}\right)$. This is precisely equivalent to the curvature-related extra production term in Eq. (11.23), thus supporting the use the flux-Richardson number:

$$Ri_f = \frac{2\nu_t U_\theta \dfrac{\partial U_\theta / r}{\partial r}}{P_k}, \tag{11.28}$$

to express the effects of swirl-related curvature. By analogy, it is also reasonable to use, as an alternative, the gradient Richardson number:

$$Ri_g = \frac{k^2}{\varepsilon^2}\frac{U_\theta}{r^2}\frac{\partial (rU_\theta)}{\partial r}, \tag{11.29}$$

which corresponds to Eq. (11.22). Either Richardson number can now be used to modify the constants $C_{\varepsilon 1}$ or $C_{\varepsilon 2}$, as indicated by Eq. (11.26). Here again, it is important to keep in mind that the influence of the curvature strain on the turbulence energy is different from that exerted on the shear stress $\overline{u_r u_\theta}$. Hence, the use of the Richardson number in the dissipation-rate equation is subject to potentially serious limitations that are difficult to quantify in any particular application.

11.5 *Rotation*

When a flow (as a 'system') is subjected to rotation at the rate $\vec{\Omega}$, one consequence is the appearance of a Coriolis force in the momentum equations. Expressed in a vectorial form, rotation gives rise to the 'body force' $R_i = -2(\vec{\Omega} \times \vec{V})$. When this term is taken into account in the derivation of the Reynolds-stress equations, via the manipulations in Eq. (4.1), with the decomposition $R_i = \overline{R_i} + r_i$, the result is the appearance of the additional rotation related production terms R_{ij} in Eq. (11.3) (see also Chapter 12, Eq. (12.7)). Notwithstanding the fact that R_{ij} is distinct from P_{ij}, system rotation is part of an interconnected set of conditions involving curvature, swirl and

[7]There is no swirl-related contribution from P_{xx}.

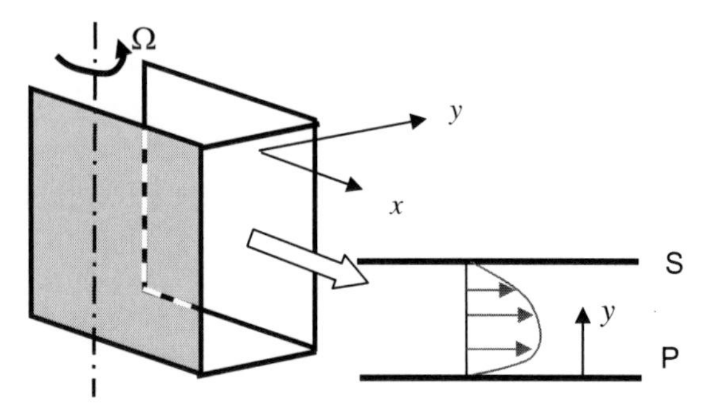

Fig. 11.5: Flow in a channel subjected to rotation (S = suction side, P = pressure side).

rotation. We have already seen that curvature and swirl are closely connected. We shall see here that rotation gives rise to extra production terms that are similar to those we have already encountered. Indeed, Durbin (2011) presents a framework which consolidates all three phenomena.

To appreciate the significance of R_{ij}, we focus on the generic situation of a channel flow subjected to orthogonal rotation, as shown in Fig. 11.5. This may be thought of as a simplified passage in a turbomachine rotor. For this particular case, in which the vector $\Omega_i = [0, 0, \Omega]$, relevant components of R_{ij} simplify to:

$$\begin{aligned} R_{yy} &= -4\Omega\overline{uv}, \\ R_{xx} &= 4\Omega\overline{uv}, \\ R_{xy} &= -2\Omega(\overline{u^2} - \overline{v^2}). \end{aligned} \tag{11.30}$$

We note immediately that a replacement $\frac{\partial U_\theta}{\partial n} = \frac{U_\theta}{R_c} = \Omega$ in the set Eq. (11.20) renders this set identical to Eq. (11.30). This is not entirely surprising, because system rotation is a form of solid body rotation. We can further identify the following physical interactions:

(i) The effects of rotation on one side of the duct are opposite to those on the other side, because the sign of $\overline{uv}$ — i.e. rotation inevitably introduces an asymmetry in the stresses, and hence in the velocity profile and the skin friction.

(ii) On the pressure side, 'P' in Fig. 11.5, the shear stress is negative, R_{22} is positive, and the effect is to enhance the turbulence intensity across the boundary layer. Consistently, since $\overline{u^2} > \overline{v^2}$, R_{xy} is negative, thus driving up the *magnitude* of the (negative) shear stress. Hence, rotation destabilizes turbulence on the pressure side. Conversely, turbulence is damped on the suction side.

(iii) Since R_{xx} is negative on the pressure side, rotation diminishes the streamwise stress. This, in conjunction with the increase in $\overline{v^2}$, reduces the anisotropy on the pressure side, thus diminishing the destabilizing effect. The reverse is the case on the suction side. Hence, stabilisation on the suction side is more pronounced than destabilisation on the pressure side.

None of the above interactions is represented by an eddy-viscosity model. In fact, since $R_{11} + R_{22} + R_{33} = 0$, rotation does not feature at all in the turbulence-energy equation. As a consequence, an eddy-viscosity model inevitably returns a symmetric velocity profile $U(y)$.

While there have been isolated efforts to address this defect by sensitising the dissipation-rate equation to rotation, via the Rossby number, $\mathrm{Ro} = \frac{U}{h\Omega}$, with h being the channel height in Fig. 11.5, there appear to have been no serious attempts to apply such corrections in practical conditions, beyond those of simple channel flow in orthogonal-mode rotation. Similarly, the effectiveness of more complex corrections (Hellsten (1998), Pettersson-Reif *et al.* (1999)) have also been examined principally for rotating channels. One likely reason for the scarcity of applications involving system-rotation corrections is that any practical flow that involves rotation — e.g. in turbomachinery — is far more complex than that shown in Fig. 11.5, involving strongly three-dimensional features. In such circumstances, the defect at issue is regarded as a relatively minor contributor to the spectrum of modelling errors arising from any eddy-viscosity framework. A much more promising route, in terms of generality and realism, is to use some form of Reynolds-stress model.

11.6 *Body forces — buoyancy*

Equation (11.3) contains the general body-force-induced stress-generation term $G_{ij} = \overline{f_i u_j} + \overline{f_j u_i}$. To understand the origin of this term, we need to accept, first, that a body force (per unit mass), whatever its origin, expresses itself by the addition of this specific force — say, a vector F_i — in the Navier–Stokes equations. As before, Eq. (4.1) gives a symbolic 'recipe' of how to derive the equations governing the Reynolds stresses, including the body forces at issue here. It is thus easy to recognise that the inclusion of $F_i = \overline{F_i} + f_i$, where f_i is the turbulent fluctuation of the specific body force, leads to the appearance of the related correlations in G_{ij}.

For the sake of clarity, we consider the case of a body force caused by density differences in a flow subjected to gravity, i.e. $F_i = (\rho - \rho_0)g_i$,[8] where ρ_0 is a reference density and g_i denotes a gravitational-acceleration vector. In the case of a simple buoyant flow, in which the gravitational acceleration is pointing downwards, in the direction $-y$, $F_1 = F_3 = 0$, $F_2 = -g(\rho - \rho_0)$. In this case,

$$\begin{aligned} G_{11} &= G_{33} = 0, \\ G_{22} &= -\frac{2g}{\rho}\overline{\rho' v}, \\ G_{12} &= -\frac{g}{\rho}\overline{\rho' u}. \end{aligned} \tag{11.31}$$

Density fluctuations can be related to temperature or mass-concentration fluctuations via volumetric expansion coefficients. Thus, in general, $\rho' = -\rho\beta\phi'$, where β can be negative, depending on the nature of the property Φ.

Equation (11.31) shows that the effect of buoyancy is highly anisotropic, increasing or decreasing only $\overline{v^2}$ among the normal stresses. Whether the stress is amplified or damped depends only on the sign of the flux $\overline{\rho' v} = -\beta\rho\overline{\phi' v}$. If we make the simplest possible assumption that the scalar flux obeys the Fourier–Fick law of diffusion, we get:

$$-\overline{v\phi'} = \Gamma_\phi \frac{\partial \Phi}{\partial y}. \tag{11.32}$$

For the case of thermal stratification, $\Phi = T$ and $\beta = -\frac{1}{\rho}\frac{\partial \rho}{\partial T}$ (which is positive, because a positive density fluctuation is associated with

[8] The density ρ is to be understood as being the time-mean value.

a negative temperature fluctuation). We thus get:

$$G_{22} = -2g\beta\Gamma_T \frac{\partial T}{\partial y} = -2g\beta \frac{\nu_t}{\sigma_T} \frac{\partial T}{\partial y}, \tag{11.33}$$

where σ_T is the Prandtl–Schmidt number. Hence, as explained in conceptual terms in Chapter 2 (by reference to the schematic in Fig. 2.15) stable thermal stratification, $\frac{\partial T}{\partial y} > 0$, results in $G_{22} < 0$, and thus a reduction of $\overline{v^2}$. This leads, in turn, to a decrease in the shear stress, because $\overline{v^2}$ drives the shear-stress production P_{12}.[9] The reverse occurs in unstable stratification, $\frac{\partial T}{\partial y} < 0$.

Models based on the eddy-viscosity hypothesis appear, at first sight, to be entirely oblivious to the above interactions, because G_{ij} does not feature in the linear stress-strain relationship. However, G_{ij} does feature in the turbulence-energy equation, because the manipulations expressed by Eq. (7.26), with the body force $F_i = g_i \frac{1}{\rho}(\rho - \rho_0)$ included in the NS equations, inevitably leads to a form of Eq. (7.27), which contains the total production terms:

$$P_k + G_k = -\overline{u_i u_j} \frac{\partial U_i}{\partial x_j} + \overline{f_i u_i}, \tag{11.34}$$

which, in the case of simple buoyancy, with $g_i = \{0, -g, 0\}$, becomes:

$$P_k + G_k = -\overline{u_i u_j} \frac{\partial U_i}{\partial x_j} + 0.5 G_{22}. \tag{11.35}$$

The inclusion of G_k provides a path, albeit one based on isotropy, for buoyancy to effect either a reduction or an increase in the turbulence energy in the correct physical sense. As a consequence, the eddy viscosity tends to increase or decrease accordingly, and hence the stresses. However, this link is rather indirect, affects all stresses uniformly, and is quantitatively incorrect. An even more serious problem is that the dissipation-rate equation also includes G_k through its consistent addition to P_k in the approximation in Eq. (9.6). This

[9] P_{12} is also affected, directly, by G_{12}, as seen from Eq. (11.31), but we do not know how, at this stage, because the eddy-diffusivity hypothesis is wholly inapplicable to $\overline{uT}'$, as explained in Chapter 6 (see Eq. (6.8)).

addition tends to drive ε in the same direction as k, which substantially counteracts the effects of including G_k in Eq. (7.27) (or (7.32)), and thus renders the model weakly sensitive, at best, to body-force-related generation.

A crude fix that is suggested by the above discussion is to simply omit G_k from the ε-equation, and this has, indeed, been done in some applications (e.g. Leschziner and Rodi (1983)). In fact, this correction can be too strong, because the eddy-diffusivity approximation in Eq. (11.32) tends to overestimate the vertical flux.

A better fix is one in which a non-dimensional buoyancy parameter — a Richardson number, akin to Eq. (11.23)[10] — is used to sensitise the dissipation-rate equation to the extra buoyancy-related turbulence-energy production G_k. Hence, the equivalent of Eq. (11.23) is here:

$$Ri_f = \frac{-g\beta \dfrac{\nu_t}{\sigma_T}\dfrac{\partial T}{\partial y}}{P_k}. \tag{11.36}$$

One of several options of using Eq. (11.36) is to modify the source term in the ε-equation as follows:

$$S_\varepsilon = C_{\varepsilon 1}\frac{\varepsilon}{k}(P_k + G_k)(1 - C''_{\varepsilon 3}R_f) - C_{\varepsilon 2}\frac{\varepsilon^2}{k}, \tag{11.37}$$

where $C''_{\varepsilon 3}$ is a constant that needs to be calibrated, but is expected to be of order '1'.

A fundamentally more sound correction is based on an approach analogous to that adopted in deriving the curvature correction (11.21). This entails, here again, the use a simplified (algebraic) second-moment closure for the stresses $\overline{v^2}$ and $\overline{uv}$, and, in addition, for the fluxes $\overline{uT'}$ and $\overline{vT'}$, subject to only $\frac{\partial U}{\partial y}$ and $\frac{\partial T}{\partial y}$ being finite, and with the buoyancy-related generation terms included. As is shown by

[10]Bradshaw (1965) discusses extensively the analogy between streamline curvature and buoyancy in turbulent shear flows.

McGuirk and Rodi (1979), the outcome of this process are expressions of the following form:

$$-\overline{uv} = \underbrace{\frac{k^2}{\varepsilon} f_1 \left(\frac{\partial T}{\partial y} \right)}_{\nu_{t,y}} \times \frac{\partial U}{\partial y}, \tag{11.38}$$

$$-\overline{vT'} = \underbrace{\frac{k^2}{\varepsilon} f_2 \left(\frac{\partial T}{\partial y} \right)}_{\Gamma_{t,y}} \times \frac{\partial T}{\partial y}. \tag{11.39}$$

where f_1 and f_2 contain only model constants, g, k and ε. These expressions thus define an eddy viscosity and a diffusivity that pertain specifically to the vertical mixing and to its dependence on the buoyancy, while all other stresses and fluxes that are pertinent to the flow in question are computed with the standard values of the eddy viscosity and diffusivity.

11.7 *Length-scale corrections*

A well-known and oft-recorded predictive defect of the $k - \varepsilon$ model is its tendency to return, in adverse pressure gradient, excessive length-scale values, as determined from $k^{3/2}/\varepsilon$, which are significantly larger than those corresponding to the equilibrium variation $\kappa C_\mu^{-3/4} y$ (see Eq. (7.47)) in the log-law layer. This defect is especially detrimental when the boundary layer is about to undergo separation, because the excessive length scale, and the excessive eddy viscosity that goes with it, causes separation to be delayed or prevented altogether — the latter, for example, on gently curved convex surfaces.

It might be supposed that this defect is yet another manifestation of the wrong response of the stresses to streamwise straining, discussed under 'Realisability' earlier in this chapter (Section 11.2). However, the observation that the $k - \omega$ model yields better near-wall behaviour suggests that one source of the defect lies in the ε-equation — although it is important to add that most improvements have been recorded when the $k - \omega$ model was applied as part of the SST model of Menter, which includes the influential shear-stress limiter, Eq. (9.71).

Efforts to correct this defect are usually based on the pragmatic introduction of extra sources terms into the dissipation-rate equation, which are simply designed to drive the dissipation rate towards a value consistent with the length-scale variation $\kappa C_\mu^{-3/4} y$. The simplest proposal is that of Yap (1987):

$$S_{\varepsilon,l} = \max\left[\left(\frac{l_\varepsilon}{\kappa C_\mu^{-\frac{3}{4}} y} - 1\right)\left(\frac{l_\varepsilon}{\kappa C_\mu^{-\frac{3}{4}} y}\right)^2 \frac{\tilde{\varepsilon}^2}{k}, 0\right]. \qquad (11.40)$$

This correction is only effective if the length scale returned by the ε-equation exceeds the equilibrium level. It acts as a source term with a magnitude that increases as l_ε departs from the equilibrium value, thus tending to increase ε and consequently decrease l_ε. The use of $\tilde{\varepsilon}$ ensures that the correction is effectively deactivated in the viscous sublayer, but the length scale in this layer should, in any event, be much lower than that implied by the linear variation, which itself deactivates the correction.

An alternative correction, proposed by Jakirlić and Hanjalić (1995), is:

$$S_{\varepsilon,l} = \max\left[\left(\frac{1}{\kappa C_\mu^{-\frac{3}{4}}} \frac{\partial l_\varepsilon}{\partial y} - 1\right)\left(\frac{1}{\kappa C_\mu^{-\frac{3}{4}}} \frac{\partial l_\varepsilon}{\partial y}\right)^2 \frac{\tilde{\varepsilon}\varepsilon}{k}, 0\right]. \qquad (11.41)$$

This form does not require the wall distance to be determined, but uses the (local) wall-normal gradient of the predicted length scale. This is only equivalent to l_ε / y if the gradient is constant across the layer within which the correction is active.

Another correction by Yakhot *et al.* (1992), introduced in the context of a renormalisation-group-(RNG-) theory-based $k - \varepsilon$ model, has the form of a strain-dependent source term in the ε-equation, resulting in a reduction of the length scale at high strain rates. However, the arguments leading to its formulation are complex and thus not pursued in this introductory text.

11.8 *Summary and lessons*

The simplicity of the linear eddy-viscosity concept has made models based upon it the default choice for predicting practical flows in the industrial environment. However, users must recognise that this simplicity comes at the cost of a potentially serious loss of physical realism in any but relatively simple two-dimensional shear flows for which the eddy-viscosity concept was originally conceived. Even for such simple conditions, any eddy-viscosity model gives entirely wrong normal stresses and anisotropy. If the shear stress is the only dynamically active component, this defect may not have a major impact on the prediction of the mean flow. A potentially more serious defect is, however, that the shear stress is predicted to be grossly excessive in strong shear, in which case the production substantially exceeds the dissipation rate. To counteract this, the coefficient C_μ in the eddy-viscosity relation has to be sensitised to the strain in such a way that the shear stress does not substantially exceed $0.3k$, and this requires C_μ to be inversely related to the strain.

In more complex conditions — featuring normal straining, curvature, separation, recirculation, swirl, rotation and body forces (buoyancy, in particular) — the linear relationship between stresses and strains becomes increasingly unrealistic and damaging to the predictive fidelity of the model in question: the normal stresses are wrongly linked to the associated normal-strain components, giving excessive levels of turbulence energy and its production, and the sensitivity of the stress components to curvature strain, rotation and body forces is substantially under-estimated. All these defects are linked, in one way or another, to the misrepresentation of the anisotropy and, consequently, to an incorrect sensitivity of the shear-stress component(s) to the normal stresses and the extra strain components associated with acceleration, deceleration and curvature. In reality, any stress is linked to all other stress components, and to all strains, as demonstrated by the exact expressions for the stress-production rates. This complex, multi-faceted linkage is not represented by the linear stress-strain relations. In the case of body forces, the stresses are additionally linked, anisotropically, to scalar fluxes associated with the property causing the body force. Here again, the eddy-viscosity concept prevents the anisotropic processes from being represented, and the consequence is a far too weak sensitivity of the turbulence and the mean flow to the body force.

In efforts to counteract the defects identified above, a range of *ad hoc* corrections have been devised, and these have been outlined in the present chapter. The wrong response to strong normal straining can be counteracted by the use of realisability constraints based on functional relations of the type $C_\mu = f(S, \Omega)$ and/or limiters introduced into the turbulence production. To improve the response to curvature, swirl and body forces, a popular practice is to use variants of the Richardson number — formed most rationally as the ratio of extra production, associated with the process to be captured, to the total value — and to introduce related *ad hoc* source/sink terms into the dissipation-rate equation. These terms are formulated in a manner that causes the eddy viscosity to increase or decrease when the extra strain or the body force in question causes turbulence to be stabilised or destabilised, respectively. Finally, the tendency of linear eddy-viscosity models to predict an excessive level of near-wall length scale, and thus turbulent viscosity, in the presence of deceleration leading up to separation, is counteracted by extra source terms in the dissipation-rate equation that are designed to steer the model towards returning a length scale that is compatible with the variation in turbulence-energy equilibrium.

All the above corrections, elaborate as they are, are rarely satisfactory in more than a narrow range of the conditions for which they were devised. To progress further, in terms of modelling realism and generality, we need to abandon the linear stress-strain relations, and embark on modelling routes that rest on a more solid fundamental foundation. This foundation is the subject of Chapter 12.

Chapter Twelve

12. Reynolds-Stress-Transport Modelling

12.1 *Rationale and motivation*

The defects of the linear eddy-viscosity framework, and the need for related *ad hoc* fixes, discussed in Chapter 11, provide a powerful incentive for attempting to formulate a closure that determines the stress components directly from the exact equations governing these components. Although these equations remain to be introduced in their general form, we have already derived and discussed simplified forms in Chapter 4. For example, Eq. (4.5) governs the behaviour of the shear-stress $\overline{uv}$ for the case of a statistically two-dimensional flow, and in Eq. (4.6) we encountered the equations for the individual normal stresses $\overline{u^2}, \overline{v^2}, \overline{w^2}$, albeit only for the case of a simple two-dimensional shear layer. In fact, much of Chapter 4, and a fair proportion of Chapters 5 and 11 may be regarded as parts of an introduction to the present chapter.

To reinforce the motivation for moving away from the linear eddy-viscosity concept, we contrast below the eddy-viscosity concept, Eq. (7.5), with the proposal that the stresses may be related linearly to their respective *rates of production*, Eq. (11.1), in which the production terms P_{ij} and P_k have been replaced by their explicit stress/strain forms:

$$\overline{u_i u_j} - \frac{2}{3}\delta_{ij} k = -\nu_t \left(\frac{\partial U_i}{\partial x_j} + \frac{\partial U_j}{\partial x_i} \right), \tag{12.1}$$

$$\overline{u_i u_j} - \frac{2}{3}\delta_{ij} k = -C \frac{k}{\varepsilon} \left(\overline{u_i u_k} \frac{\partial U_j}{\partial x_k} + \overline{u_j u_k} \frac{\partial U_i}{\partial x_k} - \frac{2}{3}\delta_{ij} \left[\overline{u_k u_l} \frac{\partial U_k}{\partial x_l} \right] \right). \tag{12.2}$$

It must be emphasised that the latter relation merely represents a notional *ansatz*, reflecting the supposition that the stress components may be linked linearly (i.e. crudely) to their respective production terms via the time-scale multiplier k/ε. While this is untenable as a quantitative model, as it ignores the important redistribution and dissipation processes (see Fig. 4.5)), it does clarify, in principle, that any stress component is linked to all other stresses and strains, and it follows that all components need to be treated on an equal footing in terms of quantitative accuracy.

A consequence of the complex inter-linkage expressed by Eq. (12.2) is that the eigenvectors of the (deviatoric-) stress and strain tensors are not co-aligned, as is wrongly implied by Eq. (12.1). This is illustrated in Fig. 12.1 by a comparison of the two sets of eigenvectors in a recirculating flow across a sub-channel of a circular rod assembly.

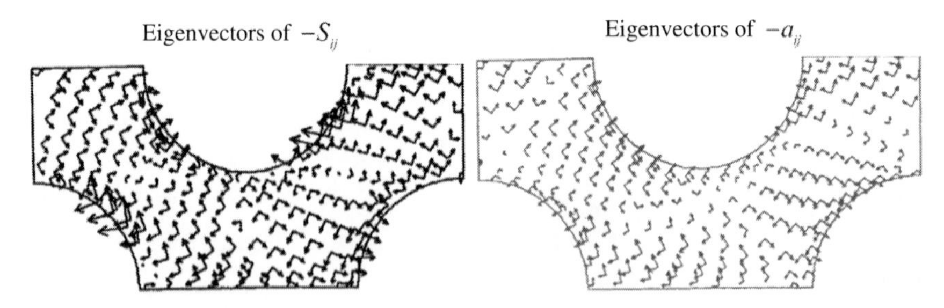

Fig. 12.1: Comparison of eigenvector fields derived from the stress and strain tensors in a flow within a sub-channel of a circular-rod assembly, computed from full Reynolds-stress-transport closure (courtesy D. Laurence). The right-hand-side plot shows $a_{ij} = \overline{u_i u_j}/k - \frac{2}{3}\delta_{ij}$.

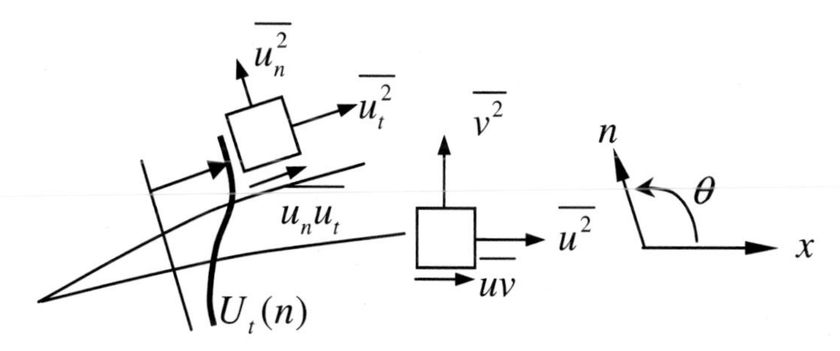

Fig. 12.2: Tilted shear layer with corresponding coordinates and stress components.

Another instructive perspective is offered by Eq. (7.8), written below with the time scale τ replaced by k/ε:

$$-\overline{uv} = C\frac{k}{\varepsilon}\overline{v^2}\frac{\partial U}{\partial y}. \tag{12.3}$$

This equation is, essentially, equivalent to Eq. (12.2) for the case of a simple x-aligned shear layer, $U(y)$. If we now imagine this shear layer to be tilted to conform with a (t,n) co-ordinate system, as shown in Fig. 12.2, and to be part of some complex flow (e.g. a shear layer bordering a recirculation zone), then Eq. (12.3) becomes[1]:

$$-\overline{u_t u_n} = C\frac{k}{\varepsilon}\overline{u_n^2}\frac{\partial U_t}{\partial n}. \tag{12.4}$$

By simple trigonometric manipulations (see Mohr's circle Fig. 12.4, e.g. S.P. Timoshenko (1983)), the stress components $(\overline{u_t^2}, \overline{u_n^2}, \overline{u_t u_n})$ can readily be expressed in terms of $(\overline{u^2}, \overline{v^2}, \overline{uv})$. This allows Eq. (12.4) to be expressed as follows:

$$\begin{aligned}&-\overline{uv}\cos 2\theta + 0.5(\overline{u^2} - \overline{v^2})\\ &\quad = C\frac{k}{\varepsilon}\left(0.5(\overline{u^2} + \overline{v^2}) + 0.5(\overline{u^2} - \overline{v^2})\cos 2\theta + \overline{uv}\sin 2\theta\right)\frac{\partial U_t}{\partial n}.\end{aligned} \tag{12.5}$$

[1]We ignore the influence of shear-layer curvature on the shear stress in this simple treatment.

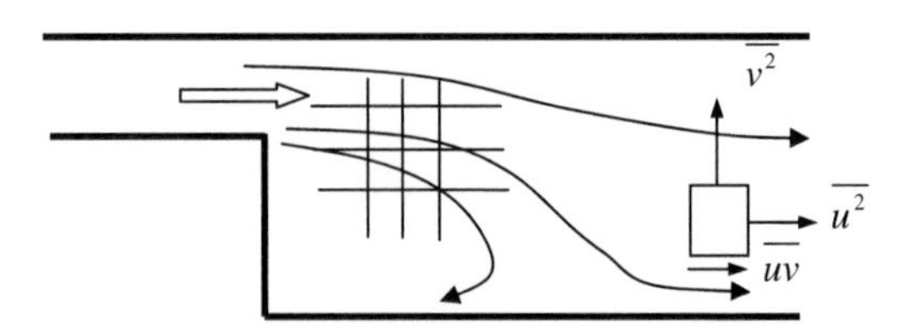

Fig. 12.3: Recirculating flow behind a backward-facing step, computed with a Cartesian mesh, illustrating the importance of the normal stresses.

Equation (12.5) can be simplified, but this form suffices to convey the main message, namely that the evaluation of the shear-stress $\overline{uv}$ relies heavily on the representation of the normal-stress anisotropy.

Apart from the above fact that the shear stress is intimately linked to the normal stresses, these latter components can themselves be dynamically influential and can exercise considerable impact on the mean-flow properties. This is readily recognised by reference to the simple recirculating flow shown in Fig. 12.3, in which it is assumed that the mean-flow equations are solved on a Cartesian grid — the most natural choice in this geometry. In this case, the momentum equation for the streamwise velocity U contains the turbulent-transport terms $-\left(\frac{\partial \overline{uv}}{\partial y} + \frac{\partial \overline{u^2}}{\partial x}\right)$. Because the flow is highly curved, streamwise gradients of all quantities are substantial, and the streamwise gradient of the streamwise stress is thus influential. Analogously, the V-velocity field is influenced by $-\frac{\partial \overline{v^2}}{\partial y}$. It follows that the normal stresses, which are poorly correlated to the respective normal strains, may need to be determined with much higher accuracy than is possible with a linear eddy-viscosity model.

It is recalled that the main closure challenge in the case of two-equation models lies in providing the correct energy-dissipation rate (or equivalent length/time scale). The only other term that requires closure is that representing the turbulent transport (see Eq. 7.27), but this is a relatively minor contributor to the turbulence-energy balance, in most circumstances, as is exemplified by Fig. 5.4, so it does not greatly affect the accuracy of the model.

The closure of the Reynolds-stress-transport equations is, unfortunately, much more demanding. This is not merely due to the fact that the model involves (in 3D flows) six equations for the stress components — rather than one for the turbulence energy — but because of the need to provide each of the six equations with its own dissipation-rate component, ε_{ij}. Even more demanding, as will transpire, is the requirement to provide accurate approximations for the pressure-velocity correlations that inter-link the stress-transport equations. This is most clearly brought out by Eq. (4.6), which pertains to a simple, two-dimensional, plane shear layer. The stress component of greatest interest is, of course, $\overline{uv}$, but its level is dictated by the production term $-\overline{v^2}\frac{\partial U}{\partial y}$. However, $\overline{v^2}$ is not generated by any strain. Rather, its value is largely governed by the balance between the gain through the pressure-strain term and the loss through the dissipation rate for that stress. The gain is affected by the transfer of energy from $\overline{u^2}$ via the pressure-strain 'bridge'. It follows that the accuracy with which this redistributive process is modelled is likely to have profound consequences for the fidelity of the model as a whole.

12.2 *The exact Reynolds-stress equations*

The manner in which the Reynolds-stress equations are derived has already been indicated by way of Eq. (4.1), reproduced below (with mean-velocity overbars omitted):

$$\overline{u_i NS(U_j + u_j) + u_j NS(U_i + u_i)} = 0, \qquad (12.6)$$

in which NS stands for the Navier–Stokes equations $NS(U_i + u_i) = 0$. Moreover, one particular equation, for $\overline{uv}$, was derived in detail, leading to Eq. (4.5). If the Navier–Stokes equations include the body-force term F_i, with turbulent components f_i, and system rotation Ω_k is also included, the sequence of the steps in Eq. (12.6) leads to:

$$\underbrace{\frac{D\overline{u_i u_j}}{Dt}}_{C_{ij}} = \underbrace{-\left(\overline{u_i u_k}\frac{\partial U_j}{\partial x_k} + \overline{u_j u_k}\frac{\partial U_i}{\partial x_k}\right)}_{P_{ij}}$$

$$+ \underbrace{\left(\overline{f_i u_j} + \overline{f_j u_i}\right)}_{G_{ij}} + \underbrace{2\Omega_k\left(\overline{u_j u_m}\varepsilon_{ikm} + \overline{u_i u_m}\varepsilon_{jkm}\right)}_{R_{ij}}$$

$$
\underbrace{-\frac{1}{\rho}\left(\overline{u_i\frac{\partial p}{\partial x_j}}+\overline{u_j\frac{\partial p}{\partial x_i}}\right)}_{\Pi_{ij}}\underbrace{-2\nu\overline{\frac{\partial u_i}{\partial x_k}\frac{\partial u_j}{\partial x_k}}}_{\varepsilon_{ij}}
$$

$$
\underbrace{+\frac{\partial}{\partial x_k}\left(\nu\frac{\partial\overline{u_i u_j}}{\partial x_k}\right)}_{D_{ij}^v}\underbrace{+\frac{\partial}{\partial x_k}\left(-\overline{u_i u_j u_k}\right)}_{D_{ij}^t}. \tag{12.7}
$$

In the above, $P_{ij}+G_{ij}+R_{ij}$ represents the total production (see also Eq. (11.3)), D_{ij}^v and D_{ij}^t represent, respectively, diffusion by viscous action and by turbulent transport, Π_{ij} represents the pressure-velocity interaction and ε_{ij} is the dissipation-rate tensor. As regards Π_{ij}, reference is made to Eq. (4.7), which conveyed the fact that this term can be, and often is, split into two contributions:

$$
\Pi_{ij}=\underbrace{\overline{\frac{p}{\rho}\left(\frac{\partial u_i}{\partial x_j}+\frac{\partial u_j}{\partial x_i}\right)}}_{\Phi_{ij}}\underbrace{-\frac{\partial}{\partial x_k}\left(\overline{\frac{p}{\rho}(u_i\delta_{jk}+u_j\delta_{ik})}\right)}_{D_{ij}^p}, \tag{12.8}
$$

referred to as the pressure-strain and the pressure-diffusion terms, respectively. As already explained in Section 4.2, the rationale of adopting this split lies in the fact that Φ_{ij} is purely redistributive among the normal stresses — the sum Φ_{kk} vanishing and thus not contributing to the turbulence energy — while D_{ij}^p is 'diffusive' in character, i.e. is *mathematically* akin to D_{ij}^t. *Prima facie*, the split appears to offer two advantages to the approximation process: first, the requirement of zero trace of Φ_{ij} is a constraint that is relatively easy to impose on its model, and this will contribute substantially to the physical realism of the model; second, the diffusive nature of D_{ij}^p may be exploited by approximating it by the same type of model as that used to represent the triple-correlation term D_{ij}^t. However, as will be argued below, there are some powerful arguments to the contrary, based on insight derived from direct numerical simulation (DNS).

The central pillar of Reynolds-stress models is that the production terms are retained in their exact form. Formally, this also applies to the correlations $\overline{f_i u_j}$, although these must be fed into Eq. (12.7) from other equations. For example, in the case of buoyancy-induced forces:

$$\overline{f_i u_j} = -g_i \beta \overline{T' u_j}, \tag{12.9}$$

where g_i is the gravitational-acceleration vector and T' is the temperature fluctuation. The correlation $\overline{T' u_j}$ represents the heat-flux vector, and this has to be derived from a separate model for the heat fluxes — for example, one based on modelled versions of the transport equations (6.6).

With stress convection and viscous diffusion requiring no approximation either, the closure problem thus amounts to approximating the processes Π_{ij}, ε_{ij} and D^t_{ij} in terms of the stresses, strains and other determinable quantities (e.g. the turbulence-energy dissipation, ε, for which we have already derived an equation, albeit in the context of two-equation modelling). There are many ways in which these terms can be, and have been, approximated, and this diversity has spawned numerous models and variants. The most important source of differences among models is the manner in which the pressure-velocity correlations have been approximated. This process is especially challenging, not only because it represents highly influential conduits for energy transfer among the normal stresses, and is the principal sink in the shear-stress equations, but also because pressure fluctuations propagate across the entire domain and are reflected at solid boundaries. Hence, the pressure-velocity-interaction process is highly non-local, and its point-wise approximation is thus extremely tenuous.

The approach adopted in this chapter parallels that taken in earlier ones: it does not aim to cover the multitude of models documented in the literature, but rather focuses on conveying the rationale underpinning the principal models, thus providing insight into modelling principles. Readers who wish to go beyond the content of this chapter, or wish to gain a broader view of the main issues pertaining to the subject, are referred to review articles by Launder (1989) and Hanjalić (1994), which summarised the position on Reynolds-stress modelling at the time the respective articles appeared.

12.3 *Closure — some basic rules*

As is generally the case in turbulence modelling, the key to successfully closing the unknown terms in Eq. (12.7) is insight into the physical implications of the terms to be modelled, and an appreciation (not least one based on intuition) of the degree to which the model is likely emulate the basic physical properties of the terms involved. Three issues need to be considered, in particular:

(i) In the case of diffusion, the model must be such that it redistributes the property in question in the spatial domain without giving rise to generation or dissipation of the property. The modelled process must, therefore, integrate to zero across the domain, subject to zero-gradient conditions of the property normal to the domain boundary.

(ii) In the case of dissipation, the model must, in all circumstances, be a sink that diminishes the property at a rate that is compatible with the exact term.

(iii) In the case of normal-stress redistribution, the model must be such as it does not generate or dissipate turbulence energy, i.e. it has to contract to zero, as does the exact term.

A second general set of rules pertain to the tensorial consistency between the exact term and its model. Most important are the following requirements:

(i) The model must have the same rank as the exact term.

(ii) The model must be Galilean-invariant and system-objective. In other words, it must not introduce a spurious sensitivity to system translation, must not depend on the orientation of the system axes, and must be applicable to an accelerating or rotating system.

Other issues relevant to the modelling process, in general, are kinematic constraints — e.g. continuity, no-slip and impermeability restrictions — and tensorial symmetry, which allows certain tensorial terms to be omitted or agglomerated with their symmetric counterparts, because of insensitivity to the order of the tensorial indices.

A third general principle pertains to the use of 'constants'. Any closure proposal will generally be in the form of an expression involving the stress tensor, i.e. the highest-order variables determinable from the model. A 'constant' determined from calibration against experimental or simulation data need not be a purely numerical value. However, if it is not a numerical value, it should be a scalar quantity and, at most, be a function of invariants of the strain and/or vorticity and/or stress tensor. Invariants include, among others, $S_{ij}S_{ji}, S_{ij}S_{jk}S_{ki}\ldots, \Omega_{ij}\Omega_{ji}\ldots, S_{ij}\Omega_{ji}\ldots, k$, and scalar representatives of the stress tensor: the second invariant $A_2 \equiv a_{ij}a_{ji}$, the third invariant $A_3 \equiv a_{ij}a_{jk}a_{ki}$ and combinations of the two, where $a_{ij} \equiv \frac{\overline{u_i u_j}}{k} - \frac{2}{3}\delta_{ij}$ is the anisotropy tensor.

12.4 *Realisability and its implications for modelling*

The subject of realisability was first discussed in Section 11.2, in relation to linear eddy-viscosity modelling. Surprisingly, perhaps, the subject is also relevant to Reynolds-stress modelling, notwithstanding the fact that this class of models is one that we inherently expect to avoid non-realisable solutions for the turbulent stresses.

It is recalled, first, that a realisable model has to ensure:

$$\left\{\overline{u^2}, \overline{v^2}, \overline{w^2}\right\} \geq 0 \quad \text{and} \quad \left\{\frac{\overline{uv}}{\sqrt{\overline{u^2}}\sqrt{\overline{v^2}}}, \frac{\overline{uw}}{\sqrt{\overline{u^2}}\sqrt{\overline{w^2}}}, \frac{\overline{vw}}{\sqrt{\overline{v^2}}\sqrt{\overline{w^2}}}\right\} < 1. \tag{12.10}$$

These conditions are satisfied if the Reynolds stresses in the principal axes, $\overline{u_\alpha^2}$, are unconditionally positive. The concept is best illustrated for two-dimensional conditions by reference to the Mohr's stress circle for solid mechanics, already mentioned in relation to Fig. 12.2 and Eq. (12.5). This is shown in Fig. 12.4 for two-dimensional conditions. Thus, given a stress state $(\tau_{xx}, \tau_{yy}, \tau_{xy})$, we seek the state — by rotating the axes — in which the shear stress vanishes. This state is $(\widehat{\tau}_{xx}, \widehat{\tau}_{yy})$. Any realisable state must preclude $\widehat{\tau}_{yy} < 0$ for whatever combination $(\tau_{xx}, \tau_{yy}, \tau_{xy})$. For any state in which this condition applies (i.e. with the circle being wholly to the right of the origin (0,0) in Fig. 12.4), $\tau_{xy}^2/(\tau_{xx}\tau_{yy}) < 1$. Applied to the turbulent stresses,

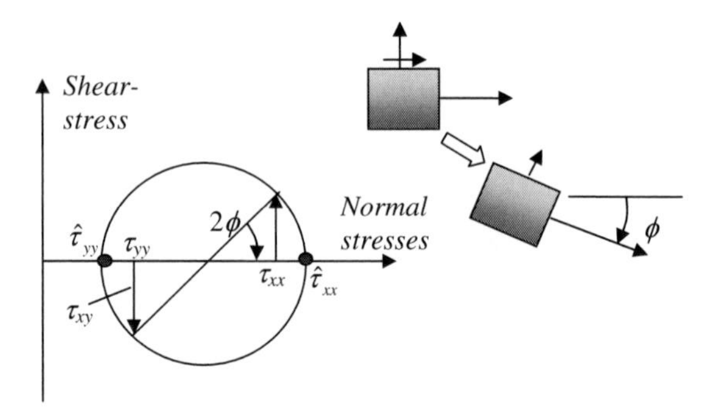

Fig. 12.4: Mohr's circle, illustrating the relationship between the stress systems in a two-dimensional solid element at two orientations, with one orientation featuring the principal stresses, in which case the shear stress is zero.

this is entirely equivalent to the relationship between $(\overline{u^2}, \overline{v^2}, \overline{uv})$ and $\overline{u_\alpha^2}(\alpha = 1, 2)$.

To secure realisability in a turbulent flow, we have to ensure that the evolution equations governing the principal stresses (i.e. the three components of the diagonalised stress tensor) satisfy the condition:

$$\frac{D\overline{u_\alpha^2}}{Dt} \to 0 \quad \text{as } \overline{u_\alpha^2} \to 0, \quad \alpha = 1, 2, 3. \tag{12.11}$$

Put in words: the (modelled) Reynolds-stress-transport equations — i.e. the right-hand side of $\frac{D\overline{u_\alpha^2}}{Dt} = \{\ldots\}$ — must not drive any of the principal stresses below zero while that stress component approaches zero. If the right-hand side of Eq. (12.7), or rather its modelled version, is denoted by $\mathbf{\Sigma}_{ij}$ then $\{\ldots\}$ comprises the diagonal elements of the transformed matrix $\mathbf{\Sigma}_{\alpha\alpha}$, derived by projecting all elements of $\mathbf{\Sigma}_{ij}$ onto the principal axes of the stress tensor. With $\mathbf{\Sigma}_{\alpha\alpha}$ obtained, realisability can be tested by reference to condition in Eq. (12.11). We shall not pursue this analysis here, in specific terms; we cannot do so at this stage, in any event, because we first need to introduce specific closure approximations for $\mathbf{\Sigma}_{ij}$. Durbin and Pettersson-Reif (2001) give a formal and fairly detailed account on the subject of realisability, examining the constraint in Eq. (12.11) for some specific closure proposals considered in sections to follow hereafter. It has to be said, however, that a formal realisability analysis of this type is difficult to undertake for any but the simplest closure approximations; there are numerous model variations that make such an analysis virtually impossible.

In the context of this book, the essential message to take forward is that realisability is an issue of which modellers and model users need to be aware, not only in relation to eddy-viscosity models, but also Reynolds-stress-transport models. There is no guarantee, *a priori*, that any particular model will satisfy condition Eq. (12.11), unless specifically designed to do so through the introduction of constraints, usually via model coefficients. This applies, in particular, to the model of the pressure-velocity interaction, because this process is responsible for the inter-normal-stress transfer of turbulence energy, and it is thus critically important to the level of the anisotropy.

A simple example of how realisability might be violated is the flow in a duct subjected to rotation around an axis orthogonal to the duct (e.g. a duct in a centrifugal fan), such as shown in Fig. 12.5. For this case, already considered in Section 11.7 (see also Eq. (11.30)), the production terms due to rotation are:

$$\begin{aligned} R_{11} &= 4\Omega\overline{uv} \\ R_{22} &= -4\Omega\overline{uv} \\ R_{12} &= -2\Omega(\overline{u^2} - \overline{v^2}). \end{aligned} \tag{12.12}$$

The rotational production of $\overline{v^2}$, R_{22}, is proportional to $\overline{uv}$. On the upper (suction) side of the duct, $\overline{uv}$ is positive, and R_{22} therefore tends to diminish $\overline{v^2}$. Now, if the modelled shear-stress equation returns a value for $\overline{uv}$ which does not decline correctly in harmony with $\overline{v^2}$ then R_{22} can drive $\overline{v^2}$ below zero.

In general, realisability is especially pertinent to turbulence close to walls and sharp fluid–fluid interfaces. While in both cases, all

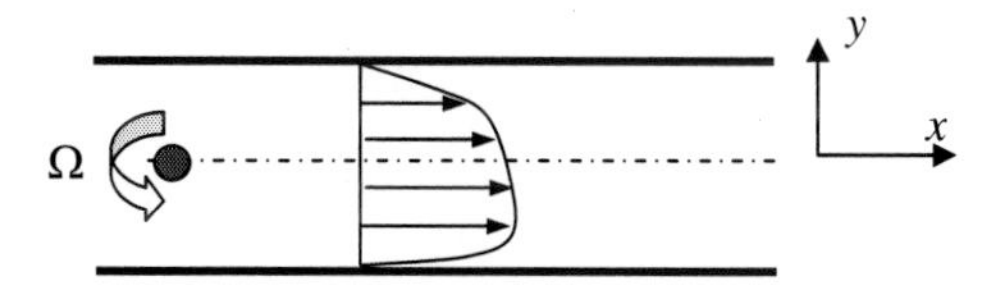

Fig. 12.5: Flow in a duct rotating normally to its axis.

Reynolds-stress components must decay to zero, the *rate* at which they do so varies, as discussed in Section 5.1. If a model fails to return the correct rates of decay for the respective stress components, this may provoke a non-realisable state.

Lumley (1978) has shown that any realisable turbulent state is characterised by pairs of values of the second and third invariants, $A_2 \equiv a_{ij}a_{ji}$ and $A_3 \equiv a_{ij}a_{jk}a_{ki}$, respectively, which fall within a triangular domain of the 'invariant map' shown in Fig. 12.6. In this figure, the origin $A_2 = A_3 = 0$ identifies isotropic turbulence, and line '1' identifies states of *two-component turbulence* — the state encountered as a wall is approached. Lumley has further shown that the combined invariant,

$$A \equiv 1 - \frac{9}{8}(A_2 - A_3), \tag{12.13}$$

often referred to as the 'flatness parameter', must always lie between 1 and 0, the former identifying isotropic turbulence and the latter two-component turbulence.

Figure 12.7 shows variations of A in a channel flow, for a range of Reynolds-number values, and in a separated flow at four down stream positions, within the separated flow shown in the stream-function inset of Fig. 12.7(b). In the latter, one position is within the boundary layer upstream of separation, two are within the separation

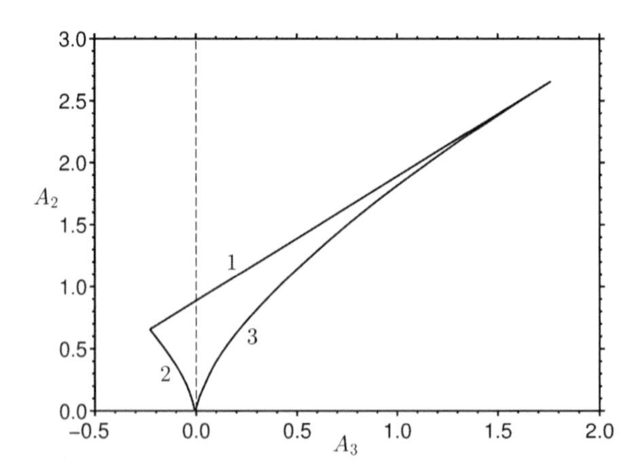

Fig. 12.6: Lumley's realisability map.

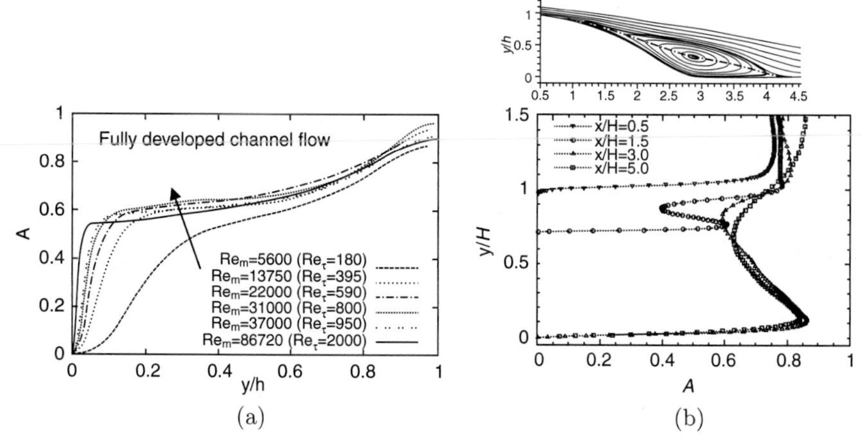

(a) (b)

Fig. 12.7: Profiles of the invariant A (Eq. (12.13)): (a) in channel flow at different Reynolds numbers (courtesy S. Jakirlić); and (b) in a statistically two-dimensional recirculating flow at $Re_H = 13700$, based on the outer velocity (U_∞) and ramp height (H); derived from DNS data (reproduced with permission of Taylor and Francis Ltd., from Bentaleb *et al.* (2012)).

zone and the last downstream of reattachment. These profiles thus show that the anisotropy increases rapidly as the wall is approached (i.e. A is declining towards the two-component limit), and is also high in the separated shear layer.

From a modelling point of view, the above invariants, especially A, can be very useful in securing the correct behaviour of model fragments — say, by introducing coefficients which are made functions of A, A_2, A_3. The exploitation of these invariants in model construction will be demonstrated in sections to follow.

12.5 *Turbulent transport*

We start the modelling process with the term D^t_{ij} in Eq. (12.7), because this is, arguably, the least difficult process to approximate,[2] its closure resting on a reasonably firm rational foundation.

DNS and wall-resolved large eddy simulation (LES) solutions for channel flow and boundary layers suggest that D^t_{ij} makes a relatively modest contribution to the balance of processes dictating the

[2]The adjective 'easiest' is avoided intentionally.

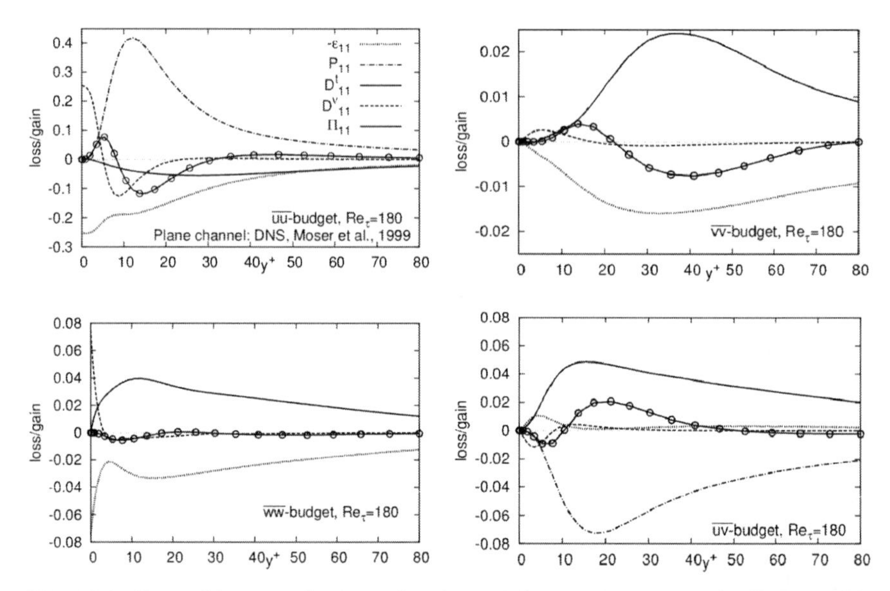

Fig. 12.8: Reynolds-stress budgets for channel flow at $Re_\tau = 180$ (bulk Reynolds number 5600). All contributions have been normalised by (u_τ^4/ν). The turbulent-diffusion profiles are accentuated by the addition of open circles; derived from DNS data of Moser *et al.* (1999) (courtesy S. Jakirlić).

stresses, and this is argued to allow more leeway, or greater level of liberty, in its approximation. This is exemplified by Fig. 12.8, which shows the contribution of turbulent transport to the stress budgets in a channel flow, determined from DNS. However, the turbulent-transport process can be a significant contributor to the stress budgets in more complex conditions, as is conveyed in Fig. 12.9 for the same separated flow considered in Fig. 12.7(b). The magnitude of the transport of both the streamwise-stress and shear-stress components peaks at around $y/H = 0.85$ (negative in the case of $\overline{u^2}$, positive in the case of $\overline{uv}$), due to shear production reaching a maximum.[3] Turbulent transport thus acts to drain the stresses from this high-production region, and to effect a transfer both towards the wall and into the outer parts of the separated shear layer. While D_{ij}^t is not the

[3]Note that diffusion opposes production in both stress components, the sign reversal from one stress to the other reflecting the opposite signs of the stress components themselves.

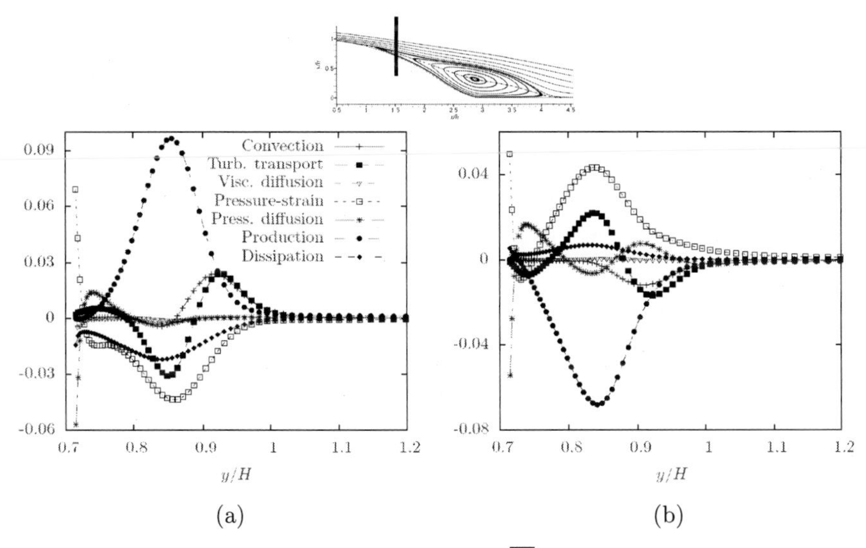

Fig. 12.9: Budgets for (a) streamwise stress $\overline{u^2}$ and (b) shear-stress $\overline{uv}$ in a separated shear layer within a recirculation zone of the flow in Fig. 12.7. All contributions have been normalised by (U_o^3/H). The turbulent-diffusion profiles are identified by full squares (reproduced with permission of Taylor and Francis Ltd., from Bentaleb *et al.* (2012)).

dominant contributor to the balance, it is clearly influential; indeed, in the case of $\overline{u^2}$, it is of the same order as that of the dissipation rate.

Major conceptual assistance to modelling D_{ij}^t is offered by the fact that the exact equations for the triple moments (not derived herein — see Amano and Goel (1986)) can be shown to feature the following production terms:

$$\underbrace{\frac{D\overline{u_i u_j u_k}}{Dt}}_{C_{ijk}} = \underbrace{-\left(\overline{u_i u_l}\frac{\partial \overline{u_j u_k}}{\partial x_l} + \overline{u_j u_l}\frac{\partial \overline{u_k u_i}}{\partial x_l} + \overline{u_k u_l}\frac{\partial \overline{u_i u_j}}{\partial x_l}\right)}_{P_{ijk}} + \cdots \tag{12.14}$$

Leaning conceptually on Eq. (12.2), we propose that the triple correlations may, here too, be approximated as their production rates, multiplied by the turbulent time scale, i.e.:

$$\overline{u_i u_j u_k} = -C_t \frac{k}{\varepsilon}\left(\overline{u_i u_l}\frac{\partial \overline{u_j u_k}}{\partial x_l} + \overline{u_j u_l}\frac{\partial \overline{u_k u_i}}{\partial x_l} + \overline{u_k u_l}\frac{\partial \overline{u_i u_j}}{\partial x_l}\right). \tag{12.15}$$

Thus[4]:

$$\frac{\partial}{\partial x_k}(-\overline{u_i u_j u_k}) = \frac{\partial}{\partial x_k}\left[C_t \frac{k}{\varepsilon}\left(\overline{u_i u_l}\frac{\partial \overline{u_j u_k}}{\partial x_l} + \overline{u_j u_l}\frac{\partial \overline{u_k u_i}}{\partial x_l} + \overline{u_k u_l}\frac{\partial \overline{u_i u_j}}{\partial x_l}\right)\right]. \quad (12.16)$$

This particular closure proposal is part of the model of Hanjalić and Launder (1972).

The use of k/ε as the time scale needs to be questioned. We have done so before, in Chapter 9 (see Eqs. (9.22) and (9.30)), where it was agued that the appropriate time scale close to the wall must involve the Kolmogorov scales. This also applies here, and several proposals have been made in the literature in this respect. For example, Durbin (1991) has proposed the replacement (see Eq. (9.30)):

$$\frac{k}{\varepsilon} \leftarrow \max\left[\frac{k}{\varepsilon}, 6\left(\frac{\nu}{\varepsilon}\right)^{\frac{1}{2}}\right], \quad (12.17)$$

the switch occurring around $y^+ \approx 10$ (based on channel-flow DNS studies).

In complex three-dimensional flows, Eq. (12.16) involves many terms and is expensive to compute. This has motivated the replacement of Eq. (12.16) by two truncated forms, both retaining system-invariance.[5] First, Mellor and Herring (1973) proposed the replacement of the pre-multiplying stresses in Eq. (12.16) by the turbulence energy k, giving:

$$\frac{\partial}{\partial x_k}(-\overline{u_i u_j u_k}) = \frac{\partial}{\partial x_k}\left[C_t \frac{k}{\varepsilon}\left(k\frac{\partial \overline{u_j u_k}}{\partial x_i} + k\frac{\partial \overline{u_k u_i}}{\partial x_j} + k\frac{\partial \overline{u_i u_j}}{\partial x_k}\right)\right]. \quad (12.18)$$

Second, an earlier proposal by Daly and Harlow (1970) retains only the last group in Eq. (12.16), namely:

$$\frac{\partial}{\partial x_k}(-\overline{u_i u_j u_k}) = \frac{\partial}{\partial x_k}\left[C_t \frac{k}{\varepsilon}\left(\overline{u_k u_l}\frac{\partial \overline{u_i u_j}}{\partial x_l}\right)\right]. \quad (12.19)$$

[4]Note the tensorial consistency: both sides are of rank 2 in (i, j).

[5]More complex forms also exist, proposed by Lumley (1978) and Magnaudet (1992), but have very rarely been used.

The most drastic simplification arises when Mellor and Herring's proposal is combined with Eq. (12.19) to give:

$$\frac{\partial}{\partial x_k}(-\overline{u_i u_j u_k}) = \frac{\partial}{\partial x_k}\left[C_t \frac{k^2}{\varepsilon}\frac{\partial \overline{u_i u_j}}{\partial x_k}\right], \tag{12.20}$$

in which case, the multiplier of the stress-gradient term is, effectively, the $k - \varepsilon$ eddy viscosity, corrected by a coefficient equivalent to a Prandtl–Schmidt number (i.e. $C_t = C_\mu/\sigma_t$).

Unsurprisingly, both Eqs. (12.19) and (12.20) are popular constituents of second-moment closures implemented in industrial and commercial codes. Both are clearly consonant with the concept of simple gradient diffusion, in which the transported property in question — here, $\overline{u_i u_j}$ — is diffused across its gradient, with the rate of diffusion dictated by a pre-multiplying diffusivity. The difference is that the diffusivity in model Eq. (12.19) is tensorial in nature, allowing each flux component ($k = 1, 2, 3$) of the stress $\overline{u_i u_j}$ to be driven by (sensitised to) the gradients of the stress in all three directions ($l = 1, 2, 3$). This more complex form is therefore referred to as the 'generalised gradient diffusion hypothesis' (GGDH).

There are very few studies in the literature that provide reliable comparisons of the predictive performance of different diffusion models in practically interesting conditions. Demuren and Sarkar (1993) present results for a plane channel flow, examined experimentally by Laufer (1951), for which they study the proposals in Eqs. (12.16), (12.18) and (12.19), with coefficients C_t = 0.22, 0.11 and 0.073, respectively. A general problem with any such study, aiming to identify the properties of any one particular closure component, is that the overall predictive performance depends on all components of the model, including those for Π_{ij} and ε_{ij}. A conclusion presented by Demuren and Sarkar is that the solutions depend little on the diffusion model in the log-law region, but are (predictably) more sensitive to the model in the outer, minimally sheared region, in which turbulence tends towards a state of isotropy. In this region, the model of Mellor and Herring, Eq. (12.18), is found to perform best.

12.6 *Dissipation*

The subject of turbulence-energy dissipation and related length-scale surrogates has occupied a substantial proportion of Chapter 7 (Section 7.5, in particular), reflecting the major impact of the length scale on the predictive properties of two-equation models. In the present framework of second-moment closure, the modelling task is considerably complicated by the fact that the dissipation rate is a tensor, ε_{ij}.

While it is possible, in principle, to start the process of modelling ε_{ij} by deriving exact equations for this tensor, along lines similar to those resulting in Eq. (7.55), this is not a practically tenable route, simply because it is too complex, and the resulting equations insufficiently transparent to form a suitable basis for a term-by-term approximation. Instead, all models currently used, in both academic and industrial codes, employ algebraic approximations that link ε_{ij} to ε. The argument underpinning this option is that the dissipative processes are associated with the smallest scales of turbulence, and these tend to be statistically isotropic, except close to walls or near discontinuities within the fluid.

A model that reflect the concept of isotropy in the smallest scales is:

$$\varepsilon_{ij} = \frac{2}{3}\varepsilon\delta_{ij}, \tag{12.21}$$

which enforces an equal level of dissipation in respect of all three normal stresses, and zero dissipation for the shear stresses. Whether this is a realistic approximation can only be answered by examining simulation data, which are unfortunately rare for any but simple shear flows at low Reynolds numbers, because it is especially difficult to obtain such data with sufficient accuracy.

We shall consider below the degree to which the dissipation rate is anisotropic and how to model this anisotropy. Before doing so, however, we note that some model originators (e.g. Lumley (1978) and Shima (1988)) have argued that this anisotropy need not be considered explicitly, but can be viewed as part of modelling the pressure-strain term. This argument is based on the observation that the dissipation anisotropy, $\varepsilon_{ij} - \frac{2}{3}\delta_{ij}\varepsilon$, like the pressure-strain term, makes no contribution to the turbulence energy and acts as a quasi-redistributive term in the normal stresses. A counter argument is, however, that we wish to distinguish carefully, on physical grounds, between different processes, and we are able to do so, at least to a limited extent, by having access to accurate simulation data that shed light on both ε_{ij} and Φ_{ij} separately.

Figure 12.10 presents a collection of budgets for the four active Reynolds-stress components in the flow to which Figs. 12.7(b) and 12.9 relate, at the location $x/H = 2$. Attention is drawn, however, to the fact that the Reynolds number, $Re_H = 13700$, is low relative to typical engineering flows. Closer examination shows

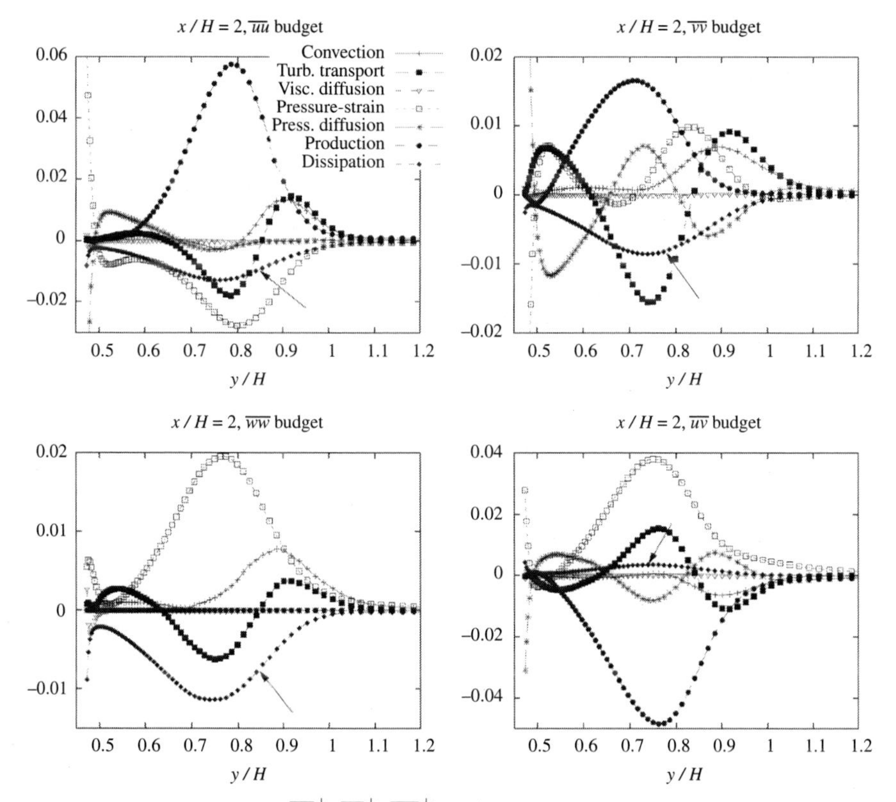

Fig. 12.10: Budgets for $\overline{u^2}^+, \overline{v^2}^+, \overline{w^2}^+, \overline{uv}^+$ for the separated flow of Fig. 12.7 at $x/H = 2$. The dissipation profiles are identified by arrows (reproduced with permission of Taylor and Francis Ltd. from Bentaleb *et al.* (2012)).

that the normal-stress-dissipation components follow very similar variations and are quite close in terms of their peak values, the respective maxima being $\varepsilon_{11}^{+} \approx 0.012$, $\varepsilon_{22}^{+} \approx 0.009$ and $\varepsilon_{33}^{+} \approx 0.011$ ($\varepsilon^{+} = \varepsilon/(u_{\tau}^{4}/\nu)$). Moreover, the maximum dissipation rate of the shear stress is $\varepsilon_{12}^{+} \approx 0.003$, which definitely makes a small contribution to the budget, the latter dominated by the production and the pressure-strain terms. These features are thus consistent with the concept of isotropy in the smallest scales.

It is important to point out that the above comments apply, primarily, to the region fairly remote from the wall, i.e. the separated shear layer and the upper part of the recirculation zone. As the wall is approached, the spectrum of the eddy-size narrows progressively, and the smallest scales are increasingly subjected to the anisotropy-promoting blocking effect of the wall. This is brought out in Fig. 12.11 for a channel flow. As the wall is approached, $\varepsilon_{22} \rightarrow 0$,

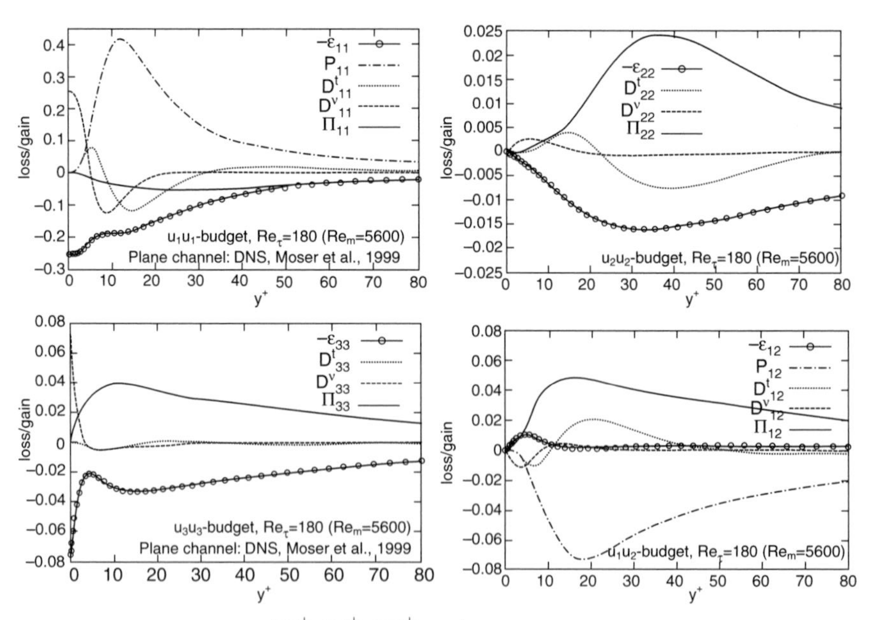

Fig. 12.11: Budgets for $\overline{u^2}^+, \overline{v^2}^+, \overline{w^2}^+, \overline{uv}^+$ for channel flow at $Re_\tau = 180$. All contributions have been normalised by (u_τ^4/ν). The dissipation profiles are accentuated by the addition of open circles; derived from DNS by Moser, Kim and Mansour (1999) (courtesy S. Jakirlić).

ε_{11} rises towards a peak wall value, while ε_{33} also rises towards a wall maximum, albeit lower than ε_{11}. All these features are characteristic of the fact that the dissipative eddies are strongly affected by the wall. In fact, it is observed that the normalised dissipation-rate components $\frac{\varepsilon_{ij}}{\varepsilon}$ behave, qualitatively, similar to the normalised stresses $\frac{\overline{u_i u_j}}{k}$ — see Fig. 5.3(b). To avoid any doubt, we emphasise that the dissipation-rate components cannot vary in harmony with the stresses themselves, because the dissipation reaches a maximum at the wall, while the stresses decay to zero.

The above observation suggests that an approximation superior to Eq. (12.21) might be:

$$\varepsilon_{ij} = \varepsilon \frac{\overline{u_i u_j}}{k}, \tag{12.22}$$

which is generally attributed to Rotta (1951). This expression is also broadly compatible with the comments made earlier in relation to the moderate dissipation anisotropy in the separated shear layer (shown in Fig. 12.10), at least at the relatively low Reynolds number of the flow considered. In this layer, Bentaleb *et al.* (2012) show that the normal-stress maxima obey $\overline{u^2}/\overline{w^2}/\overline{v^2} \approx 1.5/1.2/1$, which is not far from the corresponding ratios of peak dissipation components in the shear layer. However, Eq. (12.22) provides a poor approximation for ε_{12}, the real value of which is substantially lower.

One other problem with the model in Eq. (12.22) is that it over-estimates the dissipation-rate anisotropy in low-shear regions away from the wall, especially at high Reynolds-number values. Fig. 12.12 shows distributions of the dissipation-rate anisotropy parameter, E, which is the equivalent of the stress-anisotropy parameter, A, in Eq. (12.13), and is obtained simply by replacing $a_{ij} = \frac{\overline{u_i u_j}}{k} - \frac{2}{3}\delta_{ij}$ by $e_{ij} = \frac{\varepsilon_{ij}}{\varepsilon} - \frac{2}{3}\delta_{ij}$, and defining the anisotropy invariants as $E_2 \equiv e_{ij}e_{ji}$, $E_3 \equiv e_{ij}e_{jk}e_{ki}$.[6] A comparison of this figure with Fig. 12.7(a) demonstrates that, away from the wall, the assumption in Eq. (12.21) is more appropriate than that in Eq. (12.22), at least in channel flow.

[6] An important distinction is, however, that the realisability considerations in Section 12.4 *do not* apply to the dissipation invariants. The latter arise purely by analogy to the stress invariants.

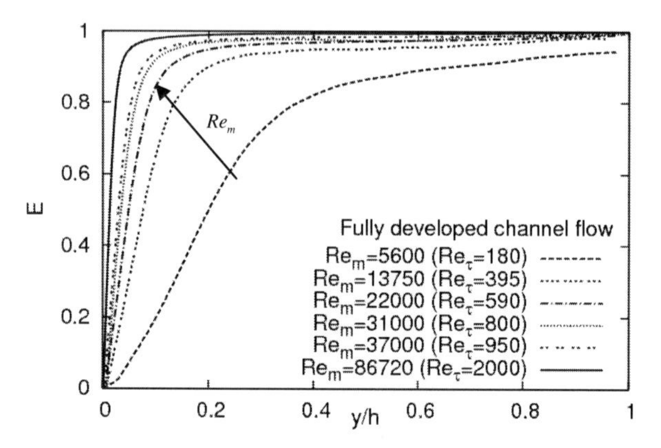

Fig. 12.12: Dissipation-rate anisotropy parameter E for channel flow at different Reynolds numbers; derived from DNS data (courtesy S. Jakirlić).

This suggests, therefore, that a model of the form,

$$\varepsilon_{ij} = f_{\varepsilon w} \varepsilon \frac{\overline{u_i u_j}}{k} + (1 - f_{\varepsilon w}) \frac{2}{3} \varepsilon \delta_{ij}, \tag{12.23}$$

is likely to offer a greater degree of generality than either Eqs. (12.21) or (12.22), if the weighting factor $f_{\varepsilon w}$ is judiciously chosen to mimic the transition from the anisotropic near-wall state to the isotropic state, although it is very unlikely that a single function will be adequate for all dissipation components. In principle, we expect $f_{\varepsilon w}$ to be a function of the turbulent Reynolds number, or a related wall-distance parameter, and/or anisotropy invariants. The question of what form $f_{\varepsilon w}$ should take needs to be addressed, in part, by reference to the wall-asymptotic variations of the components ε_{ij}, and this is an issue we consider next.

We start with the expansions in Eq. (5.1) and insert these into the exact dissipation term:

$$\varepsilon_{ij} = 2\nu \overline{\frac{\partial u_i}{\partial x_k} \frac{\partial u_j}{\partial x_k}}, \tag{12.24}$$

subject to the usual kinematic constraints (no-slip, impermeability and mass conservation). This is a somewhat laborious, but otherwise straightforward process, the result of which, truncated beyond y^2, is as follows:

$$
\begin{aligned}
\varepsilon_{11} &= 2\nu\overline{b_1 b_1} + 8\nu\overline{b_1 c_1}y \\
&\quad + 2\nu\left[4\overline{c_1^2} + 6\overline{b_1 d_1} + \overline{\left(\frac{\partial b_1}{\partial x}\right)^2} + \overline{\left(\frac{\partial b_1}{\partial z}\right)^2}\right]y^2 + HoT \\
\varepsilon_{22} &= 8\overline{c_1^2}y^2 + HOT \\
\varepsilon_{33} &= 2\nu\overline{b_3 b_3} + 8\nu\overline{b_3 c_3}y \\
&\quad + 2\nu\left[4\overline{c_3^2} + 6\overline{b_3 d_3} + \overline{\left(\frac{\partial b_3}{\partial x}\right)^2} + \overline{\left(\frac{\partial b_3}{\partial z}\right)^2}\right]y^2 + HoT \\
\varepsilon_{12} &= 4\nu\overline{b_1 c_2}y + 2\nu(\overline{b_1 d_2} + 4\overline{c_1 c_2})y^2 + HoT \\
\varepsilon &= 0.5(\varepsilon_{11} + \varepsilon_{22} + \varepsilon_{33}).
\end{aligned} \tag{12.25}
$$

As for the Reynolds-stresses, a somewhat more explicit form of that given in Eq. (5.4) is the following:

$$
\begin{aligned}
\overline{u^2} &= \overline{b_1 b_1}y^2 + 2\overline{b_1 c_1}y^3 + (2\overline{b_1 d_1} + \overline{c_1 c_1})y^4 + HoT \\
\overline{v^2} &= \overline{c_2 c_2}y^4 + HoT \\
\overline{w^2} &= \overline{b_3 b_3}y^2 + 2\overline{b_3 c_3}y^3 + (2\overline{b_3 d_3} + \overline{c_3 c_3})y^4 + HoT \\
\overline{uv} &= \overline{b_1 c_2}y^3 + (2\overline{b_1 d_2} + \overline{c_1 c_2})y^4 + HoT \\
k &= 0.5(\overline{u^2} + \overline{v^2} + \overline{w^2}).
\end{aligned} \tag{12.26}
$$

We are now able to examine the wall-asymptotic values of proposal (12.22), simply by inserting the appropriate components therein, subject to the limit $y \to 0$. The result is given in Table 12.1. These values agree closely with DNS data of Moser, Kim and Mansour (1999) for channel flow at $Re_\tau = 180$, as shown in Fig. 12.13, and immediately confirm that a single function $f_{\varepsilon w}$ in eq (12.23) will not give the requisite behaviour close to the wall. Also, while the validity of

Table 12.1: Wall-asymptotic values for the normalised dissipation-tensor components.

$i, j =$	1,1	2,2	3,3	1,2
$\dfrac{\varepsilon_{ij}}{\varepsilon} / \dfrac{\overline{u_i u_j}}{k} =$	1	4	1	2

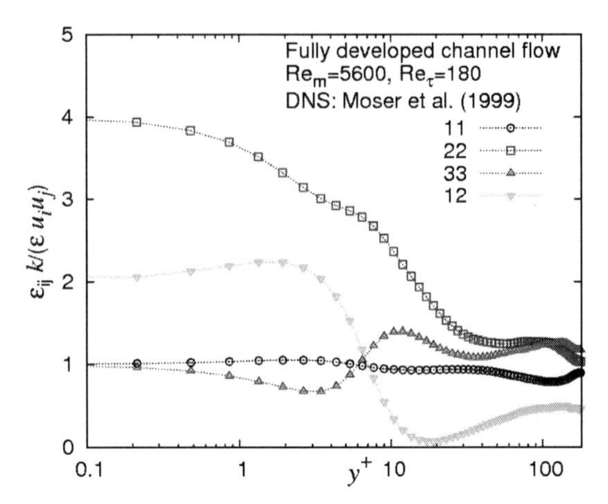

Fig. 12.13: Normalised dissipation anisotropy in channel flow at $Re_\tau = 180$; derived from DNS results by Moser, Kim and Mansour (1999) (courtesy S. Jakirlić).

Eq. (12.23) improves beyond the viscous sublayer, it is only moderately satisfactory — in fact, poor for the shear stress — and we certainly expect the approximation in Eq. (12.21) to become more appropriate away from the wall as the Reynolds number increases.

Although the model in Eq. (12.23) is incompatible with the true departure of $\frac{\varepsilon_{ij}}{\varepsilon}/\frac{\overline{u_i u_j}}{k}$ from '1', in the case of $\overline{v^2}$ and $\overline{uv}$, the significance of this difference should not be overrated. Eqs. (12.25) and (12.26) show that the origin of the defect in $\overline{v^2}$ is that ε_{22} declines in proportion to y^2, while $\overline{v^2}$ declines in proportion to y^4. However, both quantities decline rapidly as the wall is approached, and the defect for this stress diminishes quickly beyond the viscous sublayer (but not for the shear stress). This is a pertinent point to make in view of the fact that most models for ε_{ij} do not comply with the asymptotic values or, indeed, the variations shown in Fig. 12.13.

Models that use the form (12.23) and a single function $f_{\varepsilon w}$, which asymptotes to '1' at the wall, necessarily lead to the wall condition $\frac{\varepsilon_{ij}}{\varepsilon}/\frac{\overline{u_i u_j}}{k} = 1$. This is the case in the model of Hanjalić and Launder (1976), in which $f_{\varepsilon w}$ is given by:

$$f_{\varepsilon w} = \frac{1}{1 + 0.1 Re_t}. \tag{12.27}$$

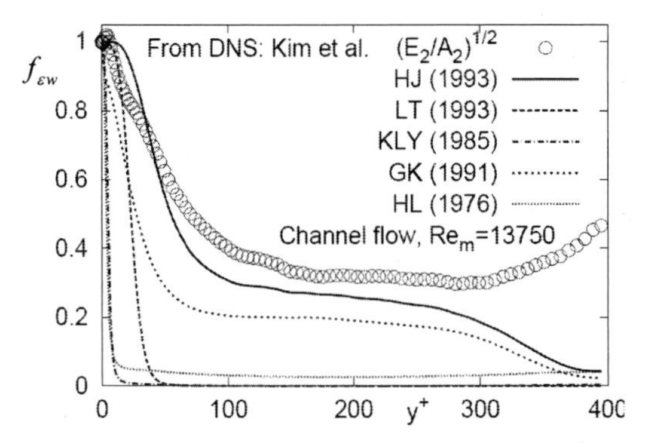

Fig. 12.14: Variations of the function $f_{\varepsilon w}$ for different models, in comparison to $(e_{ij}e_{ji}/a_{ij}a_{ji})^{1/2}$. The respective models are given in Table 12.2 (with permission from Herbert Utz Verlag GmbH from Jakirlić (1997)).

Table 12.2: Models for $f_{\varepsilon w}$ corresponding to variations in Fig. 12.14.

Model	$f_{\varepsilon w}$
HL — Hanjalić & Launder (1976)	$\dfrac{1}{1+0.1Re_t}$
KLY — Kebede *et al.* (1985)	$\exp\left(-\dfrac{R_t}{40}\right)$
LT — Launder & Tselepidakis (1993)	$\exp\left(-20A^2\right)$
GL — Gilbert & Kleiser (1991)	$1 - A^{1/2}$
HJ — Hanjalić & Jakirlić (1993)	$1 - A^{1/2}E$

Is this functional dependence realistic? Part of the answer is provided by Fig. 12.14. This shows that the Hanjalić and Launder line (see Table 12.2) gives a poor representation of the DNS-derived data for the square root of the ratio of the dissipation-anisotropy and Reynolds-stress invariants, this quantity being the nearest (scalar) representative of $f_{\varepsilon w}$. Figure 12.14 also implies that the sole use of Re_t is not promising, because $f_{\varepsilon w}$ varies substantially, well beyond the viscous sublayer, where the dependence on Re_t should be insignificant.

Any attempt to represent the variations in Fig. 12.13 with a single function $f_{\varepsilon w}$ requires the near-wall fragment in Eq. (12.23) to be made a more elaborate function of the stresses, namely:

$$\varepsilon_{ij} = f_{\varepsilon w}\varepsilon^*_{ij} + (1 - f_{\varepsilon w})\frac{2}{3}\varepsilon\delta_{ij}, \tag{12.28}$$

with $\varepsilon^*_{ij} = \varepsilon f_{ij}\left(\frac{\overline{u_i u_j}}{k}\right)$, wherein f_{ij} indicates that this has to be a second-rank tensor. We consider two specific proposals below.

The first proposal is by Kebede, Launder and Younis (1985):

$$\varepsilon^*_{ij} = \frac{\varepsilon}{k}\left[\overline{u_i u_j} + \overline{u_i u_k} n_j n_k + \overline{u_j u_k} n_i n_k + \delta_{ij}\overline{u_k u_l} n_k n_l\right], \tag{12.29}$$

together with

$$f_{\varepsilon w} = \exp\left(-\frac{R_t}{40}\right). \tag{12.30}$$

The symbol n_i in Eq. (12.29) is a unity-valued indicator of the orientation of the wall at which the approximation in this equation is applied, as illustrated by the two examples in Fig. 12.15.

To gain insight into the rationale of the model in Eq. (12.29), we consider the expression for the wall-limiting tensor $\varepsilon_{ij,w}$ at a horizontal wall, $n_1, n_2, n_3 = 0, 1, 0$. For the particular components ε^*_{22} and ε^*_{12} we have:

$$\begin{aligned} \varepsilon^*_{22} &= \frac{\varepsilon}{k}\left[\overline{u_2^2} + \overline{u_2^2} + \overline{u_2^2} + \overline{u_2^2}\right] = 4\frac{\varepsilon}{k}\overline{u_2^2}, \\ \varepsilon^*_{12} &= \frac{\varepsilon}{k}\left[\overline{u_1 u_2} + \overline{u_1 u_2} + 0 + 0\right] = 2\frac{\varepsilon}{k}\overline{uv}, \end{aligned} \tag{12.31}$$

both of which comply with the asymptotic values in Table 12.1. The correct values are also returned for the other components. As to the adequacy of Eq. (12.30), Fig. 12.14 shows that it seriously

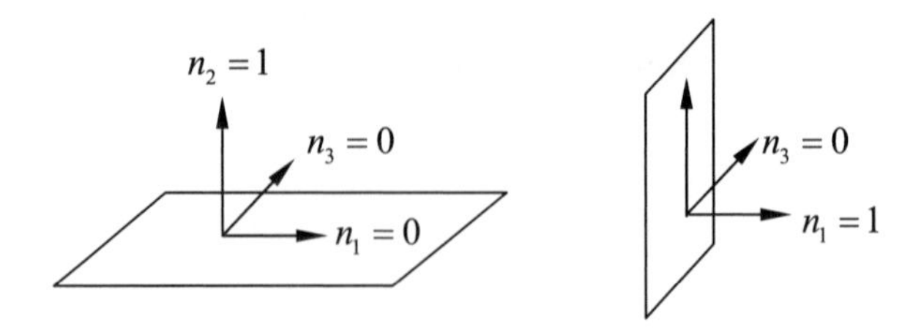

Fig. 12.15: Wall-orientation indicators in model (12.29).

underestimates the influence of the wall on the anisotropy of the dissipation — at least for the particular channel flow used here as a benchmark.

More recent models, proposed after 1990, give expression to the recognition that the near-wall anisotropy of the dissipation is likely to be greatly influenced, in common with the stress anisotropy, by inviscid features associated with wall-blocking, rather than by viscous effects. Sensitising $f_{\varepsilon w}$ to this process requires the use of a scalar quantity that represents the near-wall stress anisotropy, and which varies between '0' at the wall' and '1' in the outer region in which turbulence is close to being isotropic. This requirement is satisfied by the invariant A (Eq. (12.13)) and its dissipation-equivalent E. An example is the proposal by Launder and Tselepidakis (1993):

$$f_{\varepsilon w} = \exp(-20A^2), \tag{12.32}$$

also included in Fig. 12.14.

The most elaborate model is that of Hanjalić and Jakirlić (1993), who adopt a variation of Kebede *et al.*'s (1985) model, Eq. (12.29), which gives the correct wall-limiting dissipation anisotropy. In addition, the model sensitises the damping function $f_{\varepsilon w}$ to three scalar parameters: the anisotropy of the stresses, the anisotropy of the dissipation and the turbulent Reynolds number. In its most general form, the approximation, applicable to a wall at an arbitrary orientation, is as follows:

$$\varepsilon_{ij}^* = \frac{\frac{\varepsilon}{k}\left[\overline{u_i u_j} + \left(\overline{u_i u_k} n_j n_k + \overline{u_j u_k} n_i n_k + \overline{u_k u_l} n_k n_l n_i n_j\right) f_d\right]}{1 + \dfrac{3}{2}\dfrac{\overline{u_k u_l}}{k} n_k n_l f_d}, \tag{12.33}$$

with

$$f_{\varepsilon w} = 1 - \sqrt{A}E^2, \quad f_d = \frac{1}{1 + 0.1 Re_t}.$$

As shown in Fig. 12.12, E is close to 1.0 over 80–90% of the channel (especially at bulk Reynolds-number values larger than 20,000), declining rapidly towards zero as the wall is approached, in a manner similar to A. It follows, therefore, that this invariant is only effective,

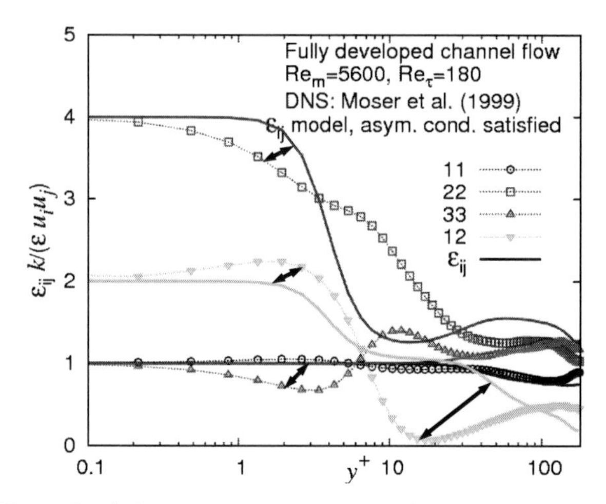

Fig. 12.16: Normalised dissipation anisotropy in channel flow at $Re_\tau = 180$, predicted by model (12.33), against DNS results by Moser, Kim and Mansour (1999) (courtesy S. Jakirlić).

as a modelling aid, in the region very close to the wall. Indeed, this is the sense in which E is exploited in Eq. (12.33), essentially as a modification of the model by Gilbert and Kleiser (1991) (see Table 12.2). The sensitisation to A, E and f_d is, essentially, a matter of trial and error, or rather a matter of attempting to achieve the best fit to DNS data. The level of fidelity thus attained is conveyed in Fig. 12.16. Here again, the principal defect is a misrepresentation of the shear-stress dissipation, although this defect is likely to diminish with increasing Reynolds number.

A final issue that needs to be covered under the heading 'Dissipation' is the equation used to determine the scalar dissipation ε. In principle, any of the dissipation-rate equations discussed in Chapter 9 qualifies. In practice, however, greater discrimination and care are needed to ensure that the form chosen is 'well suited', in terms of its predictive characteristics, to the Reynolds-stress closure used. This means, usually, that some particular corrections or modifications are introduced in conjunction with specific model forms.

In addition to the direct (explicit) use of the normal-stress productions to determine $P_k = 0.5\,(P_{11} + P_{22} + P_{33})$, a rational change (though not always adopted or effective) is to take advantage of the availability of the individual stresses in approximating the turbulent-diffusion of the dissipation rate. This is almost invariably done (e.g. Hanjalić and Launder (1972)) by analogy to the GGDH approximation in Eq. (12.19):

$$T_\varepsilon^t = \frac{\partial}{\partial x_k}\left[C_\varepsilon \frac{k}{\varepsilon}\left(\overline{u_k u_l}\frac{\partial \varepsilon}{\partial x_l}\right)\right], \tag{12.34}$$

with $C_\varepsilon = 0.15 - 0.18$. Although this modification complies, in spirit, with the second-moment-closure framework, its predictive benefits are not clearly established relative to the isotropic-diffusivity approximation. A justifiable variation of Eq. (12.34) could be the replacement of the time scale $\frac{k}{\varepsilon}$ by Eq. (12.17), so as to account for viscous effects.

A third term subject to model-specific adjustments is the destruction term. This is, essentially, analogous to the introduction of the function $f_{\varepsilon 2}$ in the context of low-Reynolds-number two-equation models (see Section 9.3). Three examples follow.

In the model of Hanjalić & Launder (1976) (and also in several other models, including that of Kebede *et al.* (1985) and Jakirlić & Hanjalić (1995)), the destruction term is approximated by:

$$D_\varepsilon = C_{\varepsilon 2} f_\varepsilon \frac{\varepsilon \widetilde{\varepsilon}}{k}, \tag{12.35}$$

with

$$f_\varepsilon = 1 - \frac{C_{\varepsilon 2} - 1.4}{C_{\varepsilon 2}} \exp\left[-\left(\frac{Re_t}{6}\right)^2\right].$$

The use of the 'isotropic dissipation' $\widetilde{\varepsilon} = \varepsilon - 2\nu\left(\frac{\partial k^{1/2}}{\partial y}\right)^2$ ensures that the destruction term vanishes at the wall, aiding the elevation of ε as the wall is approached. Figs. 5.4 and 12.11 show that the dissipation reaches a maximum at the wall. As is demonstrated by results of So *et al.* (1991) for channel flow at the bulk Reynolds-number $Re_m = 6500$, reproduced in Fig. 12.17(a), the approximation in Eq. (12.35), used as part of the Hanjalić–Launder model, does

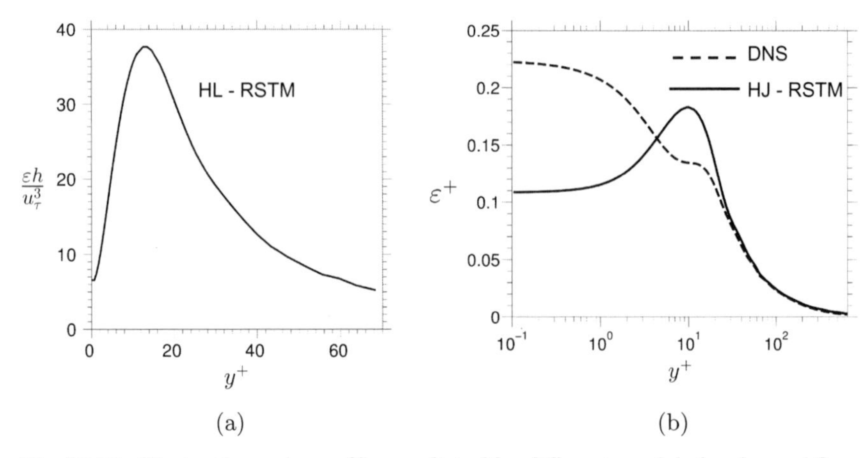

Fig. 12.17: Dissipation-rate profiles predicted by different models for channel flow, (a) Hanjalić–Launder model (1976) at bulk Reynolds number 6500 (reproduced with permission from American Institute of Aeronautics and Astronautics (AIAA) from So *et al.* (1991)); (b) Hanjalić–Jakirlić model (1993) at bulk Reynolds number 13750 (with permission from Herbert Utz Verlag GmbH from Jakirlić (1997)).

not return the dissipation profile correctly, as a strong maximum is predicted in the buffer layer, rather than at the wall.[7]

The much more recent model of Hanjalić and Jakirlić (1993), which also uses the proposal in Eq. (12.35), gives a somewhat more realistic variation, as is shown in Fig. 12.17(b), but this arises from a dissipation-rate equation that contains a further correction yet to be discussed.

The second example of model-specific modifications is the addition, by Shima (1988), of the following term, ξ, to the dissipation-rate equation:

$$D_\varepsilon + \xi = C_{\varepsilon 2}\frac{\varepsilon^2}{k} + \left[\left(\frac{7}{9}C_{\varepsilon 2} - 2\right)\frac{\varepsilon\widetilde{\varepsilon}}{k} - \frac{\widetilde{\varepsilon}'\widetilde{\varepsilon}'}{2k}\right] f_w, \qquad (12.36)$$

where $f_w = \exp\left[-\left(0.015\frac{k^{1/2}y}{\nu}\right)^4\right]$ and $\widetilde{\varepsilon}' = \varepsilon - 2\nu\frac{\partial^2 k}{\partial y^2}$ is a variation of the 'isotropic dissipation' $\widetilde{\varepsilon}$. This form reflects yet another attempt

[7] In Fig. 12.17(a), the dissipation is normalised with the channel half-height, while in Figs. 5.4, 12.11 and 12.17(b), the dissipation is normalised by u_τ^4/ν; hence, the major differences in the scales of the ordinates.

to satisfy the asymptotic wall value of ε. However, it also returns the same type of error shown in Fig. 12.17(a), although the magnitude of the error is slightly lower than that produced by the model in Eq. (12.35).

The third example is the modification of Cresswell *et al.* (1989):

$$f_\varepsilon = \frac{1.92}{(1 + 0.6A^{0.5}A_2)}. \tag{12.37}$$

This form, when compared to f_ε in Eq. (12.35), is based, again, on the notion that the effect of the wall on the dissipation rate in the sublayer is rooted, principally, in inviscid processes, rather than viscous damping. Reality is probably somewhere in-between, justifying the use of both the anisotropy invariants and the turbulent Reynolds number.

A final correction noted herein is the addition of a fragment analogous to Eq. (9.38). Its rationale and the motivation for its use are identical to those discussed in Section 9.3.1. Here, however, the availability of all the stress components allows a more general — tensorial — equivalent to be adopted:

$$E = 0.25\nu\frac{k}{\varepsilon}\overline{u_k u_l}\frac{\partial U_i}{\partial x_j \partial x_k}\frac{\partial U_i}{\partial x_j \partial x_l}. \tag{12.38}$$

This additional correction (or variations thereof) is a part of several models, among them those of Hanjalić & Launder (1976) and Hanjalić & Jakirlić (1993), the latter yielding the variation given in Fig. 12.17(b).

12.7 *Pressure-velocity interaction*

12.7.1 *Basic considerations*

We start this section by referring to the decomposition given by the identity in Eq. (12.8). The large majority of models treat these two terms separately. Indeed, a frequent assumption made is that the pressure-diffusion terms are minor and can therefore be ignored, or that these terms can be regarded as analogous to the triple-correlation terms, so that models of the type in Eqs. (12.18)–(12.20) may be regarded as approximating the sum of both.[8] We shall discuss the validity of this assumption later. Here, we focus first on the pressure-strain term Φ_{ij}.

As the modelling task is to relate Φ_{ij} to the stresses $\overline{u_i u_j}$ and strains $\partial U_i/\partial x_j$, it is desirable to attempt, first, to gain insight into the relevant dependencies, via exact manipulations. This can be done as follows:

- Derive a pressure-Poisson equation for the pressure fluctuation p.
- Multiply this by the strain fluctuations to comply with the meaning of Φ_{ij}.
- Perform an integration over a generic domain representing the flow.

A pressure-Poisson equation for p is readily derived (Rotta, 1951) by subtracting the exact Reynolds-averaged Navier–Stokes (RANS) equations from the NS equations, rearranging the result as $\frac{1}{\rho}\frac{\partial p}{\partial x_i} = RHS$, taking the derivative $\frac{1}{\rho}\frac{\partial}{\partial x_i}(\frac{\partial p}{\partial x_i}) = \frac{\partial}{\partial x_i} RHS$ and simplifying the right-hand side (RHS) by taking advantage of the mass-conservation constraint. With body forces omitted, the result is:

$$\frac{1}{\rho}\frac{\partial}{\partial x_i}\left(\frac{\partial p}{\partial x_i}\right) = \frac{\partial}{\partial x_i} RHS = -\frac{\partial^2}{\partial x_i \partial x_k}(u_i u_k - \overline{u_i u_k}) - 2\frac{\partial U_k}{\partial x_i}\frac{\partial u_i}{\partial x_k}. \tag{12.39}$$

The pressure fluctuation itself results from a formal integration of the pressure-Poisson equation over an arbitrary volume V. As shown in Fig. 12.18, x_i is the location at which the pressure fluctuation is determined, while x_i' represents any location within the volume V involved in the integration. This integration process reflects the fact that the pressure depends upon the conditions over the entire flow domain, by virtue of the elliptic nature of the equation governing p.

Next, the pressure fluctuation in the pressure-strain term, Eq. (12.8), is replaced with Eq. (12.39), and the whole group is time-averaged. With body-force terms omitted, the result of the above

[8] One notable exception is a model by Craft and Launder (1996), in which pressure-diffusion is modelled separately.

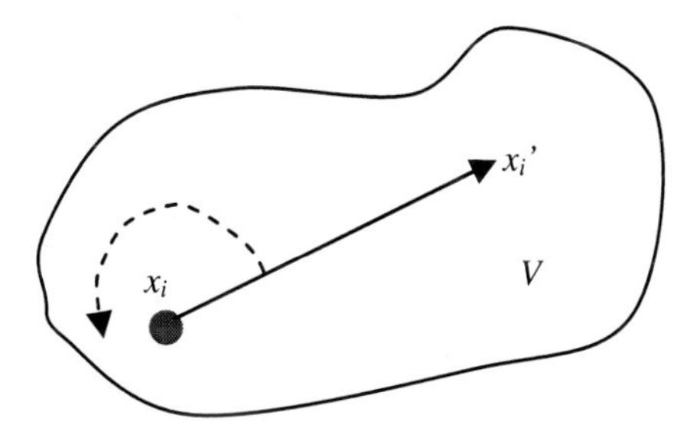

Fig. 12.18: Interpretation of the volume integration in Eq. (12.40).

process is as follows[9]:

$$\Phi_{ij} = \frac{1}{4\pi}\int\limits_V \left[\overline{\left(\frac{\partial^2 u_k u_l}{\partial x_k \partial x_l}\right)' \left(\frac{\partial u_i}{\partial x_j} + \frac{\partial u_j}{\partial x_i}\right)}\right] \frac{dV(x_i')}{|x_i - x_i'|}$$
$$+ \frac{1}{4\pi}\int\limits_V \left[2\left(\frac{\partial U_k}{\partial x_l}\right)' \overline{\left(\frac{\partial u_l}{\partial x_k}\right)' \left(\frac{\partial u_i}{\partial x_j} + \frac{\partial u_j}{\partial x_i}\right)}\right] \frac{dV(x_i')}{|x_i - x_i'|}. \tag{12.40}$$

Note that the bracketed terms contain products of quantities at the locations x_i and x_i' — i.e. the relevant products that are over-lined are *two-point correlations*, reflecting the non-locality of the processes being represented.

While the above manipulations do not directly open a transparent route to modelling, the result contains an important message — apart from the issue of non-locality already noted. There is a fundamental difference between the two terms: one involves turbulent velocity fluctuations only, while the other involves products of turbulent fluctuations and mean-strain components. Hence, any promising model can be expected to contain two types of terms: one type containing stresses only, and other involving combinations of stresses and mean strains. We distinguish these two types by defining

[9]Strictly, there should be another additive term, associated with the influence of any solid part of the boundary — a wall — on the pressure-strain process. This is a term that represents effects arising from the reflection of pressure fluctuations back into the domain. We shall consider this separately later.

the additive fragments:

$$\Phi_{ij} = \Phi_{ij,1} + \Phi_{ij,2}, \tag{12.41}$$

conventionally called the *slow* and *rapid*[10] terms, respectively. A modest improvement in insight can be derived from Eq. (12.40) by exploiting general mathematical rules that pertain to two-point-correlation derivatives (see Jovanovic *et al.* (1995)). These allow Eq. (12.40) to be re-written as follows:

$$\begin{aligned}\Phi_{ij} = &\frac{1}{4\pi}\int\limits_V \left(\frac{\partial^3 \overline{u'_k u'_l u_i}}{\partial\xi_i \partial\xi_k \partial\xi_l} + \frac{\partial^3 \overline{u'_k u'_l u_j}}{\partial\xi_k \partial\xi_l \partial\xi_i}\right)\frac{dV(x'_i)}{|x_i - x'_i|}\\ &+ \frac{1}{4\pi}\int\limits_V \left[2\left(\frac{\partial U_k}{\partial x_l}\right)'\left(\frac{\partial \overline{u'_l u_i}}{\partial\xi_l \partial\xi_j} + \frac{\partial \overline{u'_l u_j}}{\partial\xi_k \partial\xi_i}\right)\right]\frac{dV(x'_i)}{|x_i - x'_i|},\end{aligned} \tag{12.42}$$

where $\xi_i = |x_i - x'_i|$. While Eq. (12.42) still involves obscure two-point correlations, the advantage of this form is that these correlations are 'simply' products of turbulent fluctuations. This form thus suggests that valid approximations can be expected to involve higher-order (series-type) expansions in terms of quantities located at x_i, so as to represent (at least in the locality of x_i) the dependence on fluctuations at x'_i. Alternatively, one may propose correlation functions (e.g. of a decaying-exponential type) to represent the connection between fluctuations at the two neighbouring locations, thus allowing an approximate integration of Eq. (12.42).

A key property of the pressure-strain term is that it redistributes turbulence energy among the normal-stress components. In other words, it drives turbulence towards isotropy, expressed as:

$$\overline{u_i^2} \to \frac{2}{3}k \quad \text{(no summation on } i\text{)}. \tag{12.43}$$

[10]This terminology is rooted in the concept that $\Phi_{ij,2}$ describes a rapid redistribution of turbulence, driven by the mean strain, with the turbulent vorticity (strictly, the 'enstrophy' — the scalar norm of the vorticity) being conserved, and strain-induced turbulence production being zero, in the sense of the 'rapid distortion theory' (RDT) of Batchelor and Proudman (1954); see also Pope (2000), p. 404.

This is not the only effect, however. The shear-stress components are also affected, if only because Φ_{ij} also features in the shear-stress equations. Can anything be said about what this effect might be? One way to answer this question is to resort, yet again, to Mohr's circle, Fig. 12.4. Isotropy means more than is obvious from Eq. (12.43), namely that the normal stresses are identical, regardless of the orientation within the medium — i.e. not only in the Cartesian system implied by Eq. (12.43), but in any other direction. It should be obvious from Fig. 12.4 that the only way to secure identical normal stresses at all orientations is to make the circle collapse into a point on the abscissa, which means that there is no shear stress at any orientation, since the radius of the circle is zero. The implication as regards turbulence is, therefore, that any process that drives turbulence towards isotropy must also drive the shear stresses down towards zero. In other words, we expect Φ_{ij}, $i \neq j$, to be a sink term in any shear-stress equation.

In the following three sections we discuss models for the two fragments in Eq. (12.41), treated either separately or together. Although a separate exposition is convenient and appropriate in the case of the majority of models, it has to be appreciated that the two fragments interact and influence each other. For this reason, it is often necessary to determine the coefficients attached to separate model proposals as part of an examination of the performance of both fragments operating together.

An issue that requires caution, and is best highlighted here ahead of discussing specific modelling proposals for $\Phi_{ij,1}$ and/or $\Phi_{ij,2}$, is that some proposals are only valid in free flows, i.e. away from walls, while others include terms and functional corrections to coefficients that arose from a calibration of the models for near-wall flows as part and parcel of the basic modelling process. In principle, non-linear approximations (involving products of stresses) tend to be less restrictive, but still require wall-specific modifications and corrections to capture, as realistically as is possible, both the inviscid wall-reflection processes that are not included in Eqs. (12.40) and (12.42) and any effects arising from the viscosity. Unless otherwise stated, the reader should assume that a proposal does not apply to near-wall flows, and that it may require additional wall-related corrections.

12.7.2 *Modelling of the slow term* $\Phi_{ij,1}$

It is noted, first, that the slow term, the upper line in Eq. (12.42), is the sum of two second-rank tensors (in i, j). As the stress tensor is the highest-rank quantity that can be determined from the solution of a second-moment closure, any model for $\Phi_{ij,1}$ has to be of the form:

$$\Phi_{ij,1} = c_1\delta_{ij} + c_2 a_{ij} + c_3\left(a_{ik}a_{kj} - \frac{1}{3}a_{kl}a_{lk}\delta_{ij}\right) + c_4\left(a_{il}a_{lk}a_{kj} - \frac{1}{3}a_{kl}a_{lm}a_{mk}\delta_{ij}\right) + \cdots, \quad (12.44)$$

where a_{ij} is the anisotrophy tensor (see caption of Fig. 12.1 and Section 12.3). Because of the contraction constraint, $\Phi_{kk,1} = 0$, we require $c_1 = 0$. Similarly, the anisotropy invariants in the brackets are needed to achieve zero contraction.

The simplest possible model for $\Phi_{ij,1}$ that satisfies the isotropisation process described above, and the constraint that this term should involve only turbulent fluctuations, is:

$$\Phi_{ij,1} = -C_1\frac{\varepsilon}{k}\left(\overline{u_i u_j} - \frac{2}{3}\delta_{ij}k\right) = -C_1\varepsilon a_{ij}, \quad (12.45)$$

a form proposed by Rotta (1951). Note that ε/k — the inverse of the turbulent time scale — is needed for dimensional consistency.[11] The constant C_1 (usually taken as 1.8) is determined by reference to experimental data for simple canonical flows. Its value will be discussed later, because the appropriate choice of coefficients is connected to the model for $\Phi_{ij,2}$, a subject considered in the next subsection.

We note the following characteristics of the model in Eq. (12.45):

(i) The model only involves turbulent-velocity fluctuations, in the form of the Reynolds stresses themselves;

[11] Note, however, that this is only one — the most obvious — option. Other options might include the Kolmogorov scale if viscous effects are strong.

(ii) the model is tensorially consistent, of rank 2;

(iii) the sum $\Phi_{kk,1} = 0$;

(iv) $\Phi_{ij,1}, i = j$, constitutes a sink in any of the normal-stress equations when the associated stress exceeds the average $\frac{2}{3}k$, and it is a source when the stress is lower than the average — i.e. the effect is to *isotropise* the turbulence field;

(v) $\Phi_{ij,1}, i \neq j$, is always a sink in any shear-stress equation, unless that shear stress vanishes.

There is, of course, no reason to assume that a linear approximation is sufficient. Indeed, the form in Eq. (12.44) strongly suggests the appropriateness of non-linear models. One such non-linear model is the following quadratic approximation by Speziale, Sarkar and Gatski (1991):

$$\Phi_{ij,1} = -C_1 \varepsilon a_{ij} - C_1' \varepsilon \left(a_{ik} a_{kj} - \frac{1}{3} \delta_{ij} \{ a_{kl} a_{lk} \} \right). \tag{12.46}$$

As with the model in Eq. (12.45), the approximation in Eq. (12.46) contracts to zero. Again, the constants will be considered later.

No higher-order models have been proposed. One important reason for this is that the so-called Cayley–Hamilton theorem — the outcome of elaborate tensor theory, some of which will be discussed in Chapter 15 — shows that, in any relationship linking a second-rank tensor to other second-rank tensors, all tensor products of order 3 and higher can be expressed in terms of quadratic products, pre-multiplied by functions of the anisotropy invariants A_2, A_3. Thus, in principle:

$$\Phi_{ij,1} = -C_1(A_2, A_3) \varepsilon a_{ij} - C_1'(A_2, A_3) \varepsilon \left(a_{ik} a_{kj} - \frac{1}{3} \delta_{ij} \{ a_{kl} a_{lk} \} \right), \tag{12.47}$$

is the most general relationship permitted on mathematical grounds.

Remaining within the linear framework ($C_1' = 0$), Jakirlić and Hanjalić (1995) proposed, on the basis of computational calibration by reference to DNS data for channel flow at a bulk Reynolds number of 5600, the following function:

$$C_1 = E^2 A^{\frac{1}{2}} + 2.5A(\min\{0.6; A_2\})^{\frac{1}{4}} \min\left\{\left(\frac{Re_t}{150}\right)^{\frac{3}{2}}; 1\right\}, \quad (12.48)$$

in which E is the equivalent of A, determined from the anisotropy of the dissipation tensor ε_{ij} (see Section 12.6). The fact that C_1 is proportional to A, or its root, has one important consequence: as the wall is approached, turbulence tends to a state of two-component turbulence, in which case $A \to 0$. Hence, the isotropisation process is disabled, as is required by the constraint in Eq. (12.11). Apart from the use of E in the model in Eq. (12.48), which is not formally consistent with Eq. (12.47), this is a classic illustration of the fact that turbulence modelling cannot be explained purely by reference to rational principles, supported by a transparent calibration route. Rather, the function in Eq. (12.48) is a result of trial and error, designed to yield an acceptable approximation to DNS data. It is also important to point out that, in reality, C_1 should differ from stress to stress, as is shown by the DNS data. In other words, there is no unique function that correctly returns the DNS variations, even for the particular channel flow used as a basis for the calibration. The use of E, which only differs from 1.0 close to the wall, and which decays rapidly as the wall is approached (see Fig. 12.12), suggests that the rate of decline in the isotropisation process is required to be higher than the rate that is possible to achieve by using A alone close to the wall. Finally, while a Reynolds-number-dependent function might appear to be reasonable, to account for the added influence of viscosity, it can be argued that the inclusion of E, and the dependence of the ε-equation on the viscosity, might have sufficed to achieve the requisite near-wall behaviour, subject to some minor changes to the model in Eq. (12.48).

The most complex model for $\Phi_{ij,1}$ is that of Craft and Launder (1996) (see also Launder and Tselepidakis (1993) and Craft (1998)). Several variations of this model exist, the simplest being:

$$\Phi_{ij,1} = -C_1 \varepsilon a_{ij} - C_1' \varepsilon \left(a_{ik} a_{kj} - \frac{1}{3} \delta_{ij} \{a_{kl} a_{lk}\} \right) - \varepsilon A a_{ij}, \quad (12.49)$$

in which $C_1 = 3.1(A_2A)^{0.5}, C_1' = 1.2C_1$. The more elaborate variants, applicable to near-wall conditions, use the 'isotropic dissipation' $\widetilde{\varepsilon}$, include a more complex functional representation of C_1 and C_1', involving Re_t, and a function of A pre-multiplying the last term in Eq. (12.49). The rationale underpinning the inclusion of this last fragment in Eq. (12.49) deserves further comment. It was mentioned in Section 12.5 that Lumley (1978) had argued that the anisotropy of the dissipation $\varepsilon_{ij} - \frac{2}{3}\varepsilon\delta_{ij}$ need not be considered explicitly, but can be viewed as part of modelling the pressure-strain term. The last term in Eq. (12.49) reflects this argument, provided it is assumed that the anisotropy of the dissipation can be (quasi-) linearly related to the anisotropy of the stresses a_{ij}.

12.7.3 *Modelling of the rapid term* $\Phi_{ij,2}$

The starting point of modelling this term is a return to the second line of Eq. (12.42), which is the exact form of $\Phi_{ij,2}$. Upon the assumption that the mean-strain terms remain constant within the domain of integration, these terms can be taken outside the integral. Of course, strictly, this is only correct in homogeneous strain. What then remains is a fourth-rank tensor in the indices i, j, k, l, which allows the rapid term to be written, symbolically, as:

$$\Phi_{ij,2} = \frac{\partial U_k}{\partial x_l} k(A_{kj}^{li} + A_{ki}^{lj}), \tag{12.50}$$

in which the two fourth-rank tensors relate, respectively, to the two-point correlations in Eq. (12.42), divided by k to render them dimensionless.

Here again, we are constrained to using the anisotropy tensor, and lower-rank terms, but we need to ensure that any of the approximation fragments are of rank 4. The Cayley–Hamilton theorem expresses the fact that the expansion of any fourth-rank tensor in terms of second-rank tensors may include, at most, fourth-order combinations of the latter, with coefficients being, again, functions of the second and third anisotropy invariants. Fu, Launder and Tselepidakis (1987) show that a general expansion of the tensor A_{kj}^{li}, up to order 3 in a_{ij}, contains 20 independent groups:

$$
\begin{aligned}
A^{li}_{kj} = & \underbrace{c_{0,1}\delta_{li}\delta_{kj} + c_{0,2}(\delta_{lj}\delta_{ki} + \delta_{lk}\delta_{ij})}_{\text{constant terms}} + \\
& \underbrace{c_{1,1}a_{li}\delta_{kj} + c_{1,2}a_{kj}\delta_{li} + c_{1,3}(a_{lj}\delta_{ki} + a_{lk}\delta_{ij} + a_{ij}\delta_{lk} + a_{ki}\delta_{lj})}_{\text{linear term}} + \\
& \underbrace{\begin{aligned} & c_{2,1}a_{li}a_{kj} + c_{2,2}(a_{lj}a_{ki} + a_{lk}a_{ij}) + \\ & c_{2,3}a_{lm}a_{mi}\delta_{kj} + c_{2,4}a_{km}a_{mj}\delta_{li} + \\ & c_{2,5}(a_{lm}a_{mj}\delta_{ki} + a_{lm}a_{mik}\delta_{ij} + a_{im}a_{mj}\delta_{lk} + a_{km}a_{mi}\delta_{lj}) + \\ & [c_{2,6}\delta_{li}\delta_{kj} + c_{2,7}(\delta_{lj}\delta_{ki} + \delta_{lk}\delta_{ij})]\, a_{mn}a_{mn} \end{aligned}}_{\text{quadratic term}} + \\
& \underbrace{O(\boldsymbol{a}^3)}_{\text{cubic terms}} \; .
\end{aligned}
\tag{12.51}
$$

The cubic part, not given in explicit form here, comprises 7 further independent terms, each containing a cubic combination of the anisotropy tensor. The grouping of certain fragments (e.g. those premultiplied by $c_{0,2}$ and $c_{1,3}$) reflects the constraints of symmetry and zero trace of A^{li}_{kj}, simply because $\Phi_{ij,2}$ is also symmetric and traceless. This demands $A^{li}_{kj} = A^{il}_{kj} = A^{il}_{jk}$.

In principle, closure of the above form would require the 20 coefficients to be determined by a series of calibrations, in which data for a broad range of well-documented canonical flows are exploited. In this process, any of the coefficients can be made to depend on scalar parameters or invariants — in particular, the second and third invariants of the anisotropy tensor and the turbulent Reynolds number, the last accounting, primarily, for the effects of viscosity in the viscous sublayer of near-wall flows.

Unsurprisingly, in view of the complexity of the general framework, established models have adopted drastically truncated versions of Eq. (12.51). There are about 10 alternative (or related) proposals, each adopting a different set of terms, In fact, some proposals (e.g. Reynolds (1987), Shih and Lumley (1985), Johansson and Hallbaeck (1995)) include fourth-order terms. We consider, here, only relatively simple formulations that are parts of models widely used in practice.

As a preliminary exercise, it is instructive to focus on some of the terms that arise when Eq. (12.51) is inserted into Eq. (12.50). Consider, for example, the *linear* fragment $\frac{\partial U_k}{\partial x_l} k c_{1,1} a_{li} \delta_{kj}$. This may be expanded as follows:

$$\frac{\partial U_k}{\partial x_l} k c_{1,1} a_{li} \delta_{kj} = k c_{1,1} a_{li} \left(\frac{\partial U_1}{\partial x_l} \delta_{1j} + \frac{\partial U_2}{\partial x_l} \delta_{2j} + \frac{\partial U_3}{\partial x_l} \delta_{3j} \right)$$

$$= k c_{1,1} a_{li} \frac{\partial U_j}{\partial x_l}, \qquad (12.52)$$

which can also be written as:

$$\frac{\partial U_j}{\partial x_l} k c_{1,1} a_{li} = c_{1,1} \frac{\partial U_j}{\partial x_k} \left(\overline{u_i u_k} - \frac{2}{3} k \delta_{ik} \right), \qquad (12.53)$$

the first fragment of which is proportional to the first group in the Reynolds-stress-production term P_{ij} (see Eq. (12.7)). It should be clear, therefore, at least in principle, that the linear contributions to $\Phi_{ij,2}$ could, justifiably, be expressed as a linear function of P_{ij}. Of course, we also expect that the non-linear fragments would require other types of terms to be included — for example, quadratic anisotropy terms multiplied by the strain tensor.

The above considerations suggest that the simplest proposal for approximating $\Phi_{ij,2}$ is:

$$\Phi_{ij,2} = -C_2 \left(P_{ij} - \frac{2}{3} \delta_{ij} P_k \right), \qquad (12.54)$$

first proposed by Rotta (1951). As required, the bracket, representing the anisotropy of the stress production, contracts to zero. This model is, therefore, conventionally refereed to as the 'isotropization-of/by-production' (IP) model, and it is the form that is the most popular in practical implementations and applications of second-moment closure.

Before we turn our attention to more complex models, we consider the constants C_1 and C_2 in the models in Eqs. (12.45) and

(12.54), respectively. A restriction to note, first, is that both approximations are only applicable, in principle, to turbulent regions remote from walls — unless the constants are made to depend on wall-proximity parameters, which we expect to depend on anisotropy invariants and the turbulent Reynolds number, so as to represent the wall-reflection effects and the correct asymptotic variation of the stresses as any wall is approached.

In the case of a simple free mixing layer, and on the assumption of negligible convection and diffusion (DNS data show this assumption to be broadly justified), use of the models in Eqs. (12.45) and (12.54) allows the stress equations to be written:

$$P_{ij} \underbrace{-C_1\varepsilon\left(\frac{\overline{u_iu_j}}{k}-\frac{2}{3}\delta_{ij}\right)}_{\Phi_{ij,1}} \underbrace{-C_2\left(P_{ij}-\frac{2}{3}P_k\delta_{ij}\right)}_{\Phi_{ij,2}} -\frac{2}{3}\varepsilon\delta_{ij}=0, \quad (12.55)$$

or

$$\left(\frac{\overline{u_iu_j}}{k}-\frac{2}{3}\delta_{ij}\right)=\frac{1-C_2}{C_1\varepsilon}\left(P_{ij}-\frac{2}{3}\delta_{ij}P_k\right). \quad (12.56)$$

This result is insightful in two respects. First, it provides support for an earlier proposition, Eq. (11.1), of a 'primitive' second-moment closure (albeit for very simple flow conditions). Second, and more pertinent to the present discussion, it suggests it is the ratio $(1-C_2)/C_1$ that is the operative constant, rather than C_1 and C_2 themselves. That this constraint is broadly satisfied by different models is illustrated in Fig. 12.19. Although this result applies to the particular models used for $\Phi_{ij,1}$ and $\Phi_{ij,2}$, it illustrates, more generally, that the calibration to determine the respective constants has to take the link between the two fragments into account.

The simplest model included in Fig. 12.19 is that of Gibson and Launder (1978), and this combines the simplest approximations in Eqs. (12.45) and (12.54), used to derive Eq. (12.56). As seen from the figure, the constant C_2 has the value 0.6. This value arises from theoretical considerations of Crow (1968), which focus on the response of

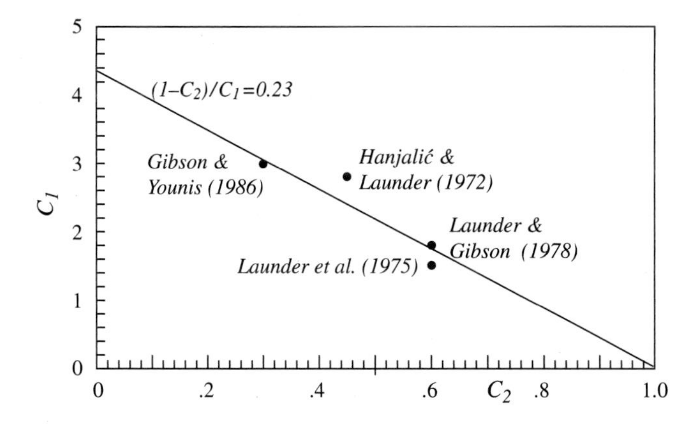

Fig. 12.19: Adherence of different models to the constraint (12.56).

isotropic turbulence (as a limiting state) to mean straining, yielding:

$$\Phi_{ij,2}|_{isotropy} = 0.4k\left(\frac{\partial U_i}{\partial x_j} + \frac{\partial U_j}{\partial x_i}\right), \tag{12.57}$$

with 0.4 having to be equal to $2/3C_2$, $C_2 = 0.6$, when the isotropic limit is imposed on Eq. (12.54). For the case of turbulence-energy equilibrium in simple shear, $P_{11} = 2P_k, P_{22} = P_{33} = 0, P_k/\varepsilon = 1.0$, and Eq. (12.56) reduces to:

$$\left(\frac{\overline{u_1^2}}{k} - \frac{2}{3}\right) = \frac{4 - 4C_2}{3C_1}; \quad \left(\frac{\overline{u_2^2}}{k} - \frac{2}{3}\right) = \frac{-2 + 2C_2}{3C_1};$$
$$\left(\frac{\overline{u_3^2}}{k} - \frac{2}{3}\right) = \frac{-2 + 2C_2}{3C_1}. \tag{12.58}$$

With $C_2 = 0.6$ and $C_1 = 1.8$, i.e. $(1 - C_2)/C_1 = 0.22$, this gives the anisotropy components $\{0.3, -0.15, -0.15\}$, respectively, agreeing reasonably well with the experimental data for homogenous shear $\{0.3, -0.18, -0.12\}$, respectively, and these are the values of the constants conventionally used in Gibson and Launder's IP model. The fact that, in reality, and especially close to a wall, the wall-normal component is lower than the spanwise component reflects a combination of the inherently unequal redistribution of turbulence energy, even in simple shear, and the influence of wall-blocking, the latter causing the wall-normal fluctuations to be depressed relative to spanwise fluctuations, thus resulting in the isotrophy rising to $\{0.55, -0.45, -0.11\}$ near a wall. In modelling terms, errors in representing the former effect are due, essentially, to the simplicity of the model, while the latter process requires wall (pressure-reflection) influences to be accounted for in the pressure-strain models in Eqs. (12.45) and (12.54).

With $\left(\frac{\overline{u_2^2}}{k} - \frac{2}{3}\right) = -0.15$, we can ask the question, for example, as to whether the shear stress in a turbulence-equilibrium layer is returned correctly. In this case, $\frac{\overline{u_1u_2}}{k} = \frac{1-C_2}{C_1\varepsilon}P_{12}$. With $\frac{P_{12}}{P_k} = \frac{\overline{u_2^2}}{\overline{u_1u_2}} = 0.51\frac{k}{\overline{u_1u_2}}$, we therefore get $-\frac{\overline{u_1u_2}}{k} = 0.33$, which is slightly too high, relative to the experimental value of approximately $0.25 - 0.3$,[12] and thus consistent with the too high level of $\overline{u_2^2}/k$ predicted by the model.

A more general (yet, still linear) form of the model in Eq. (12.54) was derived, independently and via drastically different routes, by Naot, Shavit and Wolfshtein (1973) and Launder, Reece and Rodi (1975), with all linear fragments in Eq. (12.51) being taken into consideration. In principle, this requires five coefficients to be determined. However, in any modelling process that is based on Eq. (12.51), the task of determining the coefficients can be reduced by imposing known mathematical and physical constraints. These include the following:

- The continuity equation implies $A_{ki}^{li} = 0$.
- Rotta (1951) shows, based on Green's theorem, that $A_{kk}^{li} = 2\overline{u_lu_i} = 2k\left(a_{li} + \frac{2}{3}\delta_{li}\right)$.

These constraints allow the following inter-dependence of the coefficients to be derived:

$$
\begin{aligned}
c_{0,1} &= -\frac{1}{55}(50c_{1,2} + 4); \quad c_{0,2} = \frac{1}{55}(20c_{1,2} + 6), \\
c_{1,1} &= \frac{1}{11}(4c_{1,2} + 10); \quad c_{1,3} = -\frac{1}{11}(2 + 3c_{1,2}).
\end{aligned}
\tag{12.59}
$$

[12]Recall that a value of 0.3 is consistent with $C_\mu = 0.09$ used in two-equation modelling.

Hence, the entire linear approximation of $\Phi_{ij,2}$ involves a single constant, which will be denoted by $C_2(\equiv c_{1,2})$ in the following considerations.

With Eq. (12.59) inserted into Eq. (12.51), and the result inserted into Eq. (12.50),[13] with a_{ij} etc. expanded in terms of the related Reynolds stresses, the final form for $\Phi_{ij,2}$ emerges as:

$$\Phi_{ij,2} = -\frac{C_2+8}{11}\left(P_{ij} - \frac{2}{3}\delta_{ij}P_k\right) - \frac{30C_2-8}{55}k\left(\frac{\partial U_i}{\partial x_j} + \frac{\partial U_j}{\partial x_i}\right) - \frac{8C_2-2}{11}\left(D_{ij} - \frac{2}{3}\delta_{ij}D_k\right), \tag{12.60}$$

where $D_{ij} = -\left(\overline{u_iu_k}\frac{\partial U_k}{\partial x_j} + \overline{u_ju_k}\frac{\partial U_k}{\partial x_i}\right)$ (which needs to be carefully distinguished from $P_{ij} = -\left(\overline{u_iu_k}\frac{\partial U_j}{\partial x_k} + \overline{u_ju_k}\frac{\partial U_i}{\partial x_k}\right)$).

If the assumption of stress isotropy is inserted into Eq. (12.60), then this expresses $\Phi_{ij,2}$ when isotropic turbulence is subjected to a sudden distortion. It turns out that Eq. (12.60) then reduces to Crow's result (12.57) regardless of the value of C_2.

To determine the constant C_2, a route analogous to that leading to Eq. (12.56) can be adopted: we use Eqs. (12.45) and (12.60), neglect stress transport, form a balance between production, pressure-strain and dissipation terms, assume dissipation isotropy $\varepsilon_{ij} = \frac{2}{3}\delta_{ij}\varepsilon$, and apply the system of equations to a simple shear flow. As before, for the particular case of simple shear and turbulence-energy equilibrium, $P_{11} = 2P_k, P_k/\varepsilon = 1$, the equations for the normal stresses arise as:

$$\left(\frac{\overline{u_1^2}}{k} - \frac{2}{3}\right) = \frac{8+12C_2}{33C_1}; \quad \left(\frac{\overline{u_2^2}}{k} - \frac{2}{3}\right) = \frac{2-30C_2}{33C_1};$$

$$\left(\frac{\overline{u_3^2}}{k} - \frac{2}{3}\right) = \frac{-10+18C_2}{33C_1}. \tag{12.61}$$

As noted earlier, experiments for homogeneous shear show the normal-stress-anisotropy components to be approximately

[13] A_{ki}^{lj} arises simply by interchanging the indices i, j in A_{kj}^{li}.

$\{0.3, -0.18, -0.12\}$, respectively. The choice $C_1 = 1.5$ and $C_2 = 0.4$ gives the values $\{0.26, -0.2, -0.06\}$, i.e. a reasonable fit to the streamwise and cross-flow components, but not the spanwise component. To determine the shear stress, $\Phi_{12,2}$ needs to be obtained from Eq. (12.60) and inserted into a balance equation analogous to Eq. (12.55). The result needs to be combined with $\frac{P_{12}}{P_k} = \frac{\overline{u_2^2}}{\overline{u_1 u_2}} = 0.44 \frac{k}{\overline{u_1 u_2}}$ and $\frac{D_{12}}{P_k} = \frac{\overline{u_1^2}}{\overline{u_1 u_2}} = 0.92 \frac{k}{\overline{u_1 u_2}}$, in which the numerical coefficients readily follow from Eq. (12.61). Substitution of these relations into Eq. (12.60), for $i, j = 1, 2$, noting that the dissipation rate — the last term in Eq. (12.55) — is zero, we obtain $-\frac{\overline{u_1 u_2}}{k} = 0.24$, which is lower than the experimental level.

In an earlier proposal by Hanjalić and Launder (1972) — the first attempt to construct a non-linear approximation — the first two terms in the quadratic group of Eq. (12.51) were included, but the linear term associated with $c_{1,2}$ was omitted. Following a route analogous to that leading to Eq. (12.60), the resulting model arises as:

$$\Phi_{ij,2} = -\frac{2C_2 - 8}{11}\left(P_{ij} - \frac{2}{3}\delta_{ij} P_k\right) - \frac{6C_2 - 2}{55} k \left(\frac{\partial U_i}{\partial x_j} + \frac{\partial U_j}{\partial x_i}\right) - \frac{6C_2 - 2}{11}\left(D_{ij} - \frac{2}{3}\delta_{ij} D_k\right) + 2C_2 a_{ij} P_k, \qquad (12.62)$$

with $C_2 = 0.45$, but in conjunction with $C_1 = 2.8$.

The most complex model proposed is that of Craft and Launder (1996) (see also Craft (1998)), part of which is model (12.49) for $\Phi_{ij,1}$. This model starts from the expansion in Eq. (12.51) and includes some cubic terms (the details of the process are documented in PhD Theses of Tselepidakis (1991) and Fu (1988)). This model is mathematically complex, its calibration process is elaborate, and its numerical implementation is challenging. While it has been applied to complex three-dimensional flows — including, for example, to model shock-boundary-layer interaction on realistic transonic-flight configurations (Batten *et al.* (1999)) — its use has been promulgated mainly by the group associated with the originators. Because of the its complexity and restricted adoption in practice, the model is not presented and discussed in this introductory text.

The notion that a linear approximation can be as adequate as a non-linear one, provided it includes appropriate functions of the anisotropy in the coefficients, led Jakirlić and Hanjalić (1995) to adopt Eq. (12.54), in conjunction with the linear model in Eq. (12.45) with the functional coefficient in Eq. (12.48), but with $C_2 = f(A)$:

$$\Phi_{ij,2} = -0.8A^{\frac{1}{2}}\left(P_{ij} - \frac{2}{3}\delta_{ij}P_k\right). \tag{12.63}$$

Here again, the dependence of C_2 on A is intended to ensure that the isotropisation process is disabled at walls, subject to the condition expressed by Eq. (12.11).

In Sections 12.7.5, to follow, we shall consider the inclusion of near-wall corrections to some pressure-strain models, initially formulated without reference to wall effects. The model of Jakirlić and Hanjalić, Eqs. (12.45), (12.48) and (12.63), is not among them, however, because the dependence of its coefficients on anisotropy invariants and the Reynolds number is specifically intended to secure the correct response of the model to wall-induced anisotropy alteration and viscous damping. It is, in fact, the most complex model that is based on overtly 'linear' pressure-strain approximations — although it is, of course, strictly non-linear, because of the dependence of its coefficients on the stresses. An example of its performance, relative to linear eddy-viscosity models, is given in Fig. 12.20. The figure originates from an extensive study by Apsley and Leschziner (2001), in which predictions obtained with 12 models are compared to experimental data for the flow around a generic wing-body junction. Results are given for the streamwise Reynolds stress in the symmetry plane upstream of the wing's leading edge, where the flow recirculates due to impingement, and for the turbulence energy in a plane downstream of, and normal to, the trailing edge. The flow is characterized, dynamically, by a horseshoe vortex that causes boundary-layer fluid and turbulence at the wing and plate walls to be transported outwards and lifted upwards, creating the kidney-shaped contours seen in the right-hand-side plots (b), especially that for the experimental data. Upstream of the junction, the turbulence intensity is elevated primarily in the

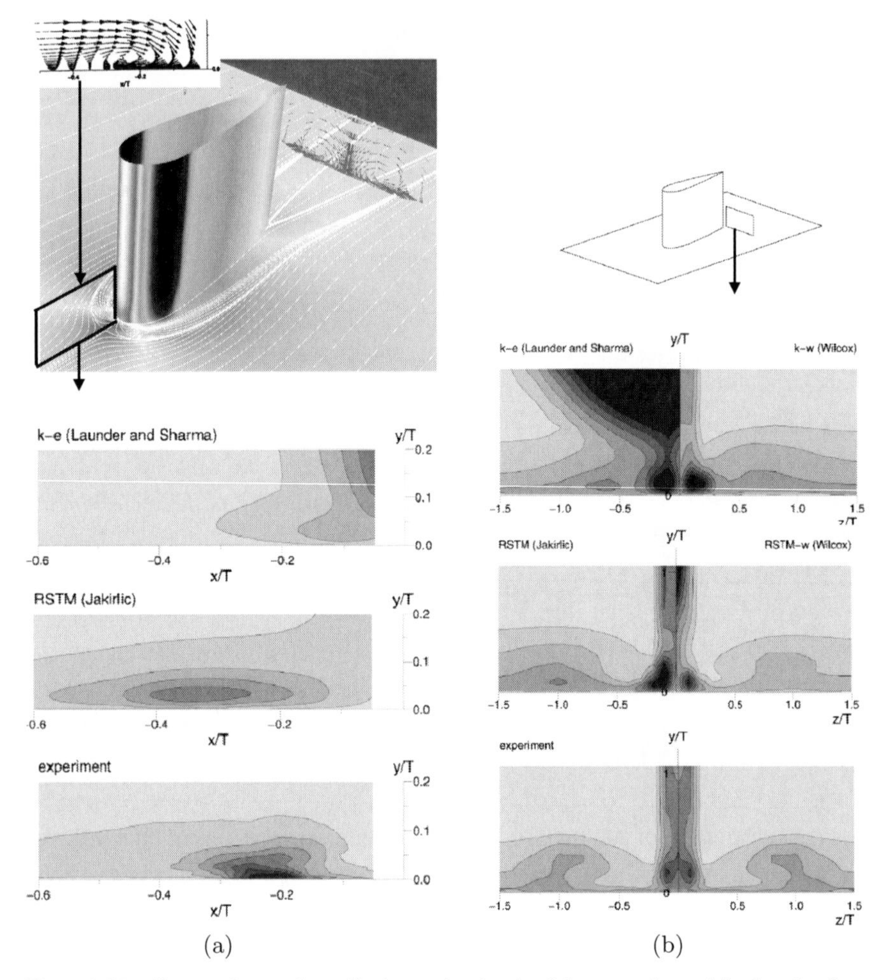

Fig. 12.20: Comparison of predictions obtained with several models for the flow around a generic wing-/body junction: (a) Streamwise normal Reynolds stress upstream of the wing leading edge; (b) Turbulence energy in the wake downstream of the wing trailing edge. The focus is on the performance of the model by Jakirlić and Hanjalić (1995) relative to that of the $k - \varepsilon$ model of Launder and Sharma (1974) (reproduced with permission of Springer from Apsley and Leschziner (2001)).

intensely sheared lower portion of the recirculation zone (see inset at the top left-hand-side corner of the figure). The Reynolds-stress model returns a much improved representation of the turbulence field, relative to the $k - \varepsilon$ model, the latter failing to capture the upwelling process explained above, and producing excessive levels of turbulence in the impingement zone and in the wake of the wing.

12.7.4 *Modelling* $\Phi_{ij,1} + \Phi_{ij,2}$ *collectively*

We consider here a modelling framework by Speziale, Sarkar and Gatski (1991), not only because it is the main representative of a general approach to modelling the pressure-strain process as a whole, but also because this model has gained a measure of popularity in practice. Moreover, this model is of relevance to Chapter 15 on algebraic approximations to second-moment transport models. Although the ultimate outcome is a model that has the form $\Phi_{ij} = \Phi_{ij,1} + \Phi_{ij,2}$ — the model fragment $\Phi_{ij,1}$ has already been stated in Eq. (12.46) — its derivation is not based on an *a priori* distinction between the two contributions, which therefore justifies its exposition in a separate section. The model involves complex matrix manipulations, which cannot be included here in detail. Rather, the principles and major steps are outlined, to an extent that allows an understanding to be gained of the rationale of the derivation of the model.

The starting point of the modelling process is, again, Eq. (12.42). Next, the whole term is assumed to be represented by a general second-rank function with arguments being anisotropy tensors, strain tensors and vorticity tensors:

$$\Phi_{ij} = \varepsilon f_{ij}(\mathbf{a}, \mathbf{S}, \boldsymbol{\Omega}), \tag{12.64}$$

where $\mathbf{S}$ and $\boldsymbol{\Omega}$ represent, respectively, various permutations of

$$S_{ij} = \frac{1}{2}\left(\frac{\partial U_i}{\partial x_j} + \frac{\partial U_j}{\partial x_i}\right), \quad \Omega_{ij} = \frac{1}{2}\left(\frac{\partial U_i}{\partial x_j} - \frac{\partial U_j}{\partial x_i}\right),$$

and $\mathbf{a}$ represents various permutations or products of the anisotropy tensor. Although $S_{ij} + \Omega_{ij} = \frac{\partial U_i}{\partial x_j}$, the use of the strain and vorticity tensors, as separate entities, is motivated by the symmetry of the former and anti-symmetry of the latter, which are desirable properties in the mathematical manipulations leading to the model. Moreover, the strain and vorticity tensors are physically meaningful variables, representing distinct processes.

A key *optional* restriction imposed on f_{ij} is that it should only include terms that are linear in the mean-velocity gradients (i.e strain and vorticity tensors). A second restriction that *must* be satisfied is that f_{ij} has to be insensitive to a change in the (Cartesian) coordinate system or its orientation — i.e. it must be system invariant. A tensor which has this property is called an 'isotropic tensor', and there are established theorems (Smith (1971)) that govern what representations of $f_{ij}(\mathbf{a}, \mathbf{S}, \mathbf{\Omega})$ are consequently permitted. The formal application of these theorems and the imposition of linearity in the strains and zero-trace of f_{ij} lead to the following general form:

$$
\begin{aligned}
f_{ij} &= \beta_1 a_{ij} + \beta_2 \left(a_{ik} a_{kj} - \frac{1}{3} a_{mn} a_{mn} \delta_{ij} \right) \\
&\quad + \beta_3 \frac{k}{\varepsilon} S_{ij} + \beta_4 \frac{k}{\varepsilon} \left(a_{ik} S_{jk} + a_{jk} S_{ik} - \frac{2}{3} a_{mn} S_{mn} \delta_{ij} \right) \\
&\quad + \beta_5 \frac{k}{\varepsilon} \left(a_{ik} a_{kl} S_{jl} + a_{jk} a_{kl} S_{il} - \frac{2}{3} a_{lm} a_{mn} S_{nl} \delta_{ij} \right) \\
&\quad + \beta_6 \frac{k}{\varepsilon} (a_{ik} \Omega_{jk} + a_{jk} \Omega_{ik}) + \beta_7 \frac{k}{\varepsilon} (a_{ik} a_{kl} \Omega_{jl} + a_{jk} a_{kl} \Omega_{il}),
\end{aligned}
\tag{12.65}
$$

where the coefficients turn out to be functions of the anisotropy invariants A_2 and A_3.

Note the following:

- The first line is closely related to Eq. (12.44), and thus may be regarded as an approximation to $\Phi_{ij,1}$ — although not treated as such in the present general framework.
- If $\beta_5 = \beta_7 = 0$, Eq. (12.65), with the first line excluded, is analogous to Launder, Reece and Rodi's (1975) linear model for $\Phi_{ij,2}$, based on the first two lines of the expansion in Eq. (12.51), provided the corresponding coefficients are constant (which they are not in Eq. (12.65)).

The task is now to determine the coefficients — one that is made more taxing by the fact that the coefficients are functions of the anisotropy invariants, as well as the turbulence-energy-production rate. Thus, strictly, the system is also non-linear in the strains, because the production rate involves the strains.

To simplify the task, Speziale *et al.* have examined the implications of applying the stress-transport equations, with Eq. (12.65) inserted, to homogeneous two-dimensional shear flow, represented by:

$$\frac{\partial U_i}{\partial x_j} = \begin{Bmatrix} 0 & S+\Omega \\ S-\Omega & 0 \end{Bmatrix}, \tag{12.66}$$

with constant value $S = \Omega = 0.5\partial U/\partial y$, and followed the evolution of the stresses and dissipation rate (say, from a state of isotropy) to the state at which the anisotropy components and the dissipation rate (normalised by the turbulence energy) reach respective equilibrium values. At this state, the values of the invariants and the production-to-dissipation ratio $A_{2,\infty}, A_{3,\infty}, (P_k/\varepsilon)_\infty$, respectively, can be determined from the solution, and are found to be universal for any homogeneous shear-strain strength. These values are then assumed as fixed (non-variable) components of the β-coefficients in the model.

A simplification of Eq. (12.65) is offered by the fact that, for fixed invariants $(.)_\infty$ and homogeneous shear, the quadratic fragments associated with β_5 and β_7 may be expressed by the following linear equivalent forms:

$$\begin{aligned} & a_{ik}a_{kl}S_{jl} + a_{jk}a_{kl}S_{il} - \frac{2}{3}a_{lm}a_{mn}S_{nl}\delta_{ij} \\ & \qquad = -a_{33}\left(a_{ik}S_{jk} + a_{jk}S_{ik} - \frac{2}{3}a_{mn}S_{mn}\delta_{ij}\right) - \frac{2}{3}\frac{A_{3,\infty}}{3a_{33}}S_{ij} \\ & a_{ik}a_{kl}\Omega_{jl} + a_{jk}a_{kl}\Omega_{il} = -a_{33}(a_{ik}\Omega_{jk} + a_{jk}\Omega_{ik}). \end{aligned} \tag{12.67}$$

This finally allows Eqs. (12.64) and (12.65) to be written as follows:

$$\begin{aligned} \Phi_{ij} = & -\varepsilon C_1 a_{ij} + \varepsilon C_2\left(a_{ik}a_{kj} - \frac{1}{3}a_{mn}a_{mn}\delta_{ij}\right) \\ & + C_3 k S_{ij} + C_4 k\left(a_{ik}S_{jk} + a_{jk}S_{ik} - \frac{2}{3}a_{mn}S_{mn}\delta_{ij}\right) \\ & + C_5 k(a_{ik}\Omega_{jk} + a_{jk}\Omega_{ik}). \end{aligned} \tag{12.68}$$

The first two fragments may thus be associated with the *slow* process, $\Phi_{ij,1}$, and the remaining fragments with the *rapid* process, $\Phi_{ij,2}$. It is emphasized, again, that the above formalism is, so far, dictated by the conditions prevailing in homogeneous two-dimensional shear.

Speziale *et al.* now introduce two, essentially heuristic, modifications:

$$\begin{aligned} C_3 &\leftarrow C_3 - C_3^* A_2^{\frac{1}{2}} \\ C_1 &\leftarrow C_1 - C_1^* \frac{P_k}{\varepsilon}. \end{aligned} \tag{12.69}$$

The former is based on the observation that the second and third strain-related fragments in the second line of Eq. (12.68) are of order $\mathbf{a}$, while the first, $C_3 k S_{ij}$, is of order '1'. This means that the term $C_3 k S_{ij}$ would vanish if the *rapid* pressure-strain part were to be expanded around the state of isotropy. Speziale *et al.* argue that the addition of the correction C_3^* avoids this inconsistency and leads to the term remaining effective, and of the correct order, in low-anisotropy states. The second replacement is based on the observation that the *slow* part is of order $\mathbf{a}^2$ and that consistency of order would be re-established by introducing P_k/ε, of order $\mathbf{a}$.

Next, the coefficients are calibrated so as to make the model perform acceptably well over a wider range of conditions. This is done by reference to known theoretical constraints and experimental observations for a range of homogeneous flows, including:

- the response of initially isotropic turbulence to rapid distortions in different types of homogenous strain, as determined from RDT;
- experimental observations on the equilibrium conditions in homogeneous shear;
- experimental observations on the decay of isotropic turbulence and the return to isotropy, starting from equilibrium conditions in homogeneous straining (e.g. plane strain, contraction strain).

Based on efforts to match the model to experimental conditions for decaying anisotropic turbulence, starting from equilibrium conditions in plane straining and then removing the strain, so that the rapid terms vanish, the constants associated with the slow terms emerge as:

$$\begin{aligned} C_1 &= 1.7 \\ C_2 &= \frac{3}{4}(2C_1 - 2) = 1.05. \end{aligned} \tag{12.70}$$

Next, Crow's limiting result, Eq. (12.57), is used to determine C_3, which is the only term remaining in rapid distortion in the isotropic limit. This immediately leads to:

$$C_3 = 0.8. \tag{12.71}$$

The remaining coefficients were determined by way of 'numerical optimisation' by reference to experimental equilibrium values for the stresses in homogeneous shear, among others. Thus, numerical calculations were performed with relevant fragments of the model comprising Eqs. (12.68) and (12.69), in conjunction with a standard form of the transport equation for the dissipation rate, yielding:

$$\begin{aligned} C_4 &= 0.62, \quad C_5 = 0.2, \\ C_1^* &= 0.9, \quad C_3^* = 0.65. \end{aligned} \tag{12.72}$$

A single example illustrating the model's performance is given in Fig. 12.21. This figure, taken from Speziale *et al.* (1991), shows the evolution of $b_{ij} \equiv 0.5a_{ij}$ for the case of two-dimensional plane strain, a process indicated by the insert in the figure and being drastically different from homogeneous shear upon which the model is founded. The computation starts with isotropic turbulence and progresses in time $t^* \equiv St$ with a particular choice of initial dissipation rate $\varepsilon_0 = 2Sk_0$. The results of Speziale *et al.*'s and Launder *et al.*'s models (the later discussed in Section 12.7.3) are compared to DNS simulations performed by Lee and Reynolds (1985).

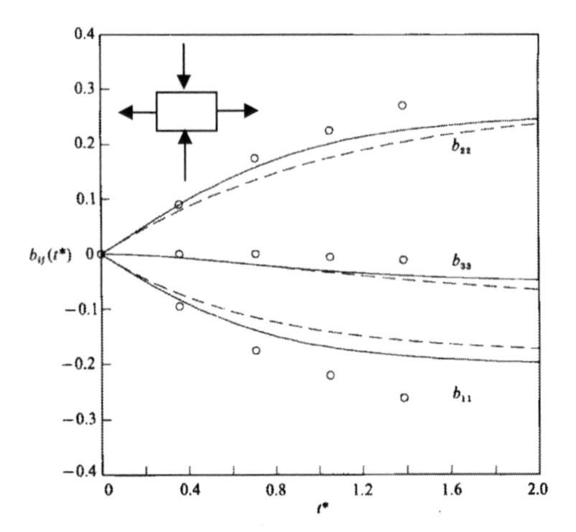

Fig. 12.21: Prediction of anisotropy evolution in homogeneous plane strain, starting from isotropic turbulence. Symbols: DNS data; solid lines Speziale, Sarkar and Gatski; dashed lines Launder, Reece and Rodi. (reproduced with permission of Cambridge University Press from Speziale, Sarkar and Gatski (1991)).

12.7.5 *Near-wall effects*

We start by recalling three facts about near-wall turbulence, based on earlier considerations in Chapter 5, as well as in this chapter:

(i) Very close to the wall, there is a viscous sublayer in which the viscosity tends to damp turbulence activity. All stresses must asymptote to zero, because of the no-slip condition.

(ii) Outside the viscous sublayer, the wall affects turbulence by inviscid mechanisms: wall-normal fluctuations are damped to a much larger extent than the wall-parallel components, because of the blocking action of the wall. In other words, different stresses asymptote to zero at different rates.

(iii) At the wall itself, all components vanish, but the ratios $\frac{\overline{u_1^2}}{k}$ and $\frac{\overline{u_3^2}}{k}$ do not, a condition termed 'two-component turbulence' (see Fig. 5.3(b)).

The basic rationale of using second-moment closure is to allow the stress anisotropy, and thus its effects on the shear-stress components, to be correctly captured. We have already observed that the models covered above do so, in principle, although the quantitative statement greatly depends on the model for the pressure-strain process. Unless the basic modelling route for the pressure-strain term includes a calibration for near-wall flows — as is the case, for example, with the model of Jakirlić and Hanjalić (Eqs. (12.48) and (12.63)) — we do not expect models that do not explicitly account for wall-proximity effects to represent correctly the behaviour very close to the wall, if only because we have not used the viscosity (or, equivalently, the Reynolds number). But do the models represent the inviscid blocking effect? The answer is *no*, in principle, and the main reason is that we have, so far, ignored an important contribution to Eq. (12.40), namely that which represents the surface integral along the boundary of the volume over which the volume integration is performed (Fig. 12.18).

In the two subsections to follow we consider how the inviscid mechanisms and the effects of viscosity are taken into account in modelling the pressure-strain process.

(i) *Inviscid wall effects*

The effect of walls on the redistribution process can be represented, in principle, by taking into account the domain boundary at which the turbulent-velocity fluctuations vanish (see Fig. 12.18) in the integration in Eq. (12.40). This has been done by Hanjalić (1984)[14] who gives the wall-related integral over the surface A as:

$$\Phi_{ij,w} = \frac{1}{4\pi}\int_A \left[\frac{1}{r}\frac{\partial}{\partial n'}\overline{p'\left(\frac{\partial u_i}{\partial x_j}+\frac{\partial u_j}{\partial x_i}\right)} - \overline{p'\left(\frac{\partial u_i}{\partial x_j}+\frac{\partial u_j}{\partial x_i}\right)}\frac{\partial}{\partial n'}\left(\frac{1}{r}\right)\right] dA, \tag{12.73}$$

where n' is the surface-normal direction. This contribution arises from the fact that the pressure fluctuations do not vanish at the surface, and neither do pressure-strain correlations. Physically, this

[14]However, considerations of boundary-related modifications of Eq. (12.40) go back to the early 1970s, as discussed in Launder, Reece and Rodi (1975).

represents a process wherein eddies impinge (or 'splat') on the surface, leading to stagnation-induced pressure peaks at the surface and thus to preferential velocity fluctuations parallel to the surface, due to mass-conservation constraints.

The need to account for the inviscid effects of walls becomes evident upon returning to the anisotropy predicted by the model in Eq. (12.58) for an equilibrium shear layer, of which the near-wall log-layer is an example:

$$\left(\frac{\overline{u_1^2}}{k}-\frac{2}{3}\right)=0.3;\quad \left(\frac{\overline{u_2^2}}{k}-\frac{2}{3}\right)=-0.15;\quad \left(\frac{\overline{u_3^2}}{k}-\frac{2}{3}\right)=-0.15. \tag{12.74}$$

However, experiments in near-wall turbulence (outside the viscous sublayer!) suggest values of order 0.55, −0.45 and −0.11, i.e. a much higher level of anisotropy than predicted by the model, with the wall-normal stress strongly depressed. In contrast, the somewhat more elaborate model of Launder *et al.* (1975), Eq. (12.60), gives the values 0.26, −0.2 and −0.06, thus also poor in wall-proximate layers.

One approach to modelling the effects of the terms in Eq. (12.73) is to include an additive term of the same form as $\Phi_{ij,1}$ and $\Phi_{ij,2}$, with different coefficients, and weighted by a function that represents the proximity to the wall. Thus, in view of Eqs. (12.46) and (12.50), we may propose:

$$\Phi_{ij,w}=\left\{\left[-C_1^w\varepsilon a_{ij}-C_1^{'w}\varepsilon\left(a_{ik}a_{kj}-\frac{1}{3}\delta_{ij}\{a_{kl}a_{lk}\}\right)\right]\right.$$
$$\left.+\,C_2^w\frac{\partial U_k}{\partial x_l}\left(B_{kj}^{li}+B_{ki}^{lj}\right)\right\}f\left(\frac{k^{\frac{3}{2}}}{\varepsilon n}\right), \tag{12.75}$$

where n is the distance normal to the wall and $k^{3/2}/\varepsilon$ is the turbulent macro-length scale, representative of the energetic eddy size. In effect, this form reflects the recognition that both the slow and rapid terms need to be modified, so as to secure the correct near-wall behaviour of the stresses — in particular, $\Phi_{nn}\to 0$ as $\overline{u_n^2}\to 0$ (no summation on n). An advantage of this approach is that the redistributive character of Φ_{ij} carries over, by default, to $\Phi_{ij,w}$. A disadvantage is that we require knowledge of the wall distance n.

An alternative approach, suggested by Eqs. (12.47) and (12.51) or (12.64),[15] is to correct Φ_{ij} itself, by sensitising the coefficients of the model fragments to anisotropy invariants, so as to secure the correct near-wall behaviour. This is the route taken, for example, by Jakirlić and Hanjalić (1995), within the linear-modelling framework, wherein Eqs. (12.48) and (12.63) are observed to be functions of A_2, A and the dissipation-rate invariant E. The main advantage of this route is that no geometric information is needed.

The proposal in Eq. (12.75) is the foundation upon which Launder *et al.* derived a correction to the model in Eq. (12.60), in conjunction with the linear model for the slow term $\Phi_{ij,1}$, Eq. (12.45). Their correction is:

$$\Phi_{ij,w} = \left\{0.125\frac{\varepsilon}{k}\left(\overline{u_i u_j} - \frac{2}{3}k\delta_{ij}\right) + 0.015(P_{ij} - D_{ij})\right\}\frac{k^{\frac{3}{2}}}{\varepsilon n}, \tag{12.76}$$

in which the two fragments are corrections relating, respectively, to $\Phi_{ij,1}$ and $\Phi_{ij,2}$. The coefficients are determined in order to return the requisite anisotropy in the near-wall layer $\{a_{11}, a_{22}, a_{33}\} = \{0.51, -0.42, -0.09\}$, so as to accord with experimental data, in contrast to the uncorrected values $\{a_{11}, a_{22}, a_{33}\} = \{0.26, -0.2, -0.06\}$.[16] Moreover, the corrected model gives the shear stress as $a_{12} = 0.24$, which is slightly too low in near-wall shear.

A restriction of Eq. (12.76) is that it is only applicable to a single wall that is normal to direction x_2. Shir (1973) proposed an intuitive correction to $\Phi_{ij,1}$ that is applicable to a wall at any orientation, followed later by an analogous proposal by Gibson and Launder (1978) for $\Phi_{ij,2}$, both implemented in conjunction with the linear models comprising Eqs. (12.45) and (12.54). Both corrections use a

---

[15]Note that the coefficients in both Eqs. (12.51) and (12.64) are functions of anisotropy invariants.

[16]As an aside, note that the spanwise component is hardly affected by the wall. The redistribution process is modified mainly by the streamwise anisotropy component being enhanced at the expense of the wall-normal component.

unit normal-to-wall vector $n_k = \{n_1, n_2, n_3\}$. Hence, in the case of a wall normal to x_2, $n_k = \{0, 1, 0\}$. Shir's correction is as follows:

$$\Phi_{ij,w,1} = C_{w1} f_w \frac{\varepsilon}{k}\left(\overline{u_k u_m} n_k n_m \delta_{ij} - \frac{3}{2}\overline{u_i u_k} n_k n_j - \frac{3}{2}\overline{u_k u_j} n_k n_i\right), \tag{12.77}$$

where $C_{w,1} = 0.5$, $f_w = \frac{k^{3/2}}{2.5\varepsilon n}$ is a wall-distance function, n being the distance normal to the wall in question. The factor 2.5 arises from $\kappa C_\mu^{-3/4}$, thus making the turbulent length scale equal to the mixing length $\ell = \kappa n$ (see Eq. (7.47)).

To gain insight into the effects of the correction in Eq. (12.77), we consider the example of a vertical wall, $n_k = \{1, 0, 0\}$, in which case:

$$\Phi_{ij,w,1} = C_{w1} f_w \frac{\varepsilon}{k}\left(\overline{u_1^2}\delta_{ij} - \frac{3}{2}\overline{u_i u_1} n_j - \frac{3}{2}\overline{u_1 u_j} n_i\right). \tag{12.78}$$

Expanding the individual components gives:

$$\begin{aligned} \Phi_{11,w,1} &= -2C_{w1} f_w \frac{\varepsilon}{k}\overline{u_1^2}; \quad \Phi_{22,w,1} = C_{w1} f_w \frac{\varepsilon}{k}\overline{u_1^2}; \\ \Phi_{33,w,1} &= C_{w1} f_w \frac{\varepsilon}{k}\overline{u_1^2} \quad \Phi_{12,w,1} = -C_{w1} f_w \frac{3}{2}\frac{\varepsilon}{k}\overline{u_1 u_2}, \end{aligned} \tag{12.79}$$

we observe the following:

- The correction is a sink term in the $\overline{u_1^2}$ -equation. It depresses this wall-normal stress in proportion to itself, and it thus acts against the isotropisation process.
- $\Phi_{11,w,1} \to 0$ as $\overline{u_1^2} \to 0$.
- The correction is redistributive: the reduction in $\overline{u_1^2}$ is matched by an increase in $\overline{u_2^2}$ and $\overline{u_3^2}$.
- The effect on the wall-parallel components is symmetric, but this is not in line with experimental evidence, which suggests that the spanwise normal stress (in the direction in which there is no shear straining) is hardly affected by the proximity of the wall.
- The correction depresses the shear stress.

- Within an equilibrium near-wall layer, $f_w = 1$, but at locations beyond this layer, away from the wall, f_w declines, and so does the correction.

Gibson and Launder's correction of $\Phi_{ij,w,2}$ follows by direct analogy with the model in Eq. (12.77):

$$\Phi_{ij,w,2} = C_{w2} f_w \frac{\varepsilon}{k} \left(\Phi_{km,2} n_k n_m \delta_{ij} - \frac{3}{2} \Phi_{ik,2} n_k n_j - \frac{3}{2} \Phi_{kj,2} n_k n_i \right), \tag{12.80}$$

where $C_{w,2} = 0.3$. This form follows rationally upon comparing the models in Eqs. (12.54) and (12.77), and upon noting that $\Phi_{ij,2}$ is simply a linear function of $P_{ij} - \frac{2}{3}\delta_{ij} P_k$.

A more complex correction has been proposed by Craft and Launder (1996) and Craft (1998) in an effort to counter the tendency of the model in Eq. (12.80) to overestimate the shutting down of the isotroprisation process in flows impinging normally on a wall. However, this is used in conjunction with a cubic model for $\Phi_{ij,2}$, and it is therefore not considered further herein.

(ii) *Viscous wall effects*

The wall-related modifications considered in the previous section do not take into account any damping exerted by the viscosity very close to the wall.[17] We assume that such explicit corrections are needed, but is this assumption justified? The answer is not clear-cut. We note that Eq. (12.73), which represents the effects of boundaries on the pressure-velocity correlation, does not contain the viscosity. Hence, theoretically, there should be no need for viscous corrections, and this argument provides a justification for the absence of any such corrections in the model of Hanjalić & Launder (1976). On the other hand, there must clearly be differences between how the set of stress components behave as a wall and, say, a shear-free water-air interface are

[17]The model by Jakirlić and Hanjalić (1995) is an exception, in so far as the coefficient (12.48) is a function of the turbulent Reynolds number. This reflects a calibration process that does not explicitly distinguishes between inviscid and viscous effects on the near-wall behaviour.

approached: in the former, all fluctuations vanish due to the no-slip condition, associated with the viscosity, while in the latter only the interface-normal component vanishes (at least theoretically). Also, from a pragmatic point of view, the need to achieve agreement with the broadest range of experimental observations and DNS data has led to the use of viscosity-containing corrections in several models, that of Jakirlić and Hanjalić being one (see Eq. (12.48)).

Consistent with earlier near-wall-modelling efforts, in the context of two-equation models, an examination of the exact asymptotic variation of the pressure-velocity-interaction terms is the appropriate starting point. In Sections 5.1 and 12.6, we have already considered the question of how the stresses and the dissipation-rate components approach the wall, and we have shown that this requires Taylor-series expansions for the velocity fluctuations to be used (Eqs. (5.1)–(5.4), Fig. 5.2, Eqs. (12.25) and (12.26)). This approach can be extended to all terms requiring modelling in Eq. (12.7).

Consider the pressure-velocity-interaction term in the Reynolds-stress transport equation (12.7):

$$\Pi_{ij} = -\frac{1}{\rho}\left(\overline{u_i\frac{\partial p}{\partial x_j}} + \overline{u_j\frac{\partial p}{\partial x_i}}\right). \tag{12.81}$$

Inserting

$$\begin{aligned} u_i &= a_i + b_i y + c_i y^2 + HOT, \\ p &= a_p + b_p y + c_p y^2 + HOT, \end{aligned} \tag{12.82}$$

into (12.81), noting the constraints introduced in relation to Eq. (5.1), and time-averaging, we obtain the following asymptotic behaviour:

$$\Pi_{ij}\big|_{y\to 0} = -\frac{1}{\rho}\left\{\underbrace{2\overline{b_1\frac{\partial a_p}{\partial x}}y}_{ij=11}, \underbrace{2\overline{c_2 b_p}y^2}_{22}, \underbrace{2\overline{b_3\frac{\partial a_p}{\partial z}}y}_{33}, \underbrace{\overline{b_1 b_p}y}_{12}\right\}. \tag{12.83}$$

The main message derived from this expression is that all components vanish at the wall, with Π_{22} decaying quadratically and the others linearly. It can further be shown that the asymptotic variation of the pressure-strain components is as follows:

$$\Phi_{ij}\big|_{y\to 0} = \frac{1}{\rho}\left\{\underbrace{2\overline{a_p\frac{\partial b_1}{\partial x}}y}_{ij=11}, \underbrace{4\overline{c_2 a_p}y}_{22}, \underbrace{2\overline{a_p\frac{\partial b_3}{\partial z}}y}_{33}, \underbrace{\overline{a_p b_1}}_{12}\right\}. \quad (12.84)$$

An important fact to highlight here is that the asymptotic behaviour of Φ_{22} and Φ_{12} differs drastically from that of Π_{22} and Π_{12}, respectively. In particular, the fact that Φ_{12} is $O(1)$ means that the shear-stress, which decays cubically towards the wall, continues to be driven by a source or sink that is finite right down to the wall. The drastic differences are due to the asymptotic variations of the pressure-diffusion terms:

$$\{D_{22}^p; D_{12}^p\}\big|_{y\to 0} = \{-4\overline{a_p c_2}y - 4\overline{c_2 c_2}y^2; \quad -\overline{a_p b_1} - 2\overline{b_1 c_2}y\}. \quad (12.85)$$

We observe, in particular, that the leading terms in both components of Eq. (12.85) are identical to the corresponding terms in Eq. (12.84), but with opposite signs. In other words, the linear components of the pairs $\{\Phi_{22}, D_{22}^p\}$ and $\{\Phi_{12}, D_{12}^p\}$ diverge as the wall is approached, so as to meet the higher-order asymptotic terms in Π_{22} and Π_{12}. Hence the pressure-diffusion terms play an important role close to the wall.

Figure 12.22 shows near-wall budgets for $\overline{v^2}$ and $\overline{uv}$ is a separated flow in a channel with hill-shaped constrictions. These provide striking confirmation of the behaviour inferred from the foregoing asymptotic analysis. Both imply that it would be futile to try to reproduce the asymptotic variations of Φ_{ij} in models in which D_{ij}^p is assumed to be zero or part of the turbulent-diffusion model. Rather, it seems more reasonable to construct a near-wall approximation that follows the asymptotic variation of Π_{ij}. A fundamental problem is, however, that Π_{ij} does not satisfy the re-distributive properties of Φ_{ij}. As most models were formulated prior to the availability of detailed DNS data, when measurements of the pressure-velocity correlations were impossible (and are still very difficult at the time of writing), there was little awareness of the issues highlighted by

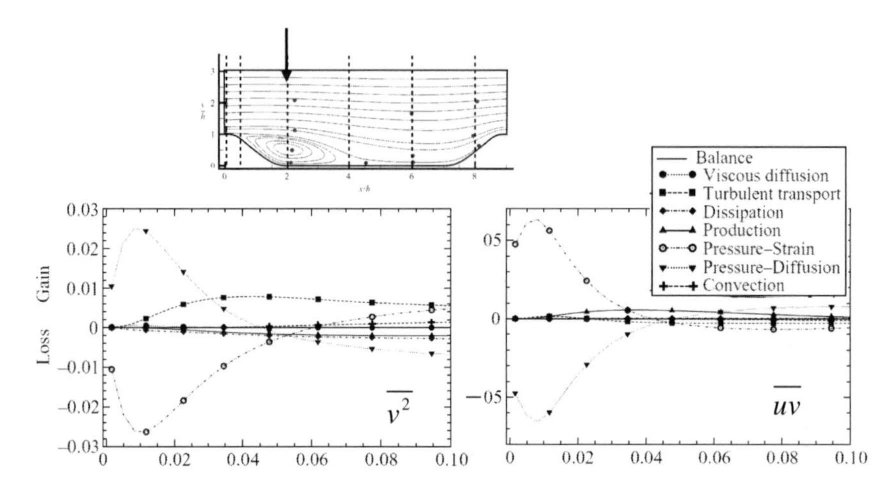

Fig. 12.22: Near-wall budgets of the wall-normal and shear-stress components in a separated channel flow at streamwise location '2' in the inset (reproduced with permission from Cambridge University Press from Froehlich *et al.* (2005)).

Eqs. (12.83)–(12.85) and Fig. 12.22. Hence, any viscous corrections were introduced, pragmatically, into Φ_{ij} in an effort to secure the correct asymptotic variation of Φ_{ij} towards zero at the wall, as implied by Π_{ij} in Eq. (12.83). This was followed by a calibration of the corrections, so as to optimise the near-wall predictive performance of the model as a whole.

A flavour of viscous corrections to Φ_{ij} is given below by reference to two examples, namely the models of Launder and Shima (1989) and Prud'homme & Elghobashi (1983). A broader review of models proposed before 1991 is given by So *et al.* (1991), and this conveys the fact that several models are closely allied variations, adopting elements from other models and augmenting these with additional corrections, modifications or adjustments. The review illustrates the fact that most models do not return the near-wall behaviour especially well — the two models selected here being close to the extreme ends of the performance range.

A difficulty with presenting the above corrections in a transparent manner is that they often go hand-in-hand with further modifications to other model elements, especially to the dissipation-rate equation. As noted earlier, this reflects efforts by the modellers concerned to 'optimize' the near-wall performance of the model as a whole. Also, because the anisotropy of the dissipation rate can be argued to be rationally linked to the pressure-strain process (Lumley (1978), Shima (1988)), some viscous corrections do not simply apply to Φ_{ij}, but to $\Phi_{ij} - (\varepsilon_{ij} - \frac{2}{3}\varepsilon\delta_{ij})$, so that corrections to the dissipation-rate equation impinge directly on the pressure-strain term.

Launder and Shima's model uses Launder *et al.*'s (1975) basic approximations in Eqs. (12.45) and (12.60) with the wall corrections Eqs. (12.77) and (12.80), but with different coefficients, namely:

$$\begin{aligned} C_1 &= 1 + 2.58AA_2^{\frac{1}{4}}\left\{1 - \exp\left[-(0.0067R_T)^2\right]\right\}, \\ C_2 &= 0.75A^{\frac{1}{2}}, \\ C_{w1} &= -\frac{2}{3}C_1 + 1.67, \\ C_{w2} &= \max\left[\left(\frac{2}{3}C_2 - \frac{1}{6}\right)\Big/C_2, 0\right]. \end{aligned} \tag{12.86}$$

These are supplemented by a correction term in the dissipation-rate equation, which also depends on A, A_2 and R_T. Hence, these modifications attempt to procure the correct near-wall behaviour (in channel flow), especially of the normal-stress components, by a computational optimisation (essentially, by trial and error), involving the anisotropy invariants and the Reynolds number.

The foundation of Prud'homme & Elghobashi's model (1983) is, again, Launder *et al.*'s (1975) approximations, Eqs. (12.45) and (12.60), to which is added the wall fragment:

$$\Phi_{ij,w} = C_3(P_{ij} - D_{ij})\exp\left(-\frac{C_4k^{\frac{1}{2}}y}{\nu}\right), \tag{12.87}$$

which is clearly associated with the second term in the correction in Eq. (12.76). This is augmented with a viscosity-related variation of the model in Eq. (12.28) for the dissipation anisotropy, and a viscous damping function multiplying the turbulent-transport term. However, as shown in Fig. 12.23, the result is far less satisfactory than that returned by the model of Launder & Shima. This result thus illustrates the general conclusion that the use of the anisotropy invariants A and A_2 is very effective in correctly capturing the near-wall variations of the stresses.

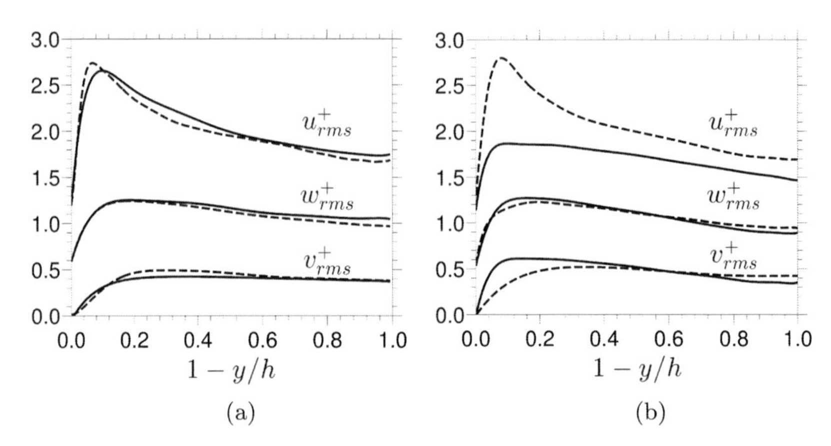

Fig. 12.23: Anisotropy of turbulence-intensity components in channel flow predicted by (a) the Launder & Shima model (1983) and (b) the Prud'homme & Elghobashi model (1983) (reproduced with permission of AIAA from So *et al.* (1991)).

12.7.6 *Effects of body forces*

An issue we have to finally consider, under the heading 'pressure-velocity interaction', is the effect of body forces on this process. We recall that this was explicitly excluded from Eq. (12.40), reflecting the omission of the body-force term F_i from the NS equations[18] when deriving Eq. (12.40). We recall, further, that the primary consequence of this body-force term is the appearance related generation terms G_{ij} in the Reynolds-stress equations (see Eq. (12.7)). If, in addition, the flow is subjected to system rotation, Ω_k, the additional body-force generation terms R_{ij} arises.

Within the framework of linear modelling of the pressure-strain process, the most straightforward approach (Launder (1975)) is to add to Eq. (12.54) fragments of the form:

$$\Phi_{ij,3} = -C_{2G}\left(G_{ij} - \frac{1}{3}\delta_{ij}G_{kk}\right) - C_{2R}\left(R_{ij} - \frac{1}{3}\delta_{ij}R_{kk}\right), \quad (12.88)$$

but this is simply based on the assumption that the effects of body-force generation can be modelled by analogy to strain-induced

[18]In the case of buoyancy-related body forces, for example, we have $F_i = -g_i\beta\rho\Delta T$.

generation. It also implies that the heat fluxes are available from, say, a separate flux-transport model of the type discussed in Chapter 10. More elaborate modelling has been pursued within the framework of non-linear approximations to the pressure-strain process, analogous to Eqs. (12.50) and (12.51) (e.g. Launder and Tselepidakis (1993) and Craft and Launder (1996)), but their discussion would go beyond the framework of this text.

12.7.7 *Elliptic relaxation of pressure-strain correlation*

A substantial proportion of our efforts in this chapter towards closing the Reynolds-stress-transport equations was devoted to the question of how to model the pressure-velocity interaction $\Pi_{ij} = -\frac{1}{\rho}\left(\overline{u_i \frac{\partial p}{\partial x_j}} + \overline{u_j \frac{\partial p}{\partial x_i}}\right)$. We recall that this term is exceptionally challenging, in so far as Eq. (12.40) reveals it to involve two-point correlations of the form $\overline{f(x_i)g(x_i')}$. This reflects the fact that pressure fluctuations propagate in all directions, and therefore link $\Phi_{ij}(x_i)$ to processes in all parts of the domain V over which the integration is performed in Eq. (12.40).

In closing this term — or rather the pressure-strain part Φ_{ij} in Π_{ij},

$$\underbrace{\overline{\frac{p}{\rho}\left(\frac{\partial u_i}{\partial x_j} + \frac{\partial u_j}{\partial x_i}\right)}}_{\Phi_{ij}} = \Pi_{ij} + \underbrace{\frac{\partial}{\partial x_k}\left(\overline{\frac{p}{\rho}(u_i \delta_{jk} + u_j \delta_{ik})}\right)}_{D_{ij}^p}, \qquad (12.89)$$

we adopted, in contravention to the non-locality of the process, various *local* (point) approximations of the form of Eqs. (12.44) and (12.51). Our hope was that the inclusion of higher-order terms would go some way towards compensating for the omission to account, explicitly, for the influence of processes at (x_i') on those at (x_i), and also for the assumption of homogeneity in the mean strain. We found that this was an inadequate approach to modelling near-wall effects, and this forced us to adopt, in Section 12.7.5, corrections that account for the influential reflection of pressure fluctuations from adjacent walls on Φ_{ij}.

Recognising the weaknesses of ignoring the non-locality of Φ_{ij}, Durbin (1991) proposed an alternative modelling framework that gives expression to the importance of approximating the two-point corrections in Φ_{ij}, the primary practical target being the avoidance of the need for *ad hoc* correction terms that account for pressure-reflection effects at walls.

The key proposition underpinning Durbin's model is that any two-point correlation in Eq. (12.40) may be approximated by:

$$\overline{f(x_i)g(x_i')} = \overline{f(x_i')g(x_i')} exp\left(-\frac{|x_i - x_i'|}{L}\right), \tag{12.90}$$

The essential logic of this approximation should be transparent: the correlation between events at (x_i') and (x_i) diminishes exponentially with the decline dictated by the length scale L, which governs the region (radius) within which the functions f and g are statistically correlated.

When the approximation in Eq. (12.90) is inserted into Eq. (12.40), and the terms in the outer square brackets of the 'slow' and 'rapid' contributions (involving velocity fluctuations and mean strains) are lumped into A_{ij} and B_{ij}, respectively, Eq. (12.40) can be written:

$$\Phi_{ij} = \int_V (A_{ij}' + B_{ij}') \left\{ \frac{e^{-|x_i - x_i'|/L}}{4\pi |x_i - x_i'|/L} \right\} \frac{1}{L} dV(x_i'), \tag{12.91}$$

Reference to the theory underpinning Green's Function shows that the expression in the curly bracket is Green's Function corresponding to the differential operator $(1/L^2 - \nabla^2)$. In other words,

$$(1/L^2 - \nabla^2)\Phi_{ij} = (A_{ij}' + B_{ij}')$$

or

$$\Phi_{ij} - L^2\nabla^2\Phi_{ij} = L^2(A_{ij}' + B_{ij}'). \tag{12.92}$$

If we imagine that the region of two-point correlation is vanishingly small, i.e. $L \to 0$, it follows that $\Phi_{ij} \to L^2(A_{ij}' + B_{ij}')$, which demonstrates that the right-hand side must be the pressure-strain term evaluated at the point $x_i = x'_i$. In fact, this represents any of the one-point models we considered in Section 12.7, but excluding any inhomogeneity corrections — wall-related corrections, in particular. We denote this by Φ^h_{ij} ('homogeneous' pressure-strain term), i.e.

$$\Phi_{ij} - L^2 \nabla^2 \Phi_{ij} = \Phi^h_{ij}. \tag{12.93}$$

One way of interpreting this equation, by reference to a familiar physical analogy, is to consider it as a *conduction* equation (a Poisson equation):

$$L^2 \nabla^2 \Phi_{ij} = (\Phi_{ij} - \Phi^h_{ij}), \tag{12.94}$$

with *conductivity* L^2 and source $(\Phi^h_{ij} - \Phi_{ij})$. Hence, the field Φ_{ij} is locally *driven* by the homogeneous term (Φ^h_{ij}), while the *conduction* term sensitises the local Φ_{ij} to the field values at neighbouring locations. Importantly, this includes conductive links to the boundary condition for Φ_{ij} at surrounding walls. The source thus tends to drive Φ_{ij} towards a weighted average of surrounding values and the value Φ^h_{ij}. This is a *relaxation* process, and the elliptic nature of the operator on the left-hand side of Eq. (12.94) is the reason for the above reference to 'elliptic relaxation'.

In most circumstances in which the flow is remote from walls, the departure from homogeneity is relatively small; hence, $\Phi_{ij} \approx \Phi^h_{ij}$. However, as a wall is approached, intense pressure reflections cause steep variations in the pressure-strain term, and its local value becomes very sensitive to the values at neighbouring locations, especially in the wall-normal direction. Hence, Eq. (12.93) is a means of sensitising Φ^h_{ij} to the proximity of a wall without the use of corrections of the type we considered in Section 12.7.5.

There are two important drawbacks to the use of the model in Eq. (12.93). First, it substantially increases the computational effort, as the number of partial differential equations to be solved is increased by six. Second, it requires boundary conditions to be specified for Φ_{ij}, and this is far from a trivial matter, because these values are unknown. Rather, what can be derived are the wall-asymptotic variations $\Phi_{ij}(y)|_{y\to 0}$. A third problem is the need to specify L.

The above description is a statement of the principles underpinning Durbin's elliptic relaxation model. In fact, Eq. (12.93) is not the form that is actually solved by Durbin. Rather, a modified equation is derived for a variable of which Φ_{ij} is a part. To appreciate the details, we return to the Reynolds-stress-transport equation, Eq. (12.7):

$$\frac{D\overline{u_i u_j}}{Dt} = P_{ij} + \Phi_{ij} + D^v_{ij} + (D^t_{ij} + D^p_{ij}) - \varepsilon_{ij}. \tag{12.95}$$

As already argued in Section 12.6, it is useful to re-write the dissipation-rate tensor in the form:

$$\varepsilon_{ij} = \left(\varepsilon_{ij} - \frac{2}{3}\delta_{ij}\varepsilon\right) + \frac{2}{3}\delta_{ij}\varepsilon. \tag{12.96}$$

The bracketed term is purely redistributive, as it contracts to zero, while the second term is the simplest closure model for ε_{ij}. An alternative decomposition is:

$$\varepsilon_{ij} = \left(\varepsilon_{ij} - \frac{\overline{u_i u_j}}{k}\varepsilon\right) + \frac{\overline{u_i u_j}}{k}\varepsilon. \tag{12.97}$$

Here again, the bracketed term is redistributive in nature, while the second term is a fairly realistic anisotropic model for ε_{ij}, proposed by Rotta (1951). We have encountered this model already in Section 12.6, Eq. (12.22), and we have shown that, while being far superior to $\varepsilon_{ij} = \frac{2}{3}\delta_{ij}\varepsilon$, it does not return the correct wall-asymptotic variation for ε_{22} and ε_{12}. Using Eq. (12.97), we can now write Eq. (12.95)

$$\frac{D\overline{u_i u_j}}{Dt} = P_{ij} + \underbrace{\left\{\Phi_{ij} - \left(\varepsilon_{ij} - \frac{\overline{u_i u_j}}{k}\varepsilon\right)\right\}}_{\varphi_{ij}} + D^v_{ij} + (D^t_{ij} + D^p_{ij}) - \frac{\overline{u_i u_j}}{k}\varepsilon. \tag{12.98}$$

We denote the group between the curly braces by φ_{ij}. To solve Eq. (12.98), we require a model for φ_{ij} and ε. The fact that we use ε, rather than ε_{ij} means that we do not need to concern ourselves with near-wall corrections for the individual dissipation-rate components that return the correct asymptotic variation for all these components. The key modification of the model in Eq. (12.94) is now that φ_{ij} should the subject of the relaxation equation, rather than Φ_{ij}, i.e.:

$$\varphi_{ij} - L^2\nabla^2\varphi_{ij} = \varphi_{ij}^h, \tag{12.99}$$

where φ_{ij}^h is the homogeneous ('uncorrected') form of φ_{ij}, i.e.:

$$\varphi_{ij}^h = \Phi_{ij}^h - \left(\frac{2}{3}\delta_{ij} - \frac{\overline{u_i u_j}}{k}\right)\varepsilon, \tag{12.100}$$

the last term being a 'homogeneous' model for $\left(\frac{\varepsilon_{ij}}{\varepsilon} - \frac{\overline{u_i u_j}}{k}\right)\varepsilon$.

As regards the boundary condition for φ_{ij}, those at the wall are most critical; far away from walls, the assumption $\varphi_{ij} = \varphi_{ij}^h$ can be expected to be entirely adequate in most circumstances. At walls, we know that φ_{ij} changes rapidly. While we do not know the actual values of φ_{ij}, we can derive statements on the near-wall behaviour from the wall-limiting form of Eq. (12.98):

$$\varphi_{ij} = -D_{ij}^v + \frac{\overline{u_i u_j}}{k}\varepsilon. \tag{12.101}$$

As $D_{ij}^v = \frac{\partial}{\partial x_k}\left(\nu\frac{\partial\overline{u_i u_j}}{\partial x_k}\right)$, it is easy to derive the asymptotic variation of these components, simply by the insertion of:

$$u_i = a_i + b_i y + c_i y^2 + O(y^3), \tag{12.102}$$

(see Eq. (12.82)) into Eq. (12.101) and setting $k = 2$. The result is:

$$D_{ij}^v = \left\{\underbrace{2\nu\overline{b_1^2} + 12\overline{b_1 c_1}y + O(y^2)}_{ij=11}, \underbrace{12\nu\overline{c_2^2}y^2 + O(y^3)}_{22}, \underbrace{2\nu\overline{b_3^2} + 12\overline{b_3 c_3}y + O(y^2)}_{33}, \underbrace{6\nu\overline{b_1 c_2}y + O(y^2)}_{12}\right\}. \tag{12.103}$$

Moreover, we have from Eqs. (12.25) and (12.26):

$$\overline{u_i u_j} = \left\{\overline{b_1^2}y^2 + O(y^3), \overline{c_2^2}y^4 + O(y^5), \overline{b_3^2}y^2 + O(y^3), \overline{b_1 c_2}y^3 + O(y^4)\right\} \tag{12.104}$$

$$k = \frac{1}{2}\left(\overline{b_1^2} + \overline{b_3^2}\right) y^2 + O(y^3), \tag{12.105}$$

$$\varepsilon = \nu\left(\overline{b_1^2} + \overline{b_3^2}\right) + O(y). \tag{12.106}$$

Substitution into Eq. (12.101), with only leading terms retained, gives:

$$\begin{aligned} \varphi_{ij}\big|_{y\to 0} &= -2\nu\left\{\overline{b_1^2} + O(y), 6\overline{c_2^2}y^2, \overline{b_3^2} + O(y), 3\overline{b_1 c_2 y}\right\} \\ &\quad + 2\nu\left\{\overline{b_1^2} + O(y), \overline{c_2^2}y^2, \overline{b_3^2} + O(y), \overline{b_1 c_2 y}\right\} \\ &= \left\{O(y), -10\nu\overline{c_2^2}y^2, O(y), -4\nu\overline{b_1 c_2 y}\right\}. \end{aligned} \tag{12.107}$$

We can now take advantage of the following wall-asymptotic results in Eqs. (12.104)–(12.106):

$$\overline{v^2} \to \overline{c_2^2}y^4, \quad \varepsilon \to 2\nu\frac{k}{y^2}, \quad \overline{uv} \to \overline{b_1 c_2}y^3. \tag{12.108}$$

This allows Eq. (12.107) to be written:

$$\varphi_{ij}\big|_{y\to 0} = \left\{O(y), -10\nu\frac{\overline{v^2}}{y^2}, O(y), -4\nu\frac{\overline{uv}}{y^2}\right\}. \tag{12.109}$$

We are especially interested to ensure the correct near-wall variation of $\overline{v^2}$ and $\overline{uv}$. If we now divide Eq. (12.109) by k, we observe that:

$$\left.\frac{\varphi_{22}}{k}\right|_{y\to 0} = -20\nu^2\frac{\overline{v^2}}{y^4\varepsilon}, \tag{12.110}$$

which is finite, because $\overline{v^2}$ declines with y^4. Hence, Eq. (12.110) is, in effect, a Dirichlet condition on φ_{22}, in the same sense as

$$\varepsilon\big|_{y\to 0} = 2\nu\frac{k}{y^2}, \tag{12.111}$$

is a Dirichlet condition on ε (see Eq. (9.2)). This result led Durbin to argue that the variable that should be the subject of the relaxation equation should be φ_{ij}/k, rather than φ_{ij} itself. This variable is conventionally denoted by f_{ij}, leading to the final relaxation equation:

$$f_{ij} - L^2\nabla^2 f_{ij} = \frac{\Phi_{ij}^h}{k} + \left(\frac{\overline{u_i u_j}}{k} - \frac{2}{3}\delta_{ij}\right)\frac{\varepsilon}{k}. \tag{12.112}$$

Unfortunately, the proposal to solve for f_{ij}, based on Eq. (12.110), poses problems for components other than f_{22}. In particular, $f_{12}|_{y\to 0} = O(1/y)$, because $\overline{uv}$ is expected to decay with $O(y^3)$. To avoid this difficulty, the shear-stress has to be allowed — or forced — to decay in accordance with $O(y^4)$. To understand how this can be done, we return to Eq. (12.103). For $i = 1, j = 2$, and using the last expression in Eq. (12.108), we can write:

$$D_{12}^v = \frac{\partial}{\partial y}\left(\nu\frac{\partial\overline{uv}}{\partial y}\right) = 6\nu\frac{\overline{uv}}{y^2} + O(y^2). \tag{12.113}$$

Self-evidently, this equation is satisfied by $\overline{uv} = Cy^3$, which is fully in accord with our expectation, of course. However, this is precisely a behaviour we wish to avoid. To enforce the variation $\overline{uv} = Cy^4$, equation (12.113) has to be modified to

$$\frac{\partial}{\partial y}\left(\nu\frac{\partial\overline{uv}}{\partial y}\right) = 12\nu\frac{\overline{uv}}{y^2} + O(y^3), \tag{12.114}$$

as is easily verified by substituting the shear stress into Eq. (12.114). Substitution of Eq. (12.114) into Eq. (12.101), using $\varepsilon/k \to 2\nu/y^2$, now leads to the condition:

$$f_{12} = \left.\frac{\varphi_{12}}{k}\right|_{y\to 0} = -10\nu^2\frac{\overline{uv}}{y^4\varepsilon}, \tag{12.115}$$

which approaches a constant value at the wall if $\overline{uv}$ is determined from Eq. (12.114), rather than Eq. (12.113).

A similar problem arises with $f_{11}|_{y\to 0}$ and $f_{33}|_{y\to 0}$. In this case, a pragmatic approach adopted in practice is to use $f_{11} = f_{33} = -0.5f_{22}$, which at least secures the correct contraction property associated with the pressure-strain process.

The final issue to address is how to determine the length scale L. In general, we can expect L to be related to the turbulent length scale, as this identifies the 'size' of the energetic eddies. However, close to the wall, L must also be related to the viscous (Kolmogorov) length scale. This argument led Durbin to propose:

$$L = \max\left\{c_L \frac{k^{\frac{3}{2}}}{\varepsilon}, c_\eta \left(\frac{\nu^3}{\varepsilon}\right)^{\frac{1}{4}}\right\}, \tag{12.116}$$

with $c_L = 0.3$ and $c_\eta = 21$.

The fact that the elliptic-relaxation strategy has to entail the solution of six additional partial differential equations is clearly a powerful disincentive for adopting it in practice, especially in general-purpose codes intended for modelling geometrically and physically complex flows. In addition to the mere increase in the number of equations, there is the significant challenge of dealing with the boundary conditions in Eq. (12.109), which turn out to be numerically destabilizing. The latter issue, in particular, has driven the evolution of a whole range of practices and model variations (some reviewed in Chapter 14), designed to achieve an acceptable compromise between fidelity and 'user-friendliness'. A review of some dozen variations can be found in the PhD thesis of Billard (2011).

As will be explained in Chapter 14, one compromise is to adapt the elliptic-relaxation concept to the eddy-viscosity framework, in which case, the method involves the solution of only two equations over and above those used to determine k and ε: one for the wall-normal Reynolds stress $\overline{v^2}$ and the other for f_{22}, which is associated with that stress.

Another route of compromise, adopted within the framework of second-moment closure, is to abandon the set of equations (12.112), and to replace it by a single equation for a scalar parameter that reflects, albeit only notionally, the elliptic-relaxation idea. This is an approach proposed by Manceau and Hanjalić (2002). The basis of this model is an elliptic equation for a scalar 'blending parameter', α:

$$\alpha - L^2 \nabla^2 \alpha = 1, \tag{12.117}$$

with boundary condition $\alpha = 0$ at walls and $= 1$ along boundaries remote from walls. This parameter is then used to 'blend' the values of φ_{ij}, derived from the homogeneous model φ_{ij}^h, with the wall values of φ_{ij}, i.e. $\varphi_{ij,w}$. With α given, the actual (blended) value of φ_{ij} is then determined from:

$$\varphi_{ij} = \alpha\varphi_{ij}^h + (1-\alpha)\varphi_{ij,w}, \tag{12.118}$$

with the wall values being:

$$\varphi_{22,w}\big|_{y\to 0} = -5\frac{\varepsilon\overline{v^2}}{k}; \quad \varphi_{11} = \varphi_{33} = -0.5\varphi_{22}; \quad \varphi_{12,w}\big|_{y\to 0} = -5\frac{\varepsilon\overline{uv}}{k}. \tag{12.119}$$

The first condition arises upon combining Eq. (12.110) and the second relation in Eq. (12.108), and the last arises from Eq. (12.115). The remaining conditions are compatible with the earlier assumption $f_{11} = f_{33} = -0.5f_{22}$. This blending practice is analogous to that used in Section 12.6 to determine the dissipation rate ε_{ij} by blending the isotropic dissipation and the wall values, according to Eq. (12.28).

Manceau and Hanjalić also provide a more general form of Eq. (12.119), applicable to any wall-orientation:

$$\varphi_{ij,w}\big|_{y\to 0} = -5\frac{\varepsilon}{k}\left[\overline{u_iu_k}n_jn_k + \overline{u_ju_k}n_in_k - \frac{1}{2}\overline{u_ku_l}n_kn_l(n_in_j - \delta_{ij})\right], \tag{12.120}$$

where $n_k = \{n_1, n_2, n_3\}$ is the unit normal-to-wall vector, determined from $n_k = \frac{\nabla\alpha}{|\nabla\alpha|}$ — i.e. the direction of steepest variation of α, thus indicating the wall-normal direction at the point closest to the wall, relative to the point at which φ_{ij} is to be determined.

12.8 *Summary and lessons*

Second-moment closure is clearly a much more elaborate framework than any linear eddy-viscosity model. In terms of mathematical complexity alone, it involves (at least) seven closely coupled, non-linear partial differential equations instead of one or two, and the equations also contain many more terms, correlations and ancillary algebraic relations. The modelling task is also far more challenging. While the stress-strain production terms are retained in their exact mathematical form, their actual magnitude during a full solution process depends critically on the ability of the model as a whole to return quantitatively accurate stress values, and these values depend most sensitively on the approximation of the pressure-velocity correlation, which is the most difficult process to model in a general manner.

A case in point is a near-wall boundary layer, in which the shear stress is dictated, principally, by a delicate balance between its generation term and the pressure-velocity term. This balance is thus dominated by two major elements, namely the wall-normal stress $\overline{v^2}$, to which the production is directly proportional, and the pressure-strain model Φ_{12}. The wall-normal stress is governed, in turn, by a balance between the pressure-strain term Φ_{22}, which feeds energy into $\overline{v^2}$, and the dissipation rate of that stress ε_{22}, which removes energy. Hence, the anisotropy of the dissipation rate contributes greatly, via the normal-stress equations, to the predictive properties of the model, particularly close to the wall, where the stress anisotropy is especially high.

The need to represent correctly the wall-asymptotic variations of the stresses can only be met by a difficult process of optimisation, wherein model coefficients are sensitised to complex combinations of anisotropy invariants and the turbulent Reynolds number. Because this process is non-rigorous, the consequence has been the emergence of a plethora of models, each reflecting a different approach to the task of achieving acceptable predictive capabilities. Hence, second-moment closure should not be viewed as a single entity, but rather as a field of models with significant fuzziness at the edges.

Apart from the complexity of the modelling task, the second-moment equations pose substantial computational challenges. This is not, primarily, a matter of the extra resources needed to solve the larger number of equations involved, but is linked, to a much greater extent, to the nature of the equations, and to the linkage between them and the RANS equations. Most importantly, the outcome of the stress-transport equations are the stresses themselves, which enter directly into the RANS equations as stress gradients. This is in stark contrast to the second-order velocity derivatives through which the effects of turbulence are represented in an eddy-viscosity model. The replacement of these terms by stress derivatives causes a radical change in the mathematical characteristics of the RANS equations, as the numerically stabilizing effect of the second-order terms is lost. The resulting drastic loss of iterative stability needs, therefore, to be counteracted by elaborate algorithmic measures (see Lien & Leschziner (1994, 1996)) that recover a sufficient level of stability, thus allowing computational solutions to be obtained without undue difficulties.

Against the above background, the critical question is whether second-moment closure offers 'value for money and effort'. This question is very difficult to answer without the addition of substantial qualifications. There is no doubt that the single most important advantage of second-moment closure, distinct from any linear eddy-viscosity models, is its ability to return the anisotropy of the normal stresses and the sensitivity of this anisotropy to different types of strain and body forces, including those arising from curvature, rotation, buoyancy and electromagnetic interactions. Second-moment closure is also much less prone to violating realisability constraints — for example, when a flow is subjected to strong compressive straining.

In many practical flows — or rather, in many numerical solutions of such flows — the crucial factor affecting predictive realism is the ability of the model to return the correct level of the shear-stress components; the normal stresses often play a dynamically subordinate role. Hence, an important issue is whether the model provides the correct linkage between the normal stresses and the shear stresses. In relatively 'simple' flows — e.g. in near-wall layers that are not subjected to strong three-dimensional skewing and/or massive deceleration/acceleration — this linkage is often (but not always) reasonably well represented. Thus, if the correct prediction of such near-wall layers is the main objective of the exercise, then second-moment closure can well offer 'good value for money'. However, in very complex flows — notably, three-dimensional separated flow, especially when separation takes place from curved surfaces — second-moment closure may give solutions that are no better than, or even worse than, eddy-viscosity models. Such flows are simply beyond the capabilities of any single-point closure, in which case a full simulation is the only realistic route to their correct prediction.

It is an inescapable fact that a great deal of experience is needed to judge which set of circumstances offers the promise of significant benefits from second-moment closure, relative to a specifically-tuned or adapted eddy-viscosity model. This is a classic example of the difficulty of choosing the right 'horses for courses'.

Chapter Thirteen

13. Scalar/Heat-Flux-Transport Modelling

13.1 *The case for flux-transport closure*

Chapter 6 introduced the exact framework that describes the relationship between the components of the scalar-flux vector $\overline{u_j\phi} = \{\overline{u\phi}, \overline{v\phi}, \overline{w\phi}\}$ and the mean values of the scalar Φ and the velocity U_i. In particular, Eq. (6.6) was derived in order to demonstrate that the scalar fluxes are generated, convected, dissipated and affected by the pressure/scalar-gradient interaction. Although the scalar-flux equations are in many respects similar to those governing the stresses, there are a number of fundamental differences between the two sets. One is that the fluxes are driven by strains, in addition to gradients of the mean scalar. Another is that pressure/scalar-gradient interaction is not redistributive in the sense that the pressure-velocity interaction is in the case of the stresses. Finally, although the scalar variance $\overline{\phi^2}$ is a property that may be regarded as the flux equivalent of the turbulence-energy equation, in the sense that both represent the intensity of the respective fluctuations, the former does not arise from the scalar-flux equations, as there are no directional 'components' of $\overline{\phi^2}$. Rather, a separate equation has to be derived for this quantity by a separate set of manipulations of the scalar-transport equation (6.2). We shall consider such an equation later in this chapter.

Chapter 6 also provided arguments as to why it makes sense to derive the fluxes from Eq. (6.6), rather than to use the eddy diffusivity approximation,

$$\overline{u_j\phi} = -\frac{\nu_t}{\sigma_\phi}\frac{\partial \Phi}{\partial x_j}. \tag{13.1}$$

To be clear, we can use this approximation either in conjunction with eddy-viscosity models or Reynolds-stress models, simply because both provide the turbulence energy and its rate of dissipation, from which we can then calculate the eddy diffusivity. However, by doing so in conjunction with a Reynolds-stress model, we would fail to take advantage of much of the information that the model yields on the anisotropy and its connection to the shear-stress components — and hence the scalar fluxes. Moreover, Eq. (13.1) requires an explicit prescription of the Prandtl–Schmidt number σ_ϕ.

As an aside, we note here that an alternative route to prescribing σ_ϕ, taken mainly within the framework of two-equation models applied to heat-transfer problems (e.g. Nagano and Kim (1988), Sommer, So and Lai (1992)), is to treat the diffusivity as being separate from the eddy viscosity, and to model the former by reference to a scalar time scale $\tau_\phi = \frac{\overline{\phi^2}}{\varepsilon_\phi}$, where $\overline{\phi^2}$ is the scalar variance and ε_ϕ is its rate of dissipation. This time scale is then used instead of, or in combination with, the fluid-dynamic time scale $\tau_t = \frac{k}{\varepsilon}$, which is part of the eddy viscosity $\nu_t = C_\mu \left(\frac{k}{\varepsilon}\right) k$. Thus, if a time-scale ratio is defined as $R \equiv \frac{\tau_\phi}{\tau_t}$, the eddy diffusivity is determined from:

$$\Gamma_\phi = C_\phi \frac{k^2}{\varepsilon} f(R), \tag{13.2}$$

where $f(R)$ depends on the particular model proposed. In effect, R dictates the ratio of the life time of scalar fluctuations relative to the life time of the energy-containing eddies. In the simplest formulation, $f(R) = R^{0.5}$, in which case the time scale associated with Γ_ϕ is simply $\left(\frac{k}{\varepsilon}\frac{\overline{\phi^2}}{\varepsilon_\phi}\right)^{0.5}$. An obvious question posed by this modelling proposal is how to determine the scalar variance and its dissipation rate. The answer will emerge below as part of the discussion of flux-transport modelling.

The next modelling level is to extend a concept adopted earlier in relation to the stresses — namely, that the fluxes are driven by their respective rates of production (see Eqs. (6.6) and (11.1)):

$$\overline{u_i\phi} = -C_\phi \frac{k}{\varepsilon}\left\{\overline{u_j\phi}\frac{\partial U_i}{\partial x_j} + \overline{u_i u_j}\frac{\partial \Phi}{\partial x_j}\right\}, \tag{13.3}$$

in which k/ε is used (validly or not) as a turbulent time scale, and C_ϕ is a constant or a function of invariants. In fact, in the case of buoyant flows, there is one more, potentially influential, source term that needs to be taken into account, namely $G_{i\phi} = -\beta g_i \overline{\phi^2}$ (see Section 13.2), in which case,

$$\overline{u_i\phi} = -C_\phi \frac{k}{\varepsilon}\left\{\overline{u_j\phi}\frac{\partial U_i}{\partial x_j} + \overline{u_i u_j}\frac{\partial \Phi}{\partial x_j} + \beta g_i \overline{\phi^2}\right\}, \tag{13.4}$$

and this is the basis of the algebraic proposal of Hanjalić, Kenjeres and Durst (1996):

$$\overline{u_i\phi} = -C_\phi \frac{k}{\varepsilon}\left\{\overline{u_j\phi}\frac{\partial U_i}{\partial x_j} + \varsigma\overline{u_i u_j}\frac{\partial \Phi}{\partial x_j} + \eta\beta g_i \overline{\phi^2}\right\}, \tag{13.5}$$

where ς and η are constants (typically 0.6). However, this requires the solution of an equation for the scalar variance, $\overline{\phi^2}$, an issue discussed in Section 13.2.

To avoid the implicit connectively among the fluxes operative in Eq. (13.4), we might choose to ignore the strain field (and the scalar variance) and use:

$$\overline{u_i\phi} = -C_\phi \frac{k}{\varepsilon}\overline{u_i u_j}\frac{\partial \Phi}{\partial x_j}. \tag{13.6}$$

The discerning reader will recognise this expression as a direct equivalent of Daly and Harlow's (1970) generalised gradient diffusion hypothesis (GGDH) used in Eqs. (12.19) and (12.34) to approximate the turbulent transport of the stresses and the dissipation rate, respectively, within the framework of Reynolds-stress modelling. Equation (13.6) is certainly better than Eq. (13.1): with an appropriate choice of the constant C_ϕ, and an accurately predicted stress field (the wall-asymptotic behaviour, in particular), it gives a fair representation of the wall-normal flux — whether in wall-parallel or wall-normal (impinging) flow — and it predicts (appropriately!) a finite streamwise flux in the absence of a streamwise mean-scalar gradient. However, the omission of the strain term in Eq. (13.3) may well cause significant errors close to the wall, especially when the fluid Prandtl–Schmidt number is low, in which case the streamwise heat flux can be large in comparison with streamwise convection of Φ. Moreover, the omission of the scalar-variance term may be detrimental in highly buoyant conditions.

Unfortunately, there are very few studies that provide reliable statements on the relative performance of the above alternative models for the scalar fluxes in flows more complex than horizontal shear layers with vertical density stratification. A recent study by Omranian, Craft, Iacovides (2014) examines the performance of the models in Eqs. (13.5) and (13.6) for a range of buoyancy-driven flows in square and tall cavities within a range of orientations relative to the gravity vector. The results are inconclusive, at best, and do not indicate with any degree of confidence that the model in Eq. (13.5) is superior to the much simpler form in Eq. (13.6).

The fact that the availability of an accurate Reynolds-stress field greatly facilitates the provision of realistic scalar fluxes, through the use of relatively simple gradient approximations (e.g. Eq. (13.6)), is exemplified by Fig. 13.1, which relates to jet impinging normally on a heated wall. The figure illustrates, in particular, the relationship between predicted levels of normal-to-the-wall turbulence intensity and the radial distribution of the Nusselt number (heat-transfer coefficient) across the impingement plate. The curves arise from four models, three of which are Reynolds-stress models of varying complexity, all operating in conjunction with the model in Eq. (13.6). The profiles furthest away from the experimental data arise from a $k - \varepsilon$ model. The essential point conveyed by the figure is that the correct prediction of the heat-transfer coefficient goes hand-in-hand with the correct prediction of the wall-normal turbulence intensity.

Figure 13.1 and results for some other flow configurations (e.g. backward-facing-step flows) suggest that modelling the full flux-transport equations may not yield benefits similar to those

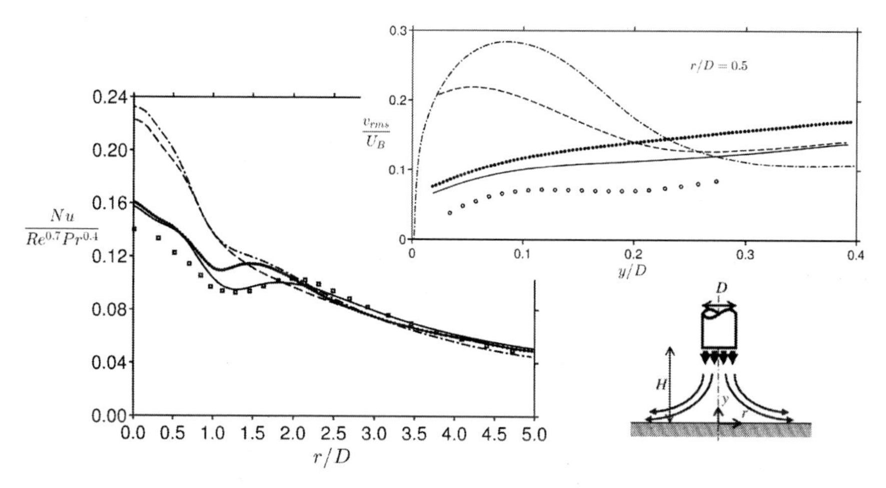

Fig. 13.1: Nusselt number and wall-normal turbulence intensity (v_{rms}) for a jet impinging on a flat plate, predicted with four different turbulence models, three of which are Reynolds-stress models (solid and dashed lines) and the forth a $k - \varepsilon$ model (chain lines). Circles represent experimental data (reproduced with permission of Elsevier from Craft, Graham and Launder (1993)).

derived from the modelling complexity pursued when approximating the stress-transport equations. Indeed, the literature on flux-transport modelling is far smaller than that on the modelling the Reynolds-stress equations, and the models are also simpler, as noted already.

Some motivation for adopting a full flux closure is provided, in principle, by the fact that the stress field can be substantially affected by the fluxes in buoyant flows. This is seen from Eqs. (12.7) and (12.9), the latter showing that the stresses are driven by the buoyancy-induced production $G_{ij} = -(g_i\beta\overline{T'u_j} + g_j\beta\overline{T'u_i})$, in which the correlations are heat fluxes and g_i is the gravity vector. Because buoyancy tends to give rise to complex motions away from walls — for example, in Rayleigh–Benard convection (Fig. 2.14) — the best possible description of the heat fluxes is highly desirable in an effort to secure a high level of overall predictive accuracy. There are very few recorded examples (if any) in the literature, however, which provide reliable evidence on the predictive benefits of full flux closure in complex flows.

The following section introduces relatively simple closure approximations for the unknown terms in the flux-transport equations.

13.2 *Closure of the flux-transport equations*

Our starting point is Eq. (6.6), re-written as follows:

$$\frac{D\overline{u_i\phi}}{Dt} = \underbrace{-\left(\overline{u_j\phi}\frac{\partial U_i}{\partial x_j} + \overline{u_iu_j}\frac{\partial \Phi}{\partial x_j}\right)}_{\text{Production}} - \Phi_{i\phi} - \varepsilon_{i\phi} + \underbrace{(T_{i\phi} + D_{i\phi})}_{\substack{\text{Turbulent \& diffusive}\\ \text{transport}}} . \tag{13.7}$$

A term that has been omitted from Eq. (6.6), as well as Eq. (13.7), is one that arises from buoyancy. In the presence of buoyancy, the Navier–Stokes (NS) equations contain the body-force terms $F_i = -\beta g_i \rho \Delta T$. As a consequence, the additional generation term $G_{iT'} = -\beta g_i \overline{T'^2}$ arises (or in terms of ϕ, $G_{i\phi} = -\beta g_i \overline{\phi^2}$), and this requires an additional transport equation for $\overline{\phi^2}$ to be derived and modelled.[1] As noted already, the buoyancy-related term $G_{iT'}$ is especially influential in free convection, such as Rayleigh–Benard (penetrative) convection, and also in double-diffusion problems (say, heated brine/fresh-water mixtures) and updrafts/downdrafts on heated/cooled vertical surfaces. In the first case, in particular, the entire dynamical system is driven by the flux $\overline{vT'}$, which is itself driven by $G_{iT'}$.

With $\overline{\phi^2}$ left aside at this stage, closure requires the approximation of the last three terms in Eq. (13.7). It is worth emphasizing, prior to modelling, that the fluxes are analogous to the shear-stress components only, because both are cross-correlations of different fluctuations. While a normal stress will always be finite in any turbulent flow, the shear stress will only be finite in the presence of straining. Similarly, the fluxes will only be finite in the presence of a mean-scalar gradient. These arguments serve to explain why, in most respects, the approximation of the terms in Eq. (6.6) follows by analogy to that

[1] This same equation is also required if the eddy-diffusivity model Eq. (13.2) is used, in which the scalar variance features in R.

adopted for the shear-stress components. However, as will be shown below, the analogy is far from perfect.

On the assumption of local isotropy in the small scales, the flux dissipation is usually approximated by:

$$\varepsilon_{i\phi} = 0. \tag{13.8}$$

This is entirely equivalent to the approximation $\varepsilon_{ij} = \frac{2}{3}\varepsilon\delta_{ij}$ for $i \neq j$ in the stress-transport equations. However, Fig. 13.2, for a heated channel flow (in which θ represents temperature fluctuations), suggests that this analogy is not necessarily justified for all flux components, and that Eq. (13.8) is thus not necessarily a satisfactory approximation. In fact, Fig. 13.2 brings to light further important pointers to the frailties of the stress-flux analogy. Thus, while the pressure-strain term balances the production in $\overline{v\theta}$ (which is analogous to the balance for $\overline{uv}$), the balance of $\overline{u\theta}$ (which has no equivalent shear-stress component) is governed by the production and dissipation, reminiscent of the conditions pertaining to the streamwise normal stress. Indeed, Eq. (13.7) shows that this flux is driven by the production $-\overline{v\theta}\frac{\partial U}{\partial y}$, which is analogous to the streamwise normal stress being driven by $-\overline{uv}\frac{\partial U}{\partial y}$. Moreover, the pressure/scalar-gradient term is non-zero at the wall, which is not compatible with the pressure-velocity correlation (see Eq. 12.83). This demonstrates the difficulty of closing the flux equations by reference to seemingly

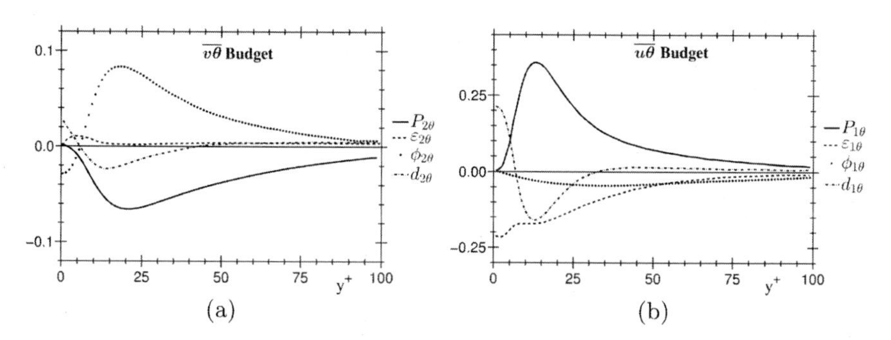

Fig. 13.2: Direct numerical simulation (DNS)-derived heat-flux budgets in a heated plane channel flow at bulk Reynolds number 4580 (reproduced with permission of ASME from Kasagi, Tomita and Kuroda (1992)).

rational extensions of the concepts used to close the Reynolds-stress-transport equations.

The diffusive flux transport is approximated in harmony with the GGDH concept by:

$$d_{i\phi} = C_\phi \frac{\partial}{\partial x_k}\left(\frac{k}{\varepsilon}\overline{u_k u_j}\frac{\partial \overline{u_i \varphi}}{\partial x_j}\right). \tag{13.9}$$

The simplest approximation of the pressure/scalar-gradient term follows the approximations of Eqs. (12.45) and (12.54) for the pressure-strain process. This is based on the observation that the exact equivalent of Φ_{ij} in Eq. (12.40) is:

$$\Phi_{\phi i} = \frac{1}{4\pi}\int_V \left(\overline{\left(\frac{\partial^2 u_k u_l}{\partial x_k \partial x_l}\right)' \frac{\partial \phi}{\partial x_i}}\right)\frac{dV(x_i')}{|x_i - x_i'|}$$
$$+ \frac{1}{4\pi}\int_V \left(2\left(\frac{\partial U_k}{\partial x_l}\right)' \overline{\left(\frac{\partial u_l}{\partial x_k}\right)' \frac{\partial \phi}{\partial x_i}}\right)\frac{dV(x_i')}{|x_i - x_i'|}, \tag{13.10}$$

hence, containing, here too, a 'slow' term and a 'rapid' term. However, an important point of difference is that the rapid fragment in Eq. (13.10) does not include one major element of the production term $P_{i\phi}$ — namely, that containing the mean-scalar gradient $\frac{\partial \Phi}{\partial x_i}$. Hence, the rapid term cannot be approximated, as done in Eq. (12.54), by way of the anisotropy of the entire production. Rather, the model for this term can only involve the production fragment containing the strain tensor. With this qualification noted, the equivalent of the linear model adopted for the pressure-strain term is:

$$\Phi_{i\phi} = -C_{1\phi}\frac{\varepsilon}{k}\overline{u_i\phi} + C_{2\phi}\overline{u_k\phi}\frac{\partial U_i}{\partial x_k}. \tag{13.11}$$

More complex, non-linear, approximations have been proposed by Craft, Ince and Launder (1996), Launder (1989), and Shih and Lumley (1985), as extensions to corresponding non-linear approximations for the pressure-strain process, but these are not discussed here.

Finally, in the case of buoyancy-affected flows, the effect of the extra generation term $G_{i\phi} = -\beta g_i \overline{\phi^2}$ on the pressure/scalar-gradient term is normally accounted for by adding a third contribution $C_3 G_{i\phi}$

to Eq. (13.11). This practice is, again, analogous to the treatment of body-force terms in the pressure-strain model (Eq. (12.89)). However, to this end, the scalar variance $\overline{\phi^2}$ needs to be determined.

As noted previously, it is readily possible to derive an exact transport equation for this quantity, by manipulating the scalar-transport equation and its time-averaged form (see Craft and Launder (1989)). This involves the substitution of Eq. (6.3) into:

$$\frac{D\phi^2}{Dt} = 2\phi\frac{D\phi}{Dt}, \tag{13.12}$$

followed by time-averaging. The result is:

$$\frac{D\overline{\phi^2}}{Dt} = \underbrace{-\left(2\overline{u_i\phi}\frac{\partial\Phi}{\partial x_i}\right)}_{P_\phi} \underbrace{- \alpha\overline{\frac{\partial\phi}{\partial x_j}\frac{\partial\phi}{\partial x_j}}}_{\varepsilon_\phi} \underbrace{- \frac{\partial\overline{u_j\phi^2}}{\partial x_j}}_{T_\phi} + \underbrace{\alpha\frac{\partial^2\overline{\phi^2}}{\partial x_j\partial x_j}}_{D_\phi}, \tag{13.13}$$

in which α is the fluid diffusivity associated with Φ. The production term P_ϕ is determinable directly from the flux-transport equations, and the diffusion term may be approximated, as before, by way of the GGDH, thus:

$$T_\phi = C_\phi\frac{\partial}{\partial x_k}\left(\frac{k}{\varepsilon}\overline{u_k u_j}\frac{\partial\overline{\phi^2}}{\partial x_j}\right). \tag{13.14}$$

This leaves the dissipation rate ε_ϕ for which an exact transport equation can also be derived, one that is equivalent to the dissipation rate of the turbulence energy ε, Eq. (9.4). The derivation involves fairly straightforward manipulations of Eq. (6.3), the result being a transport equation containing production, turbulent transport, viscous diffusion and destruction (see Nagano and Kim (1988), Craft and Launder (1989), Nagano, Tagawa and Tsuji (1991)). This equation is not given here, either in its exact or its modelled form, for it is rarely used in practical implementations of flux-closure models.[2]

[2]Nagano (2002) provides an extensive review of various alternative models, both at two-equation level, involving transport equations for the variables $\overline{\phi^2}$ and ε_ϕ, from which the turbulent diffusivity is obtained, and at second-moment-closure level, involving equations for $\overline{u_i\phi}$, $\overline{\phi^2}$ and ε_ϕ, tested mostly by reference to data for channel flow, pipe flow and flow behind a backward-facing step.

Rather, the assumption is made that the dissipation rates ε_ϕ and ε are linearly related via a time-scale ratio (see also Eq. 13.2):

$$\varepsilon_\phi = \overline{\phi^2}\left(\frac{\varepsilon}{k}\right)\frac{1}{R}. \tag{13.15}$$

In the simplest implementation, this ratio is chosen to be a constant value (of order 0.5). Alternatively, Craft and Launder (1989) have proposed an approximation in which R is made to depend on the flux invariant $A_{2\phi} = (\overline{u_i\phi}^2)/(k\overline{\phi^2})$ via $R^{-1} = 1.5(1 + A_{2\phi})$. Several other proposals exist, some involving complex non-linear relations $f(R)$, used in Eq. (13.2), with R determined directly from the solution of modelled equations for $\overline{\phi^2}$ and ε_ϕ. These models also include various damping functions, collectively designed to procure the correct wall-asymptotic behaviour of the scalar time scale (see Nagano (2002)).

13.3 *Summary and lessons*

The relative brevity of this chapter reflects the fact that the frequency of use of flux-transport models is far lower than that of Reynolds-stress-transport models. There is also a general perception that, once the flow field and the stresses are correctly captured, the use of the generalised gradient-diffusion approximation suffices to give a reasonable description of the scalar field. It is certainly true that this confidence is often misplaced in complex conditions. One situation in which flux-transport closure is highly desirable is that in which the fluid dynamics are strongly affected by buoyancy. In this case, there is a strong two-way coupling between the stresses and fluxes, so that an accurate description of the fluxes is especially important.

We have seen that flux-transport closure leans heavily, in conceptual terms, on the ideas underpinning Reynolds-stress models. Hence, diffusion is approximated by a gradient-type process, and the modelling of the pressure/scalar-gradient correlation follows, with some important differences, that used to approximate the pressure-velocity correlation with relatively simple proposals.

We have not dealt explicitly with the influence of viscosity on the modelled processes. One important reason is that viscous effects are already accounted for in the modelling of the turbulence energy, its rate of dissipation and the stresses. Hence, for example, the tensorial diffusivity in the turbulent-transport term is already sensitised to the viscosity via the modelled Reynolds stresses.

One important point to underline is that, in the presence of buoyancy, the scalar fluxes are driven by source terms that contain the scalar variance, which is a further unknown. This requires the solution of yet another modelled transport equation. One unknown process in this equation is the dissipation of the scalar variance. The simplest way of modelling this term is to use an algebraic relation that links the scalar dissipation to the dissipation of the turbulence energy, via an assumed time-scale ratio. To go further, we require the derivation and modelling of a transport equation for the scalar dissipation, equivalent to that for the turbulence-energy dissipation. This has been done, but may only be justified in extreme circumstances that involve strongly exothermic chemical reaction, multi-species systems or double-diffusion processes — the last being a combination of disparate thermal and species diffusion.

Chapter Fourteen

14. The $\overline{v^2} - f$ Model

14.1 *Relationship to second-moment closure and elliptic relaxation*

Against the background of the elliptic-relaxation method, covered in Chapter 12, (Section 12.7.7), we are now in a position to return to linear eddy-viscosity modelling in order to cover a particular group of models under the heading $\overline{v^2} - f$ (Durbin (1991)).

We recall two key facts that are especially relevant to the present discussion:

(i) Elliptic relaxation aims to account especially for the effect of wall-induced pressure reflections on the Reynolds stresses — in particular, the rapid decline in the wall-normal stress $\overline{v^2}$ as the wall is approached.

(ii) Linear eddy-viscosity models suffer from their use of the turbulence energy k as the sole velocity scale ($k^{1/2}$), as this scale does not adequately represent the turbulent mixing across shear layers, especially those close to the wall.

As regards (ii), it is instructive to review here some pertinent issues covered in Chapters 5 and 9.

- In Section 7.1, an argument is presented, on the basis of the production of the shear stress, that led to the proposal, applicable to thin shear flow:

$$-uv = C\tau\overline{v^2}\frac{\partial U}{\partial y}, \tag{14.1}$$

where τ is the turbulent time scale (later proposed to be k/ε). The implication is, therefore, that there is a need to represent correctly the wall-normal Reynolds stress, which is — as we now know — not only much smaller than $2/3k$ away from the wall, but which also declines much faster than k as the wall is approached.

- In the context of linear eddy-viscosity modelling, we opted for k to be the sole turbulent-velocity scale, and this led us, within the framework of two-equation modelling (Chapter 9), to:

$$-uv = C_\mu\left(\frac{k}{\varepsilon}\right)k\frac{\partial U}{\partial y}. \tag{14.2}$$

- While the choice $C_\mu = 0.09$ allowed us to correctly relate the shear-stress to the shear strain in regions not immediately adjacent to walls, we had to introduce influential damping functions, of the form (see Eq. (9.28)):

$$f_\mu = 1 - \alpha\exp(-\beta\, arg), \tag{14.3}$$

where *arg* is a turbulent Reynolds number or a related parameter involving the viscosity. However, we now appreciate — having covered Reynolds-stress modelling in Chapter 12 — that this is, physically, incorrect, because the rapid decline in the eddy-viscosity as the wall is approached is due, mainly, to the inviscid blocking exerted by the wall on $\overline{v^2}$, and the effect of pressure reflections on the pressure-strain process (see also the discussion in Section 12.7.5).

Against the above background it seems sensible to adopt $\overline{v^2}$ as a second turbulent-velocity scale, within the eddy-viscosity framework, and to use the elliptic relaxation strategy to represent correctly the effect of the wall on this scale. We suggested this strategy already, in principle, in Section 7.1, Eq. (7.8), and Section 9.3 Eq. (9.22).

14.2 $\overline{v^2} - f$ *model formulation and variants*

The introduction of $\overline{v^2}$ as an additional variable necessitates the addition of a third transport equation to the $k - \varepsilon$ set. This equation is, essentially, one of the set of Reynolds-stress-transport equations, namely the modelled version Eq. (12.7) for $i, j = 2, 2$, assuming we consider a wall normal to the direction x_2:

$$\frac{D\overline{v^2}}{Dt} = \underbrace{\left\{\Phi_{22} - \left(\varepsilon_{22} - \frac{\overline{v^2}}{k}\varepsilon\right)\right\}}_{\varphi_{22}=kf} + \frac{\partial}{\partial x_k}\left\{\left(\nu + \frac{\nu_t}{\sigma_k}\right)\frac{\partial \overline{v^2}}{\partial x_k}\right\} - \frac{\overline{v^2}}{k}\varepsilon, \tag{14.4}$$

with the wall condition $\overline{v^2} = 0$. This equation is also one of the set in Eq. (12.98), subject to the omission of the production term, applicable to the thin-shear flow at the wall, and to the sum of turbulent and pressure diffusion $D_{22}^t + D_{22}^p$ being represented by an eddy-viscosity-driven gradient-diffusion approximation.

A fourth equation is now also needed for φ_{22} to account for the elliptic relaxation in Eq. (14.4). This is, essentially, Eq. (12.112) in Section 12.7.7 for $i, j = 2, 2$, denoted here simply by f:

$$f - L^2\nabla^2 f = f^h, \tag{14.5}$$

where

$$f \equiv \frac{\varphi_{22}}{k}; \quad \varphi_{22} = \Phi_{22} + \left(\frac{\overline{v^2}}{k}\varepsilon - \varepsilon_{22}\right). \tag{14.6}$$

As noted in Section 12.7.7, the homogeneous form of the pressure-strain process Φ_{ij}^h can be modelled by any approximation that does not account for inhomogeneity. The simplest such model comprises Eqs. (12.45) and (12.54) for, respectively, the 'slow' and 'rapid' parts of Φ_{22}^h:

$$\Phi_{22}^h = -C_1\varepsilon\left(\frac{\overline{v^2}}{k} - \frac{2}{3}\right) + \frac{2}{3}C_2 P_k. \tag{14.7}$$

Note again that $P_{22} = 0$, which is why P_k is the only production in the 'rapid' term. As we operate within the eddy-viscosity framework, it follows that $P_k = \nu_t S^2$, where $S^2 = 2S_{ij}S_{ij}$ (see Eq. (8.43)).

As explained in Section 12.7.7, it is important to secure the correct asymptotic variation $\overline{v^2}\big|_{y\to 0} = O(y^4)$, and this is done by imposing the boundary condition (see Eq. (12.110)):

$$f\,|_{y\to 0} = -20\nu^2 \frac{\overline{v^2}}{y^4 \varepsilon}. \tag{14.8}$$

Assuming $\overline{v^2}$ and k are determined from their respective transport equations, we can calculate the eddy-viscosity from:

$$\nu_t = C_{\mu,v2}\overline{v^2}\max\left(\frac{k}{\varepsilon}, 6\sqrt{\frac{\nu}{\varepsilon}}\right), \tag{14.9}$$

in which the bracketed term is the turbulent time scale, limited by the Kolmogorov scale in the viscous sublayer, and $C_{\mu,v2} = 0.19 - 0.22$ (depending on the model variants, of which there are several — see Billard (2011)) — a value that reflects the ratio $\overline{v^2}/k$ away from the wall.

Figure 14.1 shows the results of an *a priori* study in which the proposal $\nu_t = 0.22\overline{v^2}\frac{k}{\varepsilon}$ was evaluated from direct numerical simulation (DNS) data for channel flow at five Reynolds-number values (up to $Re_\tau = 2000$), and compared to the ratio $-uv/\frac{\partial U}{\partial y}$, also evaluated from the same DNS data. This shows the proposal to give a realistic representation of the eddy-viscosity, although the coefficient 0.22 is somewhat too high at high Reynolds numbers.

In terms of wall-asymptotic properties, Eq. (14.9) is not entirely correct. We know that the decline in the eddy viscosity has to follow $O(y^3)$. However, Eq. (14.9) implies a decline according to $O(y^4)$, because the Kolmogorov time scale is $O(1)$. Fortunately, this defect is minor, as it only affects the eddy viscosity in the lower part of the viscous sublayer, where the turbulence is very low.

Figure 14.2 shows a comparison of the eddy viscosity, evaluated from the $\overline{v^2} - f$ model, relative to the conventional $k - \varepsilon$ form, but without the use of any wall-related (viscous) correction. As already

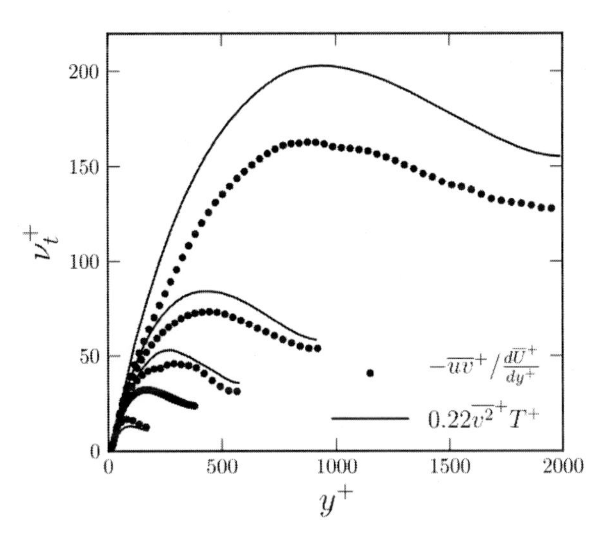

Fig. 14.1: An *a priori* study of the eddy-viscosity proposal $\nu_t = 0.22\overline{v^2}\frac{k}{\varepsilon}$, based on DNS data for channel flow at $Re_\tau = 180 - 2000$ (reproduced with permission from Billard (2011)).

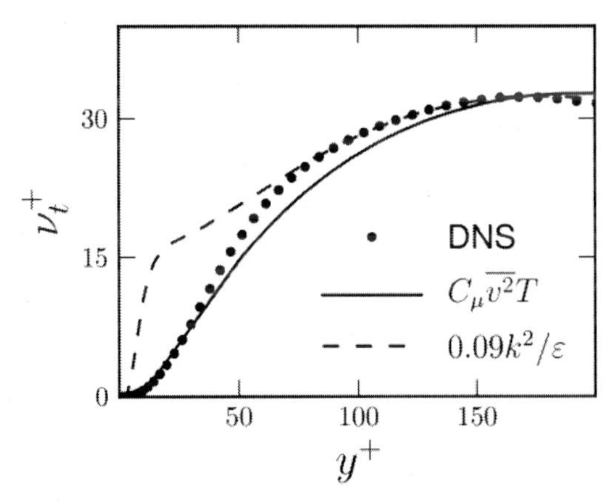

Fig. 14.2: Near-wall performance of the $\overline{v^2} - f$ model in representing the eddy viscosity, relative to the conventional $k - \varepsilon$ model without wall corrections (reproduced with permission from Billard (2011)).

observed in Chapter 7 (Fig. 7.4), the wall-induced decline in ν_t starts at $y^+ \approx 100$, which is well beyond the layer in which viscosity can possibly affect the turbulence field. This is a clear indication of the fact that the decline is driven mainly by an inviscid process. The modelled decline very close to the wall is seen to accord well with the DNS data, and this confirms the assertion made earlier that the inconsistency in the asymptotic decline of the eddy viscosity, implied by Eq. (14.9), is of little consequence.

A final point to add here, for the sake of completeness, is that the ε-equation used in the present modelling framework is (normally) the standard form, usually with coefficients $C_{\varepsilon 1} = 1.4\left(1 + c_{\varepsilon 1}\sqrt{\frac{k}{\overline{v^2}}}\right)$, $C_{\varepsilon 2} = 1.9$ and $c_{\varepsilon 2}$ being 0.045–0.06, depending on the model variant used.[1] The purpose of sensitizing $C_{\varepsilon 1}$ to $k/\overline{v^2}$ is to improve the generality of the model when used both for near-wall flows and free shear layers.

An important drawback of the above $\overline{v^2} - f$ model is that the boundary condition in Eq. (14.8) is very difficult to implement in a numerically stable fashion in general purpose codes — an issue already underlined in Section 12.7.7. Similar difficulties, although less severe, are also encountered when using the boundary condition in Eq. (5.6) for ε, and it is this difficulty that has given rise to the use of $\widetilde{\varepsilon} = \varepsilon - 2\nu\frac{k}{y^2}$ as the subject of an 'isotropic-dissipation' equation (cf. Eq. (9.1), noting that $k \propto y^2$ at the wall). This particular difficulty, and the more general wish to simplify the computational task involved in solving the additional two equations in the present framework, have motivated various attempts to derive simpler and more stable forms of the model. At the time this text is written, there are no less than 10 variants of the $\overline{v^2} - f$ model, and a comprehensive discussion of all forms is given in the PhD thesis of Billard (2011).

One of these is the so-called 'code-friendly' version proposed by Lien and Durbin (1996). The essential element of this model is the variable replacement:

$$f \leftarrow \overline{f} - 5\varepsilon\frac{\overline{v^2}}{k^2}, \tag{14.10}$$

[1] There are various alternative terms proposed, instead of $(k/\overline{v^2})^{1/2}$, and the interested reader is referred to Billard (2011) for a complete list.

which then gives rise to the modified set:

$$\frac{D\overline{v^2}}{Dt} = k\overline{f} + \frac{\partial}{\partial x_k}\left\{\left(\nu + \frac{\nu_t}{\sigma_k}\right)\frac{\partial \overline{v^2}}{\partial x_k}\right\} - 6\frac{\overline{v^2}}{k}\varepsilon, \tag{14.11}$$

$$\overline{f} - L^2\nabla^2\overline{f} = f^h + 5\varepsilon\frac{\overline{v^2}}{k^2}. \tag{14.12}$$

The advantage of this change is that the boundary condition for $\overline{f}$ becomes zero, rather than $-5\varepsilon\frac{\overline{v^2}}{k^2}$ (see Eq. (12.119)), with the consequence of weakening the linkage between the various variables involved. Equation (14.12) is obtained by substituting Eq. (14.10) into Eq. (14.5) and neglecting the term $5L^2\nabla^2(\overline{v^2}\varepsilon/k^2)$. Hence, this model is not identical to the original $\overline{v^2} - f$ parent, and it therefore displays a slightly different predictive performance, with $\overline{v^2}$ being generally overestimated.

Another 'code-friendly' version is that of Laurence, Uribe, and Utyuzhnikov (2004) (as well as Uribe (2006) and Hanjalić, Popovac, and Hadziabdić (2004)) in which an equation is solved for the ratio $\overline{v^2}/k$, rather than for $\overline{v^2}$ itself. The argument for doing so is that the wall-asymptotic variation of this ratio is $O(y^2)$, thus considerably more benign than $O(y^4)$ in terms of dealing, numerically, with the wall boundary condition.

The most recent formulations of the $\overline{v^2} - f$ ilk are Billard's $\varphi - \alpha$ and the Billard–Laurence (BL)-v^2/k models.[2] Both combine an equation for the 'code-friendly' variable $\varphi = \overline{v^2}/k$ with the blending concept of Manceau and Hanjalić (2002), explained in Section 12.7.7.

The equation for φ can be readily derived from:

$$\frac{D\varphi}{Dt} = \frac{1}{k}\frac{D\overline{v^2}}{Dt} - \frac{\overline{v^2}}{k^2}\frac{Dk}{Dt}, \tag{14.13}$$

with $\frac{D\overline{v^2}}{Dt}$ taken, in the case of the $\varphi - \alpha$ model, from Eq. (14.4), and $\frac{Dk}{Dt}$ being the standard turbulence-energy-transport equation. The

[2] Billard's PhD Thesis (Billard (2011)) provides the most comprehensive analysis of all $\overline{v^2} - f$-type models.

result is:

$$\frac{D\varphi}{Dt} = f - \frac{\varphi}{k}P_k + \frac{\partial}{\partial x_k}\left[\left(\nu + \frac{\nu_t}{\sigma_k}\right)\frac{\partial \varphi}{\partial x_k}\right] + \frac{2}{k}\left(\nu + \frac{\nu_t}{\sigma_k}\right)\frac{\partial k}{\partial x_k}\frac{\partial \varphi}{\partial x_k}. \tag{14.14}$$

The wall-asymptotic condition, equivalent to Eq. (14.8), is readily derived along the lines followed earlier and arises as:

$$f|_{y\to 0} = -10\nu\frac{\varphi}{y^2}. \tag{14.15}$$

To evaluate f, the models discussed here do not resort to the elliptic-relaxation equation (14.5), but adopt the blending formula:

$$f = (1-\alpha)f_w + \alpha f^h, \tag{14.16}$$

where (see Eq. 12.117),

$$\alpha - L^2\nabla^2\alpha = 1, \tag{14.17}$$

f_w is given by Eq. (14.15) and f^h is the homogeneous model for f, for which Billard uses the simple model in Eq. (14.7), thus yielding, upon a straightforward substitution:

$$f^h = -\frac{\varepsilon}{k}\left(C_1 - 1 + C_2\frac{P_k}{\varepsilon}\right)\left(\varphi - \frac{2}{3}\right). \tag{14.18}$$

Hence, the $\varphi - \alpha$ model consists of the equations for k, ε, φ, α, the blending formula, Eq. (14.16), and the eddy-viscosity relation, Eq. (14.9), together with the wall-boundary conditions:

$$k|_{y\to 0} = 0, \quad \varepsilon|_{y\to 0} = \frac{2\nu k}{y^2}, \quad \varphi|_{y\to 0} = 0, \quad \alpha|_{y\to 0} = 0. \tag{14.19}$$

The model constants are not given here (see Billard (2011)). Most take the standard values noted earlier, but some are adjusted, by computer optimisation, to best fit a range of DNS data for channel flow at Reynolds numbers up to $Re_\tau = 2000$.

The BL-v^2/k model involves a number of modifications to the $\varphi - \alpha$ model — without, however, changing the basic elements of the

φ and α equations and the associated boundary conditions. The main changes are the use of a particular form of the dissipation equation for the quantity $\varepsilon' = \varepsilon - \frac{1}{2}\nu\nabla^2 k$ (Jakirlić and Hanjalić (2002)); the inculsion of a viscosity-dependent correction term in the dissipation-rate equation, essentially identical to that given by Eq. (12.38); a replacement of the time scale in Eq. (14.9) — the term in brackets — by $T = \left\{\left(\frac{k}{\varepsilon}\right)^2 + 16\left(\frac{\nu}{\eta}\right)\right\}^{1/2}$; and several further adjustments to model coefficients, especially $C_{\varepsilon 2}$ in the dissipation-rate equation, mostly based on calibration by reference to DNS data. All these yield improved predictive capabilities in several test cases, as exemplified by the results in Figs. 14.3–14.5.

The comparisons for channel flow, shown in Fig. 14.3, need to be viewed with some caution, because much of the models' calibration was undertaken by reference to DNS data for this type of flow. Hence, these results should be viewed as merely demonstrating some basic capabilities, enhanced by careful calibration and optimisation. The main difference between the two models relates to the dissipation rate, and this reflects the use of different forms of dissipation variable in related equations and coefficients, as outlined above.

Much more interesting are true model predictions for two nominally two-dimensional separated flows, considered in Figs. 14.4 and 14.5, for which either large eddy simulation (LES) or experimental data are available. Perhaps the most interesting feature in both figures is the poor performance of the $\varphi - \alpha$ model, suggesting that the blending strategy is not a promising basis for approximating the elliptic relaxation based on f. In contrast, models that include an equation for f perform reasonably well. The fact that the BL-v^2/k model yields so much better results than the closely related $\varphi - \alpha$ model is intriguing. It clearly implies that the detailed modifications of the latter are extremely influential, although entirely divorced from the elliptic relaxation strategy as such, thus possibly reflecting error compensation. Whatever the reason may be, this is a sobering demonstration that model performance is extremely sensitive to many details, which are, more often than not, unrelated to the key features of the modelling strategy.

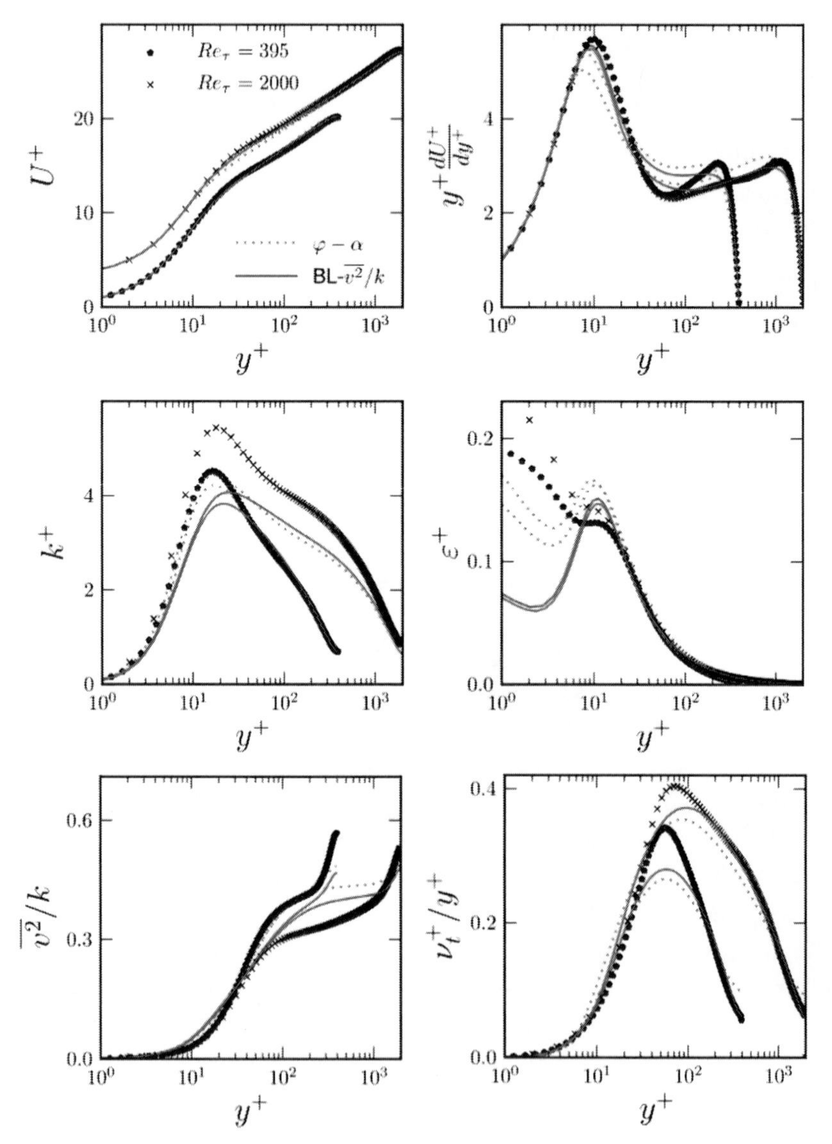

Fig. 14.3: Comparisons between $\varphi - \alpha$ and BL-v^2/k model predictions against DNS data for several quantities in channel flow at two Reynolds numbers (reproduced with permission from Billard (2011)).

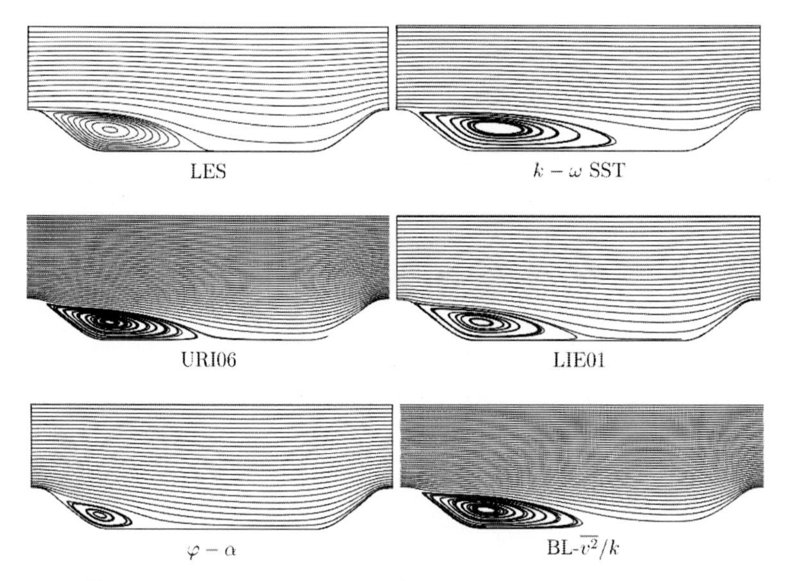

Fig. 14.4: Comparisons between stream-function predictions from several models against highly resolved LES data for a separated flow in a periodically constricted channel at bulk Reynolds number 10600, based on constriction height. All models, but $k-\omega$-SST, are of the $\overline{v^2} - f$ variety (reproduced with permission from Billard (2011)).

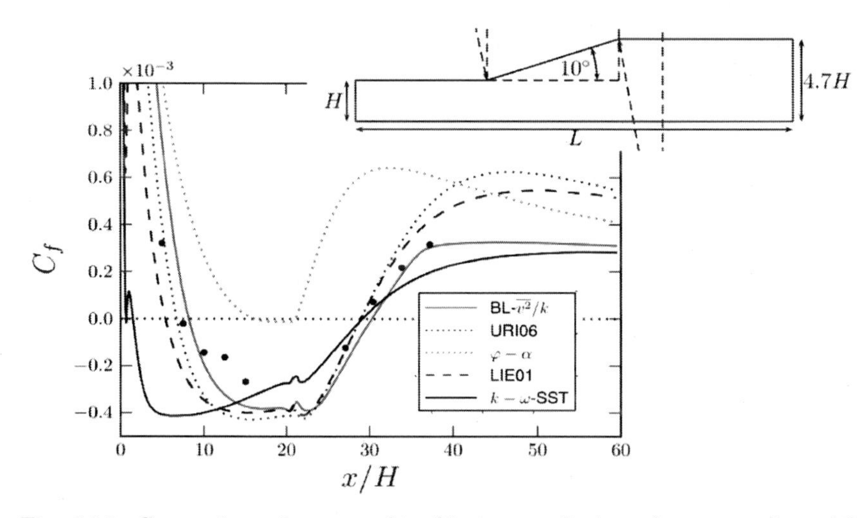

Fig. 14.5: Comparisons between skin friction predictions from several models against experimental data for a separated flow in a one-sided plane diffuser at bulk Reynolds number 18000, based on inlet passage height. All models, but $k - \omega\,$SST, are of the $\overline{v^2} - f$ variety (reproduced with permission from Billard (2011)).

14.3 *Summary and lessons*

The adoption of $\overline{v^2}$, in place of k, in the eddy-viscosity relation is an astute proposal, which has strong fundamental merits. This is so especially at walls at which $\overline{v^2}$ dictates the wall-normal mixing and the shear stress. Also, the use of the elliptic-relaxation concept provides a superior fundamental foundation for representing the influence of the wall on the wall-normal stress and the shear stress, and it obviates the need for empirical damping functions designed to represent the ratio $\overline{v^2}/k$.

Notwithstanding the above merits, the fact that the $\overline{v^2} - f$ concept is implemented within the linear eddy-viscosity framework, means that the disadvantages that have provided the motivation for embarking on second-moment closure remain firmly in place. In particular, all the complex interactions among the Reynolds stresses that are driven by the anisotropy are not resolved.

Disconcertingly, there are substantial differences in the predictive performance of different $\overline{v^2} - f$ implementations or model variations, and this leads to the conclusion that this modelling concept does not, in itself, secure superior predictive fidelity. As is almost always observed in turbulence modelling, "the devil is in the detail", and model performance is strongly influenced by fragments unrelated to the $\overline{v^2} - f$ concept. In addition, the performance depends sensitively on the precise form of the $\overline{v^2}$ equation, or a related group from which $\overline{v^2}$ is obtained, as well as on the wall boundary condition for this and the equation for f. The replacement of f by the blending factor α results in the simplest form of the $\overline{v^2} - f$ strategy, but the performance of this variant in separated flows appears to be rather poor — or at least unreliable.

Chapter Fifteen

15. Algebraic Reynolds-Stress and Non-Linear Eddy-Viscosity Models

15.1 *Rationale and motivation*

Following the exposition of Reynolds-stress-transport modelling (RSTM) in Chapter 12, there should be little doubt in readers' minds about the fundamental superiority of closure at this level, relative to linear eddy-viscosity modelling. Yet, the penetration of RSTMs in general computational fluid dynamics (CFD) for industrial applications has been remarkably weak and slow. The principal reasons are not hard to identify: RSTMs are mathematically complex, numerically challenging and computationally expensive, require the prescription of boundary conditions for all stress components, and do not display a decisive predictive superiority in all cases in which linear eddy-viscosity models fail. An especially potent computational drawback is that the Reynolds-averaged Navier–Stokes (RANS) equations lack the numerically stabilising second-order derivatives of the velocity associated with the eddy viscosity and the linear stress-strain relationships. This forces the introduction of special algorithmic measures that enhance convergence (see Lien and Leschziner (2002)). The need for significant user expertise and physical insight into modelling details, as well as the greater sensitivity of RSTM solutions to mesh quality, are yet further powerful reasons for the resistance towards adopting RSTMs in an industrial environment. This inescapable predicament has spawned many efforts to derive model forms that offer a compromise between the simplicity of eddy-viscosity modelling and the fundamental strength of RSTMs.

In the early 70s, when computer power was a tiny fraction of what it is today, it was considered impractical to apply emerging RSTMs (e.g. Launder, Reece and Rodi (1975)) to anything more complex than two-dimensional boundary layers. Specifically, the task of solving six differential transport equations for the stress components, instead of one for the turbulence energy, was considered to be too resource intensive. This prompted Rodi (1976) to make the pioneering proposal to approximate the transport of the stresses by the algebraically scaled transport of the turbulence energy, namely:

$$\frac{D\overline{u_i u_j}}{Dt} - T_{ij} \leftarrow \frac{\overline{u_i u_j}}{k}\left(\frac{Dk}{Dt} - T_k\right) = \frac{\overline{u_i u_j}}{k}\left(P_k - \varepsilon\right), \tag{15.1}$$

where T identifies turbulent (diffusive) transport. This is referred to as an 'algebraic stress model', because it only requires the solution of algebraic equations for $\overline{u_i u_j}$, given that the turbulence energy has been determined from the single transport equation for k, and that ε_{ij} has been approximated algebraically in terms of ε — for example, via $\varepsilon_{ij} = \frac{2}{3}\delta_{ij}\varepsilon$. Combining Eq. (15.1) with the definition of the anisotropy tensor, $a_{ij} \equiv \frac{\overline{u_i u_j}}{k} - \frac{2}{3}\delta_{ij}$, readily leads to the result $\frac{Da_{ij}}{Dt} = 0$. Hence, the algebraic approximation in Eq. (15.1) can be expected to be adequate when the transport of the anisotropy is weak.

Equation (15.1) is based on the expectation that the transport of the stresses occurs at the same rate as the transport of the turbulence energy. While this may be a reasonable assumption for the normal-stress components — which are, after all, components of the turbulence energy — it appears to be on weaker ground in the case of the shear-stress components. This provides some justification for an even simpler form than Eq. (15.1), namely:

$$\frac{D\overline{u_i u_j}}{Dt} - T_{ij} \leftarrow \frac{2}{3}\delta_{ij}\left(\frac{Dk}{Dt} - T_k\right), \tag{15.2}$$

which nullifies the convective transport of $\left(\overline{u_i u_j} - \frac{2}{3}\delta_{ij}k\right)$ and of $\left(T_{ij} - \frac{2}{3}\delta_{ij}T_k\right)$, hence also nullifying the transport of the shear-stress component themselves.

The ultimate simplification is achieved by introducing local Reynolds-stress equilibrium, i.e. zero transport of all stress components, leading (in the absence of body forces, rotation and viscous effects) to:

$$-\left(\overline{u_i u_k}\frac{\partial U_j}{\partial x_k} + \overline{u_j u_k}\frac{\partial U_i}{\partial x_k}\right) + \Pi_{ij} = \varepsilon_{ij}, \tag{15.3}$$

i.e. a balance between generation, pressure-velocity interaction and dissipation. As an aside, and without any details being given here, we note that a subset of Eq. (15.3) has been used in the derivation the curvature correction in Eq. (11.21).

The assumption that stress transport can be simplified in the above manner, or ignored altogether, may be justified in relatively simple flows, but is much less secure in complex flows in which streamline curvature is high. As will be shown below, the transformation of the stress-transport equations in terms of streamline-aligned co-ordinates gives rise to curvature-related convective contributions that are identical in form to some important stress-production fragments. Hence, the omission of stress transport in Eq. (15.3), or the approximation in Eq. (15.1), leads to a loss of these fragments and hence to adverse effects on the validity of the resulting model in curved flows. For example, in a strongly swirling flow, in which curvature is governed by the motion $W(r)$ (or $U_\theta(r)$), the stress-transport equations, expressed in the cylindrical-polar coordinate system (x, r, θ), feature convective contributions of the form $\overline{vw}\frac{W}{r}, \overline{uw}\frac{W}{r}, \overline{v^2}\frac{W}{r}, \overline{w^2}\frac{W}{r}$ (see Hogg and Leschziner (1989), Fu, Launder, and Tselepidakis (1987)). But Eq. (11.27) shows that these are precisely the terms that identify the impact of the swirl in the production of the stresses, which then forms the basis for the use of Richardson-number curvature corrections to represent the influence of curvature or swirl. This dichotomy of the convection/production terms has led to the proposal to treat the convection tensor C_{ij} in the same way as the production tensor P_{ij} in models for the pressure-velocity correlation Π_{ij} (see Fu, Launder and Tselepidakis (1987)).

To enhance the arguments made in relation to the algebraic approximation in Eq. (15.1), it is instructive to consider the example of a RSTM in which the pressure-velocity interaction has been modelled with the general form of a (quasi-) linear approximation, the models of Launder, Reece and Rodi and Speziale, Sarkar and Gatski (1991) being of that ilk. Using the definitions $a_{ij} \equiv \frac{\overline{u_i u_j}}{k} - \frac{2}{3}\delta_{ij}$ and $T_{ij}^{(a)} \equiv \frac{T_{ij}}{k} - \frac{\overline{u_i u_j}}{k^2}T_k$, the model may be written, following some straightforward algebra, as follows:

$$\frac{k}{\varepsilon}\left(\frac{Da_{ij}}{Dt} - T_{ij}^{(a)}\right) = \alpha_0\left[\left(\alpha_1 + \alpha_2\frac{P_k}{\varepsilon}\right)a_{ij} + \alpha_3 S_{ij}^* - \alpha_4(a_{ik}\Omega_{kj}^* - \Omega_{ik}^* a_{kj}) + \alpha_5\left(a_{ik}S_{kj}^* + S_{ik}^* a_{kj} - \frac{2}{3}a_{kl}S_{lk}^*\delta_{ij}\right)\right], \tag{15.4}$$

where $S_{ij}^* \equiv \frac{k}{\varepsilon}S_{ij}, \Omega_{ij}^* \equiv \frac{k}{\varepsilon}\Omega_{ij}$ and $\alpha_0 - \alpha_5$ are model constants, the values of which are immaterial at this stage. Imposition of the approximation in Eq. (15.1) now nullifies the left hand side, rendering the equation set algebraic. This is thus the *algebraic* (ARSM) equivalent form of the *differential* parent (RSTM). An obstacle to a straightforward solution of Eq. (15.4) is, however, that the equation set is implicit and coupled in the stresses (or the anisotropy tensor). Moreover, strictly, the set is non-linear (despite the earlier reference to a 'quasi-linear' approximation), because the product $P_k a_{ij}$ is quadratic in the stresses.

It is important to pose and discuss the question as to what advantages are derived from any one of the above simplifications in Eqs. (15.1)–(15.3). The answer is: relatively few, if any at all, unless further major manipulations are introduced to address the implicitness in, and coupling among, the stresses. In fact, penalties noted below can easily outweigh any savings due to the omission of transport. Yet, the concept of approximating the stress-governing equations by an algebraic set is important to discuss, for it forms the conceptual foundation for far more effective formulations than those arising from merely omitting the transport terms.

First, on the plus side, as the Reynolds-stress equations no longer contain differential terms, no boundary conditions are needed for the stresses. This is because the algebraic equations contain only locally determined quantities, as the stress values at any one point are no longer linked to neighbouring locations. This is helpful, in principle — although, in most practical circumstances, the assumption of local isotropy of the stresses at inlet and entrainment boundaries, of simple zero-gradient conditions at outflow boundaries, and of vanishing stresses at walls, provide adequate approximations in the context of full RSTMs, because stress transport is generally subordinate to other mechanisms. After all, if this were not so, the algebraic modelling framework would be intrinsically poor. Second, again on the positive side, the omission of the transport terms obviously reduces the number of numerical operations during the solution process. However, the savings involved are rather modest in comparison to the computational task as a whole.

On the other hand, there are a number of drawbacks to consider. First, the omission of the transport terms can have seriously detrimental effects on the iterative stability of the numerical solution. These arise over and above the challenges that are posed, with any second-moment closure, by the absence of the numerically stabilizing eddy-viscosity terms in the RANS equations, as noted earlier. The absence of transport stiffens the implicitly coupled, non-linear set of algebraic equations. The transport terms tend to 'anchor' the stresses at any one location to neighbouring stresses, thus diminishing the sensitivity of the stresses to local perturbations in the strain. The absence of transport also tends to favour the propagation and amplification of perturbations across the set of algebraic stress equations, requiring the introduction of further stability promoting algorithmic aids, and forcing the use of artificial under-relaxation measures (see Huang and Leschziner (1985) and Huang (1986)) to slow the rate of change in the solution. Second, the algebraic set, being non-linear, can have more than one solution, with the solution emerging being dependent on the initial conditions specified. Third, to actually solve the system, the turbulence energy must be derived from its own transport equation, so as to provide P_k, k and ε to the set. However, k also follows from $\frac{1}{2}\overline{u_k u_k}$, which arises from the algebraic set of equations for the stress components, and which may not comply with the turbulence energy derived from its own transport equation, unless the algebraic set contracts exactly to the k-equation. In this case, an ambiguity will also arise in respect of P_k.

The recognition that the only promising path to a broader acceptance of models that are more general than linear eddy-viscosity models has to follow the derivation of forms that are, essentially, expanded variations of the eddy-viscosity framework, has led to the emergence of two related groups of models, namely

- explicit algebraic Reynolds-stress models (EARSMs), and
- non-linear eddy-viscosity models NLEVMs.

The former is linked closely to the approximation in Eq. (15.1), and the latter is formally divorced from RSTMs and based on non-linear expansions of the stresses in terms of the strain and vorticity tensors, coupled with extensive calibration. While the routes adopted in deriving models within the above two categories differ materially, the final result is, essentially, common: a model that relates, explicitly, the stresses to derivative of the strains, products of strains of order 2 and possibly higher, the turbulence energy, and the dissipation rate (or another length-scale surrogate), with the leading term of the relationship being proportional to the strain tensor and thus interpretable as an eddy-viscosity fragment. A complicating issue, especially in efforts to discuss models under relevant model-type headings, is that the boundary between EARSMs and NLEVMs is fuzzy. Thus, several models combine elements of both, to the extent that their membership of either category is ambiguous. This will become clear in the exposition that follows.

15.2 *Explicit algebraic Reynolds-stress models (EARSMs)*

The nature of deriving an EARSM can be described very simply as: *a purely mathematical process of inverting the implicit algebraic set to arrive at explicit relationships that link the stresses to the strains and other known or determinable quantities.* Crucially, no new physics is involved in the process. Indeed, the algebraic form of the parent RSTM is inevitably less realistic than the RSTM, a point made earlier, especially in relation to streamline curvature. The purpose is purely to procure simplicity and stability.

It is instructive to consider a 'first-order solution' of the ARSM system in Eq. (15.4), with the left hand side set to zero, derived as a first step of an iterative approach. To this end, we assume that $\frac{P_k}{\varepsilon}$ is known, as determined from the solution of the k-equation. In this case, Eq. (15.4) can be written:

$$a_{ij} = -\frac{1}{\left(\alpha_1 + \alpha_2 \frac{P_k}{\varepsilon}\right)} \left[\alpha_3 S^*_{ij} + \alpha_4 (a_{ik}\Omega^*_{kj} - \Omega^*_{ik}a_{kj}) + \alpha_5 \left(a_{ik}S^*_{kj} + S^*_{ik}a_{kj} - \frac{2}{3} a_{kl}S^*_{lk}\delta_{ij} \right) \right]. \tag{15.5}$$

We note, first, that the leading term is linear in S^*_{ij}, so the multiplier of the strain is equivalent to an eddy viscosity. Second, while a_{ij} can be expected to be a complicated function of strain and vorticity tensors, the *minimal* form, with only the first term in the square bracket retained, is obviously linear in S^*_{ij}. We may view this to be the starting 'guess' of the iterative process. If this minimal form is inserted into the right-hand side, the result is clearly a quadratic model containing products such as $S^*_{ik}S^*_{kj}$, $S^*_{ik}\Omega^*_{kj}$, etc. This demonstrates, albeit tentatively, that we can expect an explicit constitutive relationship that links the stresses to (at least) quadratic products of the strain and vorticity tensors.

In the following sub-sections, we consider, first, the rigorous mathematical formalism that yields EARSMs. The terminology 'mathematical formalism' needs to be underlined here, for the process adds no physical capabilities to the baseline RSTM. It is 'merely' a set of manipulations, based on tensor algebra, that is designed to transform the implicit set to an explicit one, with the leading terms interpretable as a linear eddy-viscosity representation. In discussing the formalism, we only expose major steps and manipulations, so as to allow readers to gain an understanding of the methodology and the underlying logic. Detailed derivations, especially in respect of basic tensor and matrix algebra, are bypassed, with references given to related sources. Following this section, we consider approximations and simplifications to the formalism, which give rise to 'non-rigorous' EARSM forms.

15.2.1 *General formalism*

It should be clear from Eqs. (15.4) and (15.5) that the general solution of Eq. (15.4) must be of the form[1]:

$$a_{ij} = a_{ij}(S^*_{kl}, \Omega^*_{kl}, P_k/\varepsilon). \tag{15.6}$$

Assuming P_k/ε is known, this might suggest that the right-hand side should be an infinite series of ever higher order products of the strain and vorticity tensors, that would need to be arbitrarily truncated. This is not the case, however, because the Cayley–Hamilton theorem (see Spencer and Rivlin (1959)) states that the highest-order independent products of the tensors in Eq. (15.6) cannot exceed 5, and that products of order higher than 5 can be expressed in terms of lower-order products, so that only a finite number of linearly independent products arise.

Consider the simpler case in which a general function is sought of a single tensor, namely:

$$\phi_{ij} = \phi_{ij}(a_{kl}) = \sum_{\lambda=0}^{\infty} \beta_\lambda a^\lambda_{kl}, \tag{15.7}$$

the Cayley–Hamilton theorem states the following:

$$a_{ik}a_{kl}a_{lj} = I_a a_{ik}a_{kj} + \frac{1}{2}(II_a - I_a^2)a_{ij} + \frac{1}{6}(2III_a - 3I_a II_a + I_a^3)\delta_{ij}, \tag{15.8}$$

where $I_a = a_{kk}$, $II_a = a_{kl}a_{lk}$, $III_a = a_{kl}a_{lm}a_{mk}$ are the first, second and third invariants of the anisotropy tensor, respectively. This means that all products of order higher than 3 can also be expressed in terms of second-order products. Thus, the most general

[1]In fact, this *Ansatz* is the *starting point* of the formulation of non-linear eddy-viscosity models.

relationship is of the form

$$\phi_{ij} = \beta_1 \delta_{ij} + \beta_2 a_{ij} + \beta_3 a_{ik} a_{kj}, \tag{15.9}$$

in which the coefficients are functions of the invariants.

The theorem extends to functions of the type in Eq. (15.6), in which a tensor depends on two other tensors. In this case, Pope (1975) shows that the Cayley–Hamilton theorem dictates that the general — i.e. complete — expression contains ten independent groups of tensor products of $\mathbf{S} \equiv S^*_{kl}$ and $\boldsymbol{\Omega} \equiv \Omega^*_{kl}$, the highest-order product being 5. This expression may be written as follows:

$$\begin{aligned} a_{ij} = {} & \beta_1 \underbrace{\mathbf{S}}_{\mathbf{T}^{(1)}} + \beta_2 \underbrace{\left(\mathbf{S}^2 - \frac{1}{3}\mathbf{I}\{\mathbf{S}^2\}\right)}_{\mathbf{T}^{(2)}} + \beta_3 \underbrace{\left(\boldsymbol{\Omega}^2 - \frac{1}{3}\mathbf{I}\{\boldsymbol{\Omega}^2\}\right)}_{\mathbf{T}^{(3)}} \\ & + \beta_4 \underbrace{(\mathbf{S}\boldsymbol{\Omega} - \boldsymbol{\Omega}\mathbf{S})}_{\mathbf{T}^{(4)}} + \beta_5 \underbrace{(\mathbf{S}^2\boldsymbol{\Omega} - \boldsymbol{\Omega}\mathbf{S}^2)}_{\mathbf{T}^{(5)}} \\ & + \beta_6 \underbrace{\left(\mathbf{S}\boldsymbol{\Omega}^2 + \boldsymbol{\Omega}^2\mathbf{S} - \frac{2}{3}\mathbf{I}\{\mathbf{S}\boldsymbol{\Omega}^2\}\right)}_{\mathbf{T}^{(6)}} \\ & + \beta_7 \underbrace{\left(\mathbf{S}^2\boldsymbol{\Omega}^2 + \boldsymbol{\Omega}^2\mathbf{S}^2 - \frac{2}{3}\mathbf{I}\{\mathbf{S}^2\boldsymbol{\Omega}^2\}\right)}_{\mathbf{T}^{(7)}} + \beta_8 \underbrace{(\mathbf{S}\boldsymbol{\Omega}\mathbf{S}^2 - \mathbf{S}^2\boldsymbol{\Omega}\mathbf{S})}_{\mathbf{T}^{(8)}} \\ & + \beta_9 \underbrace{(\boldsymbol{\Omega}\mathbf{S}\boldsymbol{\Omega}^2 - \boldsymbol{\Omega}^2\mathbf{S}\boldsymbol{\Omega})}_{\mathbf{T}^{(9)}} + \beta_{10} \underbrace{(\boldsymbol{\Omega}\mathbf{S}^2\boldsymbol{\Omega}^2 - \boldsymbol{\Omega}^2\mathbf{S}^2\boldsymbol{\Omega})}_{\mathbf{T}^{(10)}} \end{aligned} \tag{15.10}$$

or

$$a_{ij} = \sum_{\lambda} \beta_\lambda \mathbf{T}^{(\lambda)}, \tag{15.11}$$

where $\mathbf{I}$ is the Kronecker delta and $\{\ldots\}$ denotes the invariant of the product in the braces. Thus, for example: $\mathbf{S}^2 = S^*_{ik} S^*_{kj}$, $\mathbf{S}\boldsymbol{\Omega} = S^*_{ik}\Omega^*_{kj}$, $\mathbf{S}^2\boldsymbol{\Omega} = S^*_{ik}\Omega^*_{kl}\Omega^*_{lj}$, $\mathbf{S}^2\boldsymbol{\Omega}^2 = S^*_{ik}S^*_{kl}\Omega^*_{lm}\Omega^*_{mj}$, $\boldsymbol{\Omega}\mathbf{S}^2\boldsymbol{\Omega}^2 = \Omega^*_{ik}$ $S^*_{kl}S^*_{lm}\Omega^*_{mn}\Omega^*_{nj}$, $\{\mathbf{S}^2\} = S^*_{kl}S^*_{lk}$, $\{\mathbf{S}\boldsymbol{\Omega}^2\} = S^*_{ik}\Omega^*_{kl}\Omega^*_{li}$, $\{\mathbf{S}^2\boldsymbol{\Omega}^2\} = S^*_{ik}$ $S^*_{kl}\Omega^*_{lm}\Omega^*_{mi}$. The scalar coefficients $\beta_1 - \beta_{10}$ are unknown and remain to be determined, but are functions of the five independent scalar invariants:

$$\beta_\lambda = \beta_\lambda \left(\{\mathbf{S}^2\}, \{\mathbf{\Omega}^2\} \{\mathbf{S}^3\}, \{\mathbf{S}\mathbf{\Omega}^2\} \{\mathbf{S}^2\mathbf{\Omega}^2\} \right), \tag{15.12}$$

which is an outcome of the application of the Cayley–Hamilton theorem to the derivation of Eq. (15.10).

The task is now to obtain the coefficients β_λ. One way of doing so is by calibration against a range of flows, possibly subject to formal constraints (e.g. realisability, as done by Shih *et al.* (1995)). This is the approach taken in the formulation of non-linear eddy-viscosity models (see Section 15.4), in which case the modelling path does not entail the use of a second-moment closure. Here, in contrast, the coefficients β_λ are determined, in principle, by inserting Eq. (15.10) into the implicit ARSM representation in Eq. (15.5), or another form that arises from a different RSTM.

Assuming a 'linearized' representation, in which $\frac{P_k}{\varepsilon}$ is known, and in which all coefficients, if functions of P_k and stress invariants (as is the case with the RSTM of Speziale, Sarkar and Gatski (1991)), are treated as constants (i.e. temporarily 'frozen'), we can write Eq. (15.5), compactly, as follows:

$$\begin{aligned} a_{ij} &= \alpha_1^* S_{ij}^* + \alpha_2^* (a_{ik}\Omega_{kj}^* - \Omega_{ik}^* a_{kj}) \\ &\quad + \alpha_3^* \left(a_{ik} S_{kj}^* + S_{ik}^* a_{kj} - \frac{2}{3} a_{kl} S_{lk}^* \delta_{ij} \right). \end{aligned} \tag{15.13}$$

Substitution of Eq. (15.10) into Eq. (15.13) gives:

$$\begin{aligned} \sum_\lambda \beta_\lambda \mathbf{T}^{(\lambda)} &= \alpha_1^* \mathbf{T}^{(\lambda)} \delta_{1\lambda} + \alpha_2^* \left(\mathbf{T}^{(\lambda)} \Omega_{kj}^* - \Omega_{ik}^* \mathbf{T}^{(\lambda)} \right) \\ &\quad + \alpha_3^* \sum_\lambda \beta_\lambda \left(\mathbf{T}^{(\lambda)} S_{kj}^* + S_{ik}^* \mathbf{T}^{(\lambda)} - \frac{2}{3} \mathbf{T}^{(\lambda)} S_{lk}^* \delta_{ij} \right). \end{aligned} \tag{15.14}$$

Because the tensors $\mathbf{T}^{(\lambda)}$ are linearly independent functions of S_{ij}^* and Ω_{ij}^*, it is possible to express the last two groups on the right-hand side, unambiguously and uniquely, as functions of $\mathbf{T}^{(\lambda)}$. This allows the following substitutions:

$$\left(\mathbf{T}^{(\lambda)}\Omega^*_{kj} - \Omega^*_{ik}\mathbf{T}^{(\lambda)}\right) = \sum_\gamma J_{\lambda\gamma}\mathbf{T}^{(\gamma)}$$

$$\left(\mathbf{T}^{(\lambda)}S^*_{kj} + S^*_{ik}\mathbf{T}^{(\lambda)} - \frac{2}{3}\mathbf{T}^{(\lambda)}S^*_{lk}\delta_{ij}\right) = \sum_\gamma H_{\lambda\gamma}\mathbf{T}^{(\gamma)}, \quad (15.15)$$

where $H_{\lambda\gamma}$ and $J_{\lambda\gamma}$ are 10×10 matrices with elements that contain only invariants of the strain and vorticity tensors, and these may be determined by a repeated application of the Cayley–Hamilton theorem (see Pope (1975)).

Eq. (15.14) can now be written:

$$\sum_\lambda \beta_\lambda \mathbf{T}^{(\lambda)} = \alpha^*_1\mathbf{T}^{(\lambda)}\delta_{1\lambda} + \alpha^*_2\sum_\lambda \beta_\lambda \sum_\gamma J_{\lambda\gamma}\mathbf{T}^{(\gamma)}$$
$$+ \alpha^*_3\sum_\lambda \beta_\lambda \sum_\gamma H_{\lambda\gamma}\mathbf{T}^{(\lambda)}. \quad (15.16)$$

Because $\mathbf{T}^{(\lambda)}$ are independent, their respective coefficients on either side of Eq. (15.16) may be equated, i.e.

$$\beta_\lambda = \alpha^*_1\delta_{1\lambda} + \alpha^*_2\sum_\gamma \beta_\gamma J_{\gamma\lambda} + \alpha^*_3\sum_\gamma \beta_\gamma H_{\gamma\lambda}. \quad (15.17)$$

This is a set of 10 equations in 10 unknowns — the vector β_λ, with coefficient matrices $H_{\lambda\gamma}$ and $J_{\lambda\gamma}$. Eq. (15.17) can be written in the following matrix form:

$$\underbrace{[\alpha^*_1\delta_{\gamma\lambda} + \alpha^*_2 J_{\gamma\lambda} + \alpha^*_3 H_{\gamma\lambda}]}_{\mathbf{A}}\,\underbrace{\beta_\lambda}_{\mathbf{B}} = \underbrace{-\alpha^*_1\delta_{1\lambda}}_{\mathbf{C}}. \quad (15.18)$$

To obtain β_λ — the solution — matrix $\mathbf{A} \equiv [\ldots]$ needs to be inverted, a task that is not normally possible to perform analytically in three-dimensional conditions, although much depends on the precise nature of the ARSM equation and the underlying RSTM. In any event, the above explanation should suffice to make clear, in principle, how the coefficients β_λ can be determined, and that they are dictated by the particular RSTM that formed the foundation of the process.

15.2.2 *The model of Pope (1975)*

Pope was the first to derive an EARSM along the formalism summarised in Section 15.2.1, but restricted himself to two-dimensional flows. In this case, Eq. (15.10) simplifies drastically to:

$$a_{ij} = \beta_1 \mathbf{S} + \beta_2 \left(\mathbf{S}^2 - \frac{1}{3}\mathbf{I}\left\{\mathbf{S}^2\right\} \right) + \beta_4 \left(\mathbf{S}\boldsymbol{\Omega} - \boldsymbol{\Omega}\mathbf{S} \right), \qquad (15.19)$$

the tensors containing only three independent elements ($i, j = 1, 2$), and the anisotropy tensor having an extra non-zero diagonal element due to the constraint $a_{33} = -(a_{11} + a_{22})$. The third quadratic term in Eq. (15.10), associated with $\boldsymbol{\Omega}^2$, reduces to zero, along with all other terms following. In this case, an analytic solution for $\beta_1, \beta_2, \beta_4$ is possible.

Pope's EARSM is based on the RSTM of Launder, Reece and Rodi (1975) (see Section 12.7). This model can be written:

$$a_{ij} = -b_0 \left[b_1 S^*_{ij} + b_2 \left(a_{ik} S^*_{kj} + S^*_{ik} a_{jk} - \frac{2}{3}\delta_{ij} a_{kl} S^*_{lk} \right) \right.$$
$$\left. - b_3 (a_{ik}\Omega^*_{kj} + \Omega^*_{ik} a_{jk}) \right], \qquad (15.20)$$

where

$$b_0 = \left(\frac{P_k}{\varepsilon} + C_1 - 1 \right)^{-1}, \quad b_1 = \frac{8}{15}, \quad b_2 = \frac{1}{11}(5 - 9C_2),$$
$$b_3 = \frac{1}{11}(7C_2 + 1), \quad C_1 = 1.5, \quad C_2 = 0.4.$$

The substitution of Eq. (15.19) into Eq. (15.20), eventually leads to the solution:

$$\beta_1 = -\frac{b_1 b_0}{\left(1 - 2b_0^2 b_3^2 \left\{\boldsymbol{\Omega}^2\right\} - \frac{2}{3} b_0^2 b_2^2 \left\{\mathbf{S}^2\right\}\right)}, \quad \beta_2 = \beta_1 b_0 b_2, \quad \beta_4 = \beta_1 b_0 b_3. \qquad (15.21)$$

Hence, β_1 may be interpreted as a strain- and vorticity-sensitised eddy-viscosity coefficient analogous to $2C_\mu$. Here, however, the coefficient is a function of the second strain and vorticity invariants. For the particular case of simple shear, $\{\mathbf{\Omega}^2\} = -\{\mathbf{S}^2\}$.[2] Also, if we make the (incorrect) assumption that $P_k/\varepsilon \approx 1$, regardless of the strain rate, we obtain from Eq. (15.21), after substitution of the model constants:

$$C_\mu = 0.177\frac{1}{(1+0.102\,\{\mathbf{S}^2\})}. \qquad (15.22)$$

In equilibrium conditions, $S^*_{eq} \equiv \sqrt{2S^*_{kl}S^*_{kl}}|_{eq} \approx 3.33$. Hence, $\{\mathbf{S}^2\}|_{eq} = 5.55$, and $C_\mu = 0.113$ which is fairly close to 0.09. However, the more important point is that Eq. (15.22) demonstrates that the EARSM returns a coefficient the value of which diminishes steadily with increasing strain rate, i.e. of the form needed to render the eddy-viscosity model realisable, as discussed in Section 11.2 (see Eq. (11.12), for example).

A further characteristic of Eq. (15.19) is that it predicts an anisotropic state in simple shear, in contrast to any linear formulation. Thus, the substitution of $S_{12} = \Omega_{12} = \frac{1}{2}\frac{\partial U_1}{\partial x_2}$ into Eq. (15.19), all other components being zero, yields:

$$\begin{aligned} a_{11} &= \frac{1}{12}\beta_2\left(\frac{k}{\varepsilon}\frac{\partial U_1}{\partial x_2}\right)^2 - \frac{1}{2}\beta_4\left(\frac{k}{\varepsilon}\frac{\partial U_1}{\partial x_2}\right)^2 \\ a_{22} &= \frac{1}{12}\beta_2\left(\frac{k}{\varepsilon}\frac{\partial U_1}{\partial x_2}\right)^2 + \frac{1}{2}\beta_4\left(\frac{k}{\varepsilon}\frac{\partial U_1}{\partial x_2}\right)^2. \end{aligned} \qquad (15.23)$$

Hence, the right-most quadratic term in Eq. (15.19) causes an increase in $\overline{u_1^2}$ and an equal and opposite decrease in $\overline{u_2^2}$. As the sum of the normal-stress components of the anisotropy tensor must be zero, it follows that $a_{33} = -\frac{1}{6}\beta_2\left(\frac{k}{\varepsilon}\frac{\partial U_1}{\partial x_2}\right)^2$.

15.2.3 *The model of Gatski and Speziale (1993)*

The extension of the above formalism to three-dimensional conditions was first undertaken by Taulbee (1992), who adopted a simplified form of the RSTM by Launder, Reece and Rodi — Eq. (15.20), with

[2] Note that the minus sign is due to the fact that, while $\{\mathbf{S}^2\} = S^*_{kl}S^*_{lk} = S^*_{kl}S^*_{kl}$, $\{\mathbf{\Omega}^2\} = \Omega^*_{kl}\Omega^*_{lk} = -\Omega^*_{kl}\Omega^*_{kl}$.

the assumption (or approximation) $C_2 = 5/9$, instead of 0.4, thus giving $b_2 = 0$, and with $\frac{P_k}{\varepsilon}$ retained as an separate entity in the EARSM derivation, as done by Pope. That same model was also used by Wallin & Johansson (2000), but with $\frac{P_k}{\varepsilon}$ treated explicitly, as will be explained in Section 15.2.4. Previous to this, Gatski and Speziale derived a three-dimensional EARSM based on the general quasi-linear RSTM of Speziale, Sarkar and Gatski (1991). This is the most complex model underpinning any EARSM (see also Jongen and Gatski (1998)). It is rarely used in its three-dimensional form, however, and it also requires a 'regularisation' to obviate pathological solutions.

The algebraic approximation of the RSTM may be written in the following form, corresponding to Eq. (15.5)[3]:

$$a_{ij} = -b_0 \left[b_1 S^*_{ij} + b_2 \left(a_{ik} S^*_{kj} + S^*_{ik} a_{kj} - \frac{2}{3} a_{kl} S^*_{lk} \delta_{ij} \right) + b_3 (a_{ik} \Omega^*_{kj} - \Omega^*_{ik} a_{kj}) \right], \tag{15.24}$$

with

$$b_0 = \left(1 - \frac{P_k}{\varepsilon} - \frac{1}{2} C_1\right)^{-1}, \quad b_1 = C_2 - \frac{4}{3}, \quad b_2 = \frac{1}{2} C_3 - 1,$$

$$b_3 = 1 - \frac{1}{2} C_4, \quad C_1 = 3.4 + 1.8 \frac{P_k}{\varepsilon}, \quad C_2 = 0.8 - 0.65 (a_{kl} a_{kl})^{\frac{1}{2}},$$

$$C_3 = 1.25, \quad C_4 = 0.4.$$

Here again, the extraction of the EARSM is based on the assumption of quasi-linearity, i.e. the model coefficients are assumed frozen. In particular, $\frac{P_k}{\varepsilon}$ is assumed to be at the equilibrium value ≈ 1.9 (not 1, as applicable to the log-law region of a boundary layer!), which corresponds to the condition in which the turbulent time scale k/ε achieves a constant level in homogeneous shear (see Eqs. (9.19)–(9.21)). The

[3]The analysis of Gatski and Speziale includes system rotation Ω_k, in which case the body-force term $\varepsilon_{lji} \Omega_l$ is added to Ω_{ij}. We exclude system rotation in the present exposition.

imposition of a constant value for this ratio necessarily results in an internal inconsistency in any condition other than equilibrium in the above sense.

As noted in Section 15.2.1, the process of deriving the EARSM for three-dimensional conditions is much more laborious than for two-dimensional ones. In particular all tensorial groups in Eq. (15.10) are active, and the result of the process of inserting Eq. (15.10) into Eq. (15.24) is the 10×10 system of the form in Eq. (15.18) for the coefficients $\beta_1 - \beta_{10}$. These coefficients are lengthy functions of the various invariants of the strain and vorticity tensors as well as the model constants. The expressions are not given here, and may be found in the reference given above.

In practice, the application of this three-dimensional form is extremely rare, because of the complexity of Eq. (15.10), which is substantially aggravated by the complexity of the β-coefficients. For two-dimensional conditions, the model is formally identical to Eqs. (15.20) and (15.21), provided the model coefficients $b_0 - b_3$ in $\beta_1, \beta_2, \beta_4$ are replaced by those applicable to the model in Eq. (15.24).

Gatski and Speziale point out that the coefficients $\beta_1 - \beta_{10}$ may become singular, as a consequence of the possibility of the common denominator in all coefficients becoming zero. This is because the terms appearing in the coefficients (a combination of invariants and model constants) have variously positive and negative signs. To avoid this possibility, Gatski and Speziale introduced a synthetic *regularisation.* To illustrate the nature of the regularisation, we consider the two dimensional case for which (see Eq. (15.21));

$$\begin{aligned}\beta_1 &= -\frac{b_1 b_0}{\left(1 - 2b_0^2 b_3^2 \left\{\boldsymbol{\Omega}^2\right\} - \frac{2}{3} b_0^2 b_2^2 \left\{\mathbf{S}^2\right\}\right)} \\ &= -\frac{b_1 b_0}{\left[1 + 2b_0^2 b_3^2 (\Omega_{kl}^* \Omega_{kl}^*) - \frac{2}{3} b_0^2 b_2^2 (S_{kl}^* S_{kl}^*)\right]}. \end{aligned} \tag{15.25}$$

It is thus clear that large strains can result in $\beta_1 = 0$. One way to avoid this condition is to introduce the following replacement:

$$b_0^2 b_2^2 S_{kl}^* S_{kl}^* \leftarrow 1 - \frac{1}{1 + b_0^2 b_2^2 S_{kl}^* S_{kl}^*}. \tag{15.26}$$

For $S^*_{kl}S^*_{kl} \ll 1$, this substitution is quite accurate. As the strain increases, however, the right-hand side will become progressively smaller than the left-hand side, the limiting value being 1. In simple shear, we have noted, in Section 15.2.2, that $S^*_{kl}S^*_{kl}|_{eq} = 5.55$. For this state, $b_0^2 b_2^2 S^*_{kl}S^*_{kl} \approx 0.2$, and the difference between the two sides of Eq. (15.26) is around 15%. It is only for substantially larger strain rates that the difference becomes large. The approximation in Eq. (15.26), when inserted into Eq. (15.25), is found to avoid a singular behaviour.

15.2.4 *The model of Wallin and Johansson (2000)*

This is the most recent (formal) EARSM, and it has become the most popular, at least in academic work targeting (in a broad sense) industrial flows. Its merits include relative simplicity, partly due to the use (in one version) of the simplified Launder *et al.* RSTM ($C_2 = 5/9$, instead of 0.4, thus rendering $b_2 = 0$ in Eq. (15.20)), and the use of a formalism that allows the explicit treatment of $P_k/\varepsilon = -\{a_{ij}\mathbf{S}\}$, leading to an internally consistent formulation.

The starting point follows directly from Eq. (15.20):

$$b_0^{-1} a_{ij} = -b_1 S^*_{ij} + b_3(a_{ik}\Omega^*_{kj} + \Omega^*_{ik}a_{jk}), \tag{15.27}$$

with the coefficients given after Eq. (15.20), except for $C_2 = 5/9$ in b_3 and $C_1 = 1.8$, instead of 1.5.

With $b_0 = \left(C_1 - 1 - \frac{P_k}{\varepsilon}\right)$, Eq. (15.27) may be written:

$$Na_{ij} = -\frac{6}{5}\mathbf{S} + (a_{ik}\mathbf{\Omega} - \mathbf{\Omega}a_{kj})$$

or

$$N\mathbf{a} = -\frac{6}{5}\mathbf{S} + (\mathbf{a\Omega} - \mathbf{\Omega a}), \tag{15.28}$$

where $N = C_1' + \frac{9}{4}\frac{P_k}{\varepsilon}, C_1' = \frac{9}{4}(C_1 - 1)$.

For two-dimensional conditions, there are only two non-zero terms in Eq. (15.10):

$$a_{ij} = \beta_1 \mathbf{S} + \beta_4 (\mathbf{S\Omega} - \mathbf{\Omega S}), \tag{15.29}$$

which is even simpler than Pope's version, Eq. (15.19). The solution, following the formalism in Section 15.2.1 (but much simplified), is:

$$\beta_1 = -\frac{6}{5}\frac{N}{N^2 - 2\{\mathbf{\Omega}^2\}}, \quad \beta_4 = -\frac{6}{5}\frac{1}{N^2 - 2\{\mathbf{\Omega}^2\}}. \tag{15.30}$$

With this solution, Eq. (15.29) can now be substituted into the definition of N, in which $\frac{P_k}{\varepsilon} = -\{\mathbf{aS}\}$, the result being:

$$N^3 - C_1' N^2 - \left(\frac{27}{10}\{\mathbf{S}^2\} + 2\{\mathbf{\Omega}^2\}\right) N + 2C_1'\{\mathbf{\Omega}^2\} = 0. \tag{15.31}$$

Wallin and Johansson provide the only positive and real root to this equation, which is not reproduced here, but is, self-evidently, of the form:

$$N = N\left(C_1', \{\mathbf{S}^2\}\{\mathbf{\Omega}^2\}\right), \tag{15.32}$$

yielding P_k/ε that is consistent with the algebraic model equations.

For the particular case of an equilibrium shear layer, $P_k/\varepsilon = 1$, $N = 3.37$. From Eq. (15.30), $\beta_1 = -\frac{4.05}{11.4+2\{\mathbf{S}^2\}}$; $\beta_2 = -\frac{1.2}{11.4+2\{\mathbf{S}^2\}}$. Hence, the implied linear eddy-viscosity coefficient (in the vicinity of the equilibrium point) is:

$$C_\mu = 0.5\frac{4.05}{11.4 + 2\{\mathbf{S}^2\}}. \tag{15.33}$$

For $\{\mathbf{S}^2\}|_{eq} = 5.55$, $C_\mu = 0.09$, which complies with the value expected. Here again, the form of C_μ, equivalent to Eq. (15.22), is such that the implied C_μ drops steadily with the strain. Figure 15.1 shows the actual variation of P_k/ε with S^* for a simple shear layer in which $S^*\left(= 2\sqrt{0.5\{\mathbf{S}^2\}}\right) = \Omega^*\left(= 2\sqrt{-0.5\{\mathbf{\Omega}^2\}}\right)$. For this same condition, the model predicts the anisotropy levels given in Table 15.1.

To appreciate the relevance of Fig. 15.1(a), we recall that $\frac{P_k}{\varepsilon} = \frac{-\overline{uv}}{k}S^*$. Hence, Fig. 15.1(a) implies that the slope of the (almost)

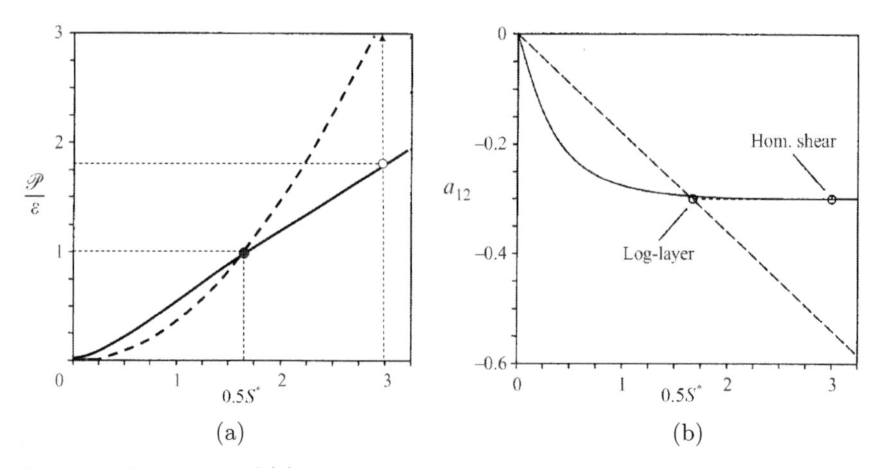

Fig. 15.1: Variations of (a) turbulence-energy-production-to-dissipation ratio and (b) ratio of shear stress to the turbulence energy, as a function of strain rate for simple shear. Solid lines: EARSM; dashed lines: linear eddy-viscosity model. The full circles identifies the log-region equilibrium condition; the open circle identifies the homogeneous-shear equilibrium state (reproduced with permission of Cambridge University Press from Wallin and Johansson (2000)).

Table 15.1: comparison of model-predicted anisotropy with DNS data for channel flow at $Re_\tau = 395$.

	a_{12}	a_{11}	a_{22}	a_{33}
Model	−0.3	0.25	−0.25	0
DNS (channel log layer)	−0.29	0.34	−0.26	−0.08

straight line of the EARSM is compatible with $\frac{-\overline{uv}}{k} \approx 0.3$, regardless of the strain rate. This is confirmed in Fig 15.1(b). It is recalled (see Section 9.4.3) that Menter's shear-stress transport (SST) limiter imposes the ratio $\frac{-\overline{uv}}{k} = 0.3$ beyond the equilibrium point (the solid circle in Fig. 15.1). In contrast, the eddy-viscosity model predicts the ratio to rise non-linearly with strain, $\frac{-\overline{uv}}{k} = 0.3\sqrt{\frac{P_k}{\varepsilon}}$ (see Eq. (9.76)). This is precisely the wrong behaviour that needed to be counteracted by the shear-stress limiter in Menter's SST model — the realisability correction which sensitises the coefficient C_μ to the strain rate (Eqs. (9.72)–(9.74)). The fact that a_{12} tends to a constant value at large strain rates is a consequence of the consistent (explicit) treatment of the production rate (or N, via Eq. (15.31)). In contrast, Gatski and Speziale's model yields a decline of a_{12} towards zero at large strain rates, due to the fixing of the production rate by reference to the equilibrium value $\frac{P_k}{\varepsilon} \approx 1.9$ in homogeneous strain.

For three-dimensional conditions, the use of the RSTM of Launder *et al.* (1975), in preference to the model of Speziale *et al.* (1991), in conjunction with the simplification $b_2 = 0$, results in a formulation that is substantially simpler than that of Gatski and Speziale. An added advantage is the avoidance of singularities of the type explained by reference to Eqs. (15.25) and (15.26)). As a consequence, the general expression in Eq. (15.10) reduces to one having only five finite non-zero terms:

$$a_{ij} = \beta_1 \mathbf{S} + \beta_3 \left(\mathbf{\Omega}^2 - \frac{1}{3}\mathbf{I}\left\{\mathbf{\Omega}^2\right\} \right) + \beta_4(\mathbf{S\Omega} - \mathbf{\Omega S}) + \beta_6 \left(\mathbf{S\Omega}^2 + \mathbf{\Omega}^2\mathbf{S} - \frac{2}{3}\mathbf{I}\left\{\mathbf{S\Omega}^2\right\} \right) + \beta_9(\mathbf{\Omega S \Omega}^2 - \mathbf{\Omega}^2\mathbf{S\Omega}), \tag{15.34}$$

where

$$\beta_1 = -\frac{N(2N^2 - 7\left\{\mathbf{\Omega}^2\right\})}{Q}, \quad \beta_3 = -\frac{12N^{-1}\left\{\mathbf{S\Omega}^2\right\}}{Q},$$

$$\beta_4 = -\frac{2(N^2 - 2\left\{\mathbf{\Omega}^2\right\})}{Q}, \quad \beta_6 = \frac{6N}{Q}, \beta_9 = \frac{6}{Q},$$

$$Q = \frac{5}{6}(N^2 - 2\left\{\mathbf{\Omega}^2\right\})(2N^2 - \left\{\mathbf{\Omega}^2\right\}).$$

As in the two-dimensional case, this solution can now be inserted into the definition of N (see Eq. (15.28)), and this leads to the sixth-order equation:

$$N^6 - C_1' N^5 - \left(\frac{27}{10}\left\{\mathbf{S}^2\right\} + \frac{5}{2}\left\{\mathbf{\Omega}^2\right\} \right) N^4 + \frac{5}{2} C_1' \left\{\mathbf{\Omega}^2\right\} N^3 + \left(\left\{\mathbf{\Omega}^2\right\} + \frac{189}{20}\left\{\mathbf{S}^2\right\}\left\{\mathbf{\Omega}^2\right\} - \frac{81}{5}\left\{\mathbf{S}^2\mathbf{\Omega}^2\right\} \right) N^2 - C_1' \left\{\mathbf{\Omega}^2\right\} N - \frac{81}{5}\left\{\mathbf{S\Omega}^2\right\} = 0. \tag{15.35}$$

This equation reduces to a two-dimensional form, upon the insertion $\{\mathbf{S\Omega}^2\} = 0$ and $\{\mathbf{S}^2\mathbf{\Omega}^2\} = 0.5\{\mathbf{S}^2\}\{\mathbf{\Omega}^2\}$, because $\{\mathbf{S}^2\}, \{\mathbf{\Omega}^2\}$ are the only independent invariants in two-dimensional conditions.

Equation (15.35) cannot be solved analytically, and Wallin and Johansson propose an approximate route. This entails the following steps:

- The starting point is the solution of the cubic equation for two-dimensional conditions, Eq. (15.31), with the solution in Eq. (15.32) denoted by N_{2d}.
- The invariants $\{\mathbf{S\Omega}^2\}$ and $\{\mathbf{S}^2\mathbf{\Omega}^2\}$ are perturbed around the two-dimensional solution: $\{\mathbf{S\Omega}^2\} = 0 + \varphi_1$, $\{\mathbf{S}^2\mathbf{\Omega}^2\} = 0.5\{\mathbf{S}^2\}\{\mathbf{\Omega}^2\} + \varphi_2$, where the leading terms in both are the respective two-dimensional values.
- Insertion of these perturbations into Eq. (15.35) leads to:

$$N = N_{2d} + \frac{162(\varphi_1 + \varphi_2 N_{2d}^2)}{D} + O(\varphi_1^2, \varphi_2^2, \varphi_1^4, \varphi_2^4), \tag{15.36}$$

  where

$$D = 20N_{2d}^4\left(N_{2d} - \frac{1}{2}C_1'\right) - \{\mathbf{\Omega}^2\}\,(10N_{2d}^3 + 15C_1'N_{2d}^2) + 10C_1'\,\{\mathbf{\Omega}^2\}^2.$$

For the model to remain fully explicit, one needs to make a choice for φ_1 and φ_2. This choice is arbitrary, but the perturbations must clearly be relatively 'small' fractions of $\{\mathbf{S\Omega}^2\}$ and $\{\mathbf{S}^2\mathbf{\Omega}^2\} - 0.5\{\mathbf{S}^2\}\{\mathbf{\Omega}^2\}$, respectively.[4] The alternatives are either to use $N = N_{2d}$ or to retain P_k/ε in its implicit form in the model.

A second, more elaborate, version of Wallin and Johansson's modelling approach involves the removal of the constraint $b_2 = 0$, in which case the implicit ARSM equations assume the form in Eqs. (15.20) and (15.24), representing Launder *et al.*'s and Speziale

[4] Initial values for φ_1, φ_2 may be obtained, in principle, from the expressions under the second bullet point preceding Eq. (15.36), but this is computationally laborious and not normally done.

et al.'s RSTMs, respectively. Thus, with Wallin and Johansson's nomenclature adopted here, Eq. (15.24) arises as:

$$N\mathbf{a} = -A_1\mathbf{S} + (\mathbf{a\Omega} - \mathbf{\Omega a}) - A_2\left(\mathbf{aS} + \mathbf{Sa} - \frac{2}{3}\{\mathbf{aS}\}\mathbf{I}\right), \quad (15.37)$$

where $N = A_3 + A_4\frac{P_k}{\varepsilon}$. In the case of Launder *et al.*'s RSTM, N is identical to the definition used in Eq. (15.28). For Speziale *et al.*'s model (see Eq. (15.24)), $N = 1/(b_0 b_3) = \frac{1}{(1-\frac{1}{2}C_4)}\left(1 - \frac{1}{2}C_1 - \frac{P_k}{\varepsilon}\right)$. However, it is recalled that the Gatski–Speziale EARSM derived therefrom assumed $\frac{P_k}{\varepsilon}$ to have a constant value, corresponding to equilibrium in homogeneous shear. In contrast, Wallin and Johansson account for the variation in this term by treating it explicitly as part of the model derivation.

A second difference relative to the Gatski–Speziale derivation is that the underlying RSTM is not precisely the version given by Eq. (15.24). Rather, a 'linearised' form is used, in which the coefficients are purely numerical, rather than dependent implicitly upon the solution. Thus, in the case of $C_2 = 0.8 - 0.65(a_{kl}a_{kl})^{1/2}$, the anisotropy invariant is replaced by its equilibrium value in homogeneous turbulence — that is, Eq. (15.24) is used with the coefficients:

$$b_0 = \left(1 - \frac{1}{2}C_1^{(0)} - \left(1 + \frac{1}{2}C_1^{(1)}\right)\frac{P_k}{\varepsilon}\right)^{-1}, \quad b_1 = C_2 - \frac{4}{3},$$
$$b_2 = \frac{1}{2}C_3 - 1, \quad b_3 = 1 - \frac{1}{2}C_4, \quad C_1^{(0)} = 3.4, \quad C_1^{(1)} = 1.8,$$
$$C_2 = 0.36, \quad C_3 = 1.25, \quad C_4 = 0.4.$$

As a consequence Eq. (15.24), with the above coefficients, and written in the form of Eq. (15.37), gives the coefficients, $N = 1/(b_0 b_3)$, $A_1 = -b_1/b_3$, $A_2 = b_2/b_3$.

The solution of Eq. (15.10), which reduces to Eq. (15.19), is then:

$$\beta_1 = -\frac{A_1 N}{Q}, \quad \beta_2 = 2\frac{A_1 A_2}{Q}, \quad \beta_4 = -\frac{A_1}{Q}$$
$$Q = N^2 - 2\{\mathbf{\Omega}^2\} - \frac{2}{3}A_2^2\{\mathbf{S}^2\}. \quad (15.38)$$

Wallin and Johansson show, for the case of simple shear, that the differences in the predicted anisotropy levels between the EARSMs based on Launder *et al.*'s and Speziale's *et al.*'s RSTMS are small, except for the value of a_{33}, which is zero in the case of the former model, with the approximation $b_2 = 0$ in Eq. (15.24). In contrast, the model based on the linearized Speziale *et al.* RSTM predicts $a_{33} \approx -0.1$, and this results in a commensurate reduction in the value of a_{11}. However, the more important value of a_{12} is not very sensitive to the model version used.

The full three-dimensional version, described by Eqs. (15.34)–(15.36), is substantially more complex than that outlined by reference to Eqs. (15.29)–(15.31). In practice, it is often the case that the two-dimensional version is used even for three-dimensional conditions. To avoid any misinterpretation, it is important to point out here, however, that 'two-dimensional' merely means that certain model fragments are omitted from the full model, and that the reduced model still yields a three-dimensional solution $a_{ij}(i, j = 1, 3)$. In this case, all components of the velocity field need, of course, to be used in $\mathbf{S}$ and $\mathbf{\Omega}$ to secure a valid objective solution. Experiences reported by Naji, Mompean and Yahyaoi (2004) and Weinmann and Sandberg (2009), for example, suggest that the differences are modest, but this must depend on the flow details, and care is needed to avoid important flow features being missed. For example, the two-dimensional version will not resolve the streamwise corner vortices in a rectangular duct, caused by wall-induced stress anisotropy (referred to as 'transverse motion of the second kind'). This feature can be influential, as is exemplified by Menter, Garburak and Egorov's (2009) application of the EARSM to a three-dimensional diffuser preceded by a straight rectangular duct.

15.2.5 *Near-wall behaviour*

The physical realism and range of applicability of the EARSMs discussed above is self-evidently dictated by the validity of the parent RSTMs and of the transport equations for k and ε (or another length-scale surrogate). Neither, the model of Launder *et al.* nor that of Speziale *et al.* has been designed to represent correctly the flow in the viscosity-affected near-wall region. It is necessary, therefore, to introduce into the EARSMs near-wall corrections analogous to those discussed in Section 9.3 in relation to two-equation models. An important difference is, however, that the EARSMs can be expected to represent, at least to some degree, the inviscid blocking effect of the wall on the anisotropy, especially the strong reduction in the wall-normal stress $\overline{v^2}$ as the wall is approached. The implication is, therefore, that the corrections required can be expected to genuinely reflect the influence of only the viscosity on the stresses. Neither the EARSM of Pope (1975) nor that of Gatski and Speziale (1993) include such corrections. As will be recognized below, the reality is that the viscous corrections need to be such that their influence extends beyond the viscosity-affected region — a fact that is indicative of weaknesses in the models, rooted both in defects in the pressure-velocity-correlation model of the baseline RSTM and the weak-equilibrium constraint applied via Eq. (15.1).

In principle, the approach to adjusting the near-wall behaviour of EARSMs is to introduce van Driest-type damping functions into the expressions for a_{ij}, which parallels, essentially, the practice applied in linear eddy-viscosity modelling.

Close to a wall, a two-dimensional boundary layer is characterised by:

$$S_{12}^* = \frac{1}{2}\frac{k}{\varepsilon}\frac{\partial U}{\partial y}, \quad \{\mathbf{S}^2\} = -\{\boldsymbol{\Omega}^2\} = 2(S_{12}^*)^2, \tag{15.39}$$

in which case Eq. (15.19) becomes:

$$a_{12} = \beta_1 S_{12}^*, \quad a_{11} = \left(\frac{1}{3}\beta_2 - 2\beta_4\right)(S_{12}^*)^2,$$
$$a_{22} = \left(\frac{1}{3}\beta_2 + 2\beta_4\right)(S_{12}^*)^2. \tag{15.40}$$

The simplest approach is to pre-multiply a_{12} by

$$f_\mu = 1 - exp(-A^+ y^+), \tag{15.41}$$

with $A^+ = 0.038$, chosen by calibration against direct numerical simulation (DNS) data. This is the practice adopted by Wallin and Johansson.

Next, the normal-stress terms need to be modified. We recall first (see Fig. 5.3) that $\overline{v^2}/k \to 0$ and $\overline{u^2}/k \to 1.44$, i.e. $a_{22} \to -\frac{2}{3}, a_{11} \to 0.78$, as the wall is approached. In contrast, Wallin and Johansson show that their model gives $a_{22} \to -0.4, a_{11} \to 0.4$ or $\overline{u^2}/k \to 1.07$, $\overline{v^2}/k \to 0.266$. Hence, while the model captures the anisotropic state, it does not return the correct asymptotic behaviour. To procure the correct behaviour, Wallin and Johansson propose a blending between the unmodified values and the correct asymptotic conditions. The latter is described by Eqs. (5.4) and (5.5), which may be re-written, in terms of a_{ij} as follows:

$$a_{11} = \frac{\overline{u^2}}{k} - \frac{2}{3} = \left(\frac{2\alpha_2}{(\alpha_2+\alpha_3)} - \frac{2}{3}\right) + O(y)$$

$$a_{22} = \frac{\overline{v^2}}{k} - \frac{2}{3} = \left(\frac{2\alpha_4}{(\alpha_2+\alpha_3)}y^2 - \frac{2}{3}\right) + O(y^3) = -\frac{2}{3} + O(y^2). \tag{15.42}$$

The blending proposed is thus as follows:

$$\begin{aligned} a_{11} &\leftarrow f_\mu^2 a_{11} + (1-f_\mu^2)\left(\frac{2\alpha_2}{(\alpha_2+\alpha_3)} - \frac{2}{3}\right) \\ a_{22} &\leftarrow f_\mu^2 a_{22} + (1-f_\mu^2)\left(-\frac{2}{3}\right), \end{aligned} \tag{15.43}$$

where a_{11}, a_{22} on the right-hand side are the values arising from the EARSM.

A problem with Eq. (15.43) is that the α-coefficients are not known, and these need to be determined from DNS or experimental data. Wallin and Johansson state that the choice $\frac{2\alpha_2}{(\alpha_2+\alpha_3)} = 1.8$ provides an adequate fit, which is broadly in line with the value 1.4–1.5 suggested by Fig. 5.3.

Substitution of Eq. (15.43) into Eq. (15.40) is readily shown to result in:

$$\beta_2 \leftarrow \frac{3\alpha^* - 4}{2(S_{12}^*)^2}(1-f_\mu^2), \quad \beta_4 \leftarrow f_\mu^2\beta_4 - \frac{\alpha^*}{4(S_{12}^*)^2}(1-f_\mu^2), \tag{15.44}$$

where $\alpha^* = \frac{2\alpha_2}{(\alpha_2+\alpha_3)}$.

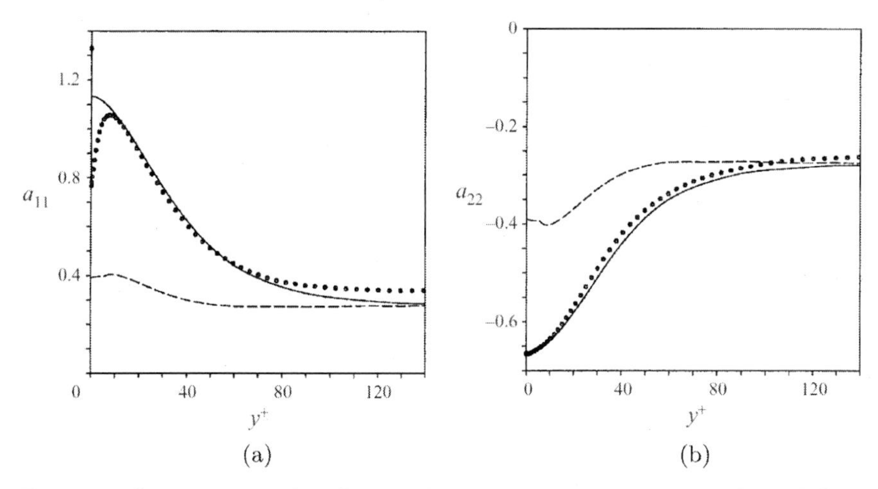

Fig. 15.2: Streamwise and wall-normal anisotropy components in channel flow, compared to DNS data at $Re_\tau = 395$ predicted by the model of Wallin and Johansson (2000) with and without viscous damping (reproduced with permission of Cambridge University Press from Wallin and Johansson (2000)).

A disconcerting aspect of the above viscosity-related correction is that its effect is required to extend well beyond the viscous layer, pointing to defects in the parent RSTM in respect of the near-wall decay of the stresses. This is illustrated in Fig. 15.2. As seen, the effects of the damping extends to $y^+ \approx 80$, which is much further than the distance within which viscous effects are expected to be influential. This is also why the constant A^+ in Eq. (15.41) is considerably smaller than it should be if the applicability of the damping function were to be confined to $y^+ \approx 30$.

The above discussion conveys the rationale of the approach to near-wall modelling. However, in practice, the model needs to be generalised to accommodate stagnation points (e.g. separation and reattachment) and three-dimensionality, the latter requiring a more general treatment of the invariants $\{\mathbf{S}^2\}$ and $\{\mathbf{\Omega}^2\}$. The generalisation is fairly straightforward, but also involves the introduction of limiters to compensate for defects rooted in the weak-equilibrium constraint, especially in near-zero shear-strain regions (e.g. the edge of a boundary layer) where the production vanishes and the transport of anisotropy becomes significant. However, as neither introduces new conceptual facets, we do not pursue the generalisation here, and the interested reader is referred to Wallin (2000), Wallin and Johansson (2000) and references therein.

15.2.6 *Scale-governing equations*

With any ARSM, there is a need to determine the turbulence energy and its dissipation rate, for these ultimately allow the stresses to be determined from the anisotropy tensor. Sections 7.4 and 9.1 have conveyed the importance of the length-scale-governing equation and its influence on the predictive properties of two-equation models. This influence is also pertinent to the present topic.

The equation governing the turbulence energy requires little discussion beyond a reference to Eq. (7.32), subject to the following modifications:

- The production is replaced by its generic form: $P_k = -\overline{u_i u_j}\frac{\partial U_i}{\partial x_j}$.
- Viscous diffusion is added.
- The scalar turbulent diffusivity, consistent with the linear eddy-viscosity approximation, is replaced by the tensorial GGDH form consistent with Eq. (12.19): $D_k = \frac{\partial}{\partial x_k}\left(C_s \frac{k}{\varepsilon}\overline{u_k u_l}\frac{\partial k}{\partial x_l}\right)$ (see also Eq. 12.34), with the constant $C_s = 0.22 - 0.25$, as recommended by Launder *et al.* (1975).

In contrast, there are many options to consider and examine in relation to the length-scale equation — particular examples being the choice between ε and ω, and the inclusion of various damping functions and limiters (see Section 9.4). Gatski and Speziale (1993) have used the standard, high-Reynolds-number form of the ε-equation, in harmony with the high-Reynolds-number restriction applicable to the underlying RSTM of Speziale *et al.* In contrast, Wallin and Johansson's formulation is applicable to low Reynolds numbers, via the modifications outlined in Section 15.2.5, and they have adopted either the $k - \varepsilon$ model of Chien (1982) (see Section 9.3.1) or the $k - \omega$ model of Wilcox (1994) (see Section 9.4.1), subject to the simplification of using a scalar turbulent diffusivity (e.g. $C_\mu \frac{k^2}{\sigma_\varepsilon \varepsilon}$ in the ε-equation, $C_\mu = -f_\mu \beta_1/2$ — see Eqs. (15.40) and (15.41)), and

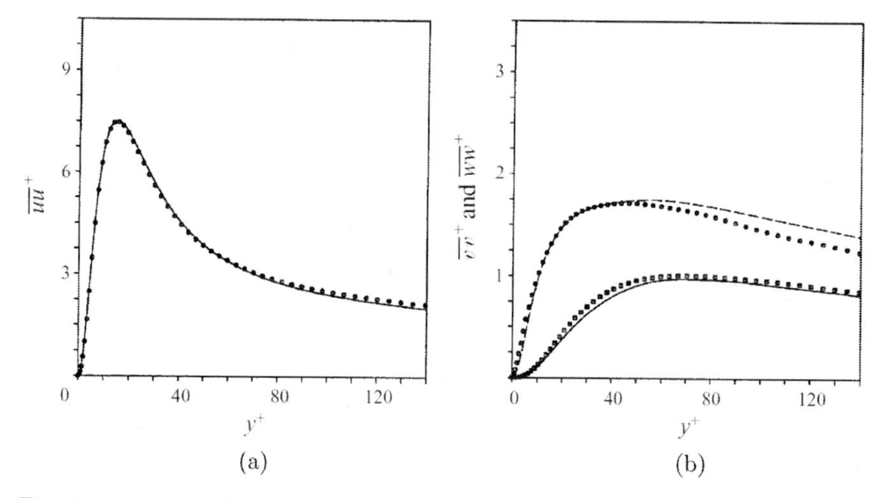

Fig. 15.3: Normal Reynolds stresses in channel flow, compared to DNS data at $Re_\tau = 395$, predicted by the model of Wallin and Johansson (2000) with viscous damping and the ω-equation (reproduced with permission of Cambridge University Press from Wallin and Johansson (2000)).

two further minor variations to a constant in the $k - \omega$ model and to the precise relationship between $\widetilde{\varepsilon}$ and ε (see Eq. 9.2).

With viscous damping introduced into the Wallin–Johansson model and the ω-equation used to determine the length scale, the normal stresses predicted by the model for channel flow are shown in Fig. 15.3. While this is an impressively accurate representation, it must be borne in mind that this is due, in no mean measure, to the empirical adjustments through Eq. (15.43).

15.3 *Approximations to rigorous EARSMs*

There are a number of models that have a close relationship to the above type of EARSMs, but have not been derived along the rigorous routes discussed in Section 15.2. Most of these precede the model of Wallin and Johansson (2000), and their formulation appears to have been motivated by the wish of their originators to avoid the complexity of earlier 'rigorous' models by Taulbee (1992) and Gatski and Speziale (1993), and/or to derive formulations that include a high-fidelity representation in the viscosity-affected near-wall layer. We cover here two representative models: those of Abe, Kondoh and Nagano (1997) and of Apsley and Leschziner (1998). A common feature of these models is that they start from, or lean on, the formal ARSM framework, either in implicit or explicit form, but then involve manipulations or approximations of the baseline form, to yield a result that does not comply with the EARSM formalism. In Section 15.4 we discuss non-linear eddy-viscosity models (NLEVMs) that do not involve the use of Reynolds-stress models at all. While this seems to present a clear line of distinction from the models considered in this section, it has to be acknowledged that the categorisation of a model under the present heading is not entirely clear-cut, because the degree to which models are based on the ARSM equations can be modest, fragmentary and obscured by the approximation process. This uncertainty is reflected by the fact that the designations EARSM and NLEVM are occasionally viewed and used interchangeably, even in the research literature.

While the exposition of the models introduced below is quite detailed, elements and steps that are not essential to conveying a basic understanding of the underlying principles and rationale are only sketchily discussed, or omitted altogether. Often, the lack of rigour in proposing certain model terms (usually referred to as 'corrections'), on the basis of intuition and calibration, defies a rational exposition and can only be stated as given by the models' originators. These model fragments need to be taken into account carefully, of course, if the reader wishes to incorporate any of the models into a code.

15.3.1 *The Model of Abe et al. (1997, 2003)*

Starting with the implicit algebraic form in Eq. (15.24),

$$a_{ij} = -b_0 \left[b_1 S^*_{ij} + b_2 \left(a_{ik} S^*_{kj} + S^*_{ik} a_{kj} - \frac{2}{3} a_{kl} S^*_{lk} \delta_{ij} \right) \right. \\ \left. + \; b_3 (a_{ik} \Omega^*_{kj} - \Omega^*_{ik} a_{kj}) \right], \tag{15.45}$$

Abe *et al.* first introduce the normalisations:

$$S^{\wedge}_{ij} \equiv b_0 b_2 S^*_{ij}; \quad \Omega^{\wedge}_{ij} \equiv b_0 b_3 \Omega^*_{ij}; \quad a^{\wedge}_{ij} \equiv -\frac{b_2}{b_1} a_{ij}. \tag{15.46}$$

With the above inserted, the equation may be solved, in two-dimensional conditions, for $a_{ij}^{\wedge}$, to give[5]:

$$a_{ij}^{\wedge} = \frac{6}{3 + 6\{\Omega_{mn}^{\wedge}\Omega_{mn}^{\wedge}\} - 2\{S_{mn}^{\wedge}S_{mn}^{\wedge}\}} \times \left[-S_{ij}^{\wedge} - (S_{ik}^{\wedge}\Omega_{kj}^{\wedge} - \Omega_{ik}^{\wedge}S_{kj}^{\wedge}) + 2\left(S_{ik}^{\wedge}S_{kj}^{\wedge} - \frac{1}{3}S_{mn}^{\wedge}S_{mn}^{\wedge}\delta_{ij}\right)\right], \tag{15.47}$$

which is entirely equivalent to Eqs. (15.19) and (15.21) arising from Pope's model. Hence, at this stage, the model contains no new features.

Abe *et al.* argue, correctly, that the model in Eq. (15.45) is not applicable to the near-wall region, especially in respect of viscous effects. This weakness has previously been the subject of Section 15.2.5, which addressed the provision of influential near-wall corrections to the basic Wallin–Johansson model, that aim to yield the correct wall-asymptotic variations of the stresses. Here, a different route is adopted, by reference to an early non-linear eddy-viscosity model proposed by Speziale (1987):

$$a_{ij} = -2\nu_t S_{ij} - 4C_D\frac{\nu_t^2}{k}(S_{ik}\Omega_{kj} - \Omega_{ki}S_{kj}) + 4C_D\frac{\nu_t^2}{k}\left(S_{ik}S_{kj} - \frac{1}{3}\{\mathbf{S}^2\}\delta_{ij}\right), \tag{15.48}$$

with $\nu_t = C_\mu f_\mu \frac{k^2}{\varepsilon}$. This model thus offers, in principle, a means of representing viscous effects via an appropriately calibrated, fluid-viscosity-damped, eddy viscosity.

With the substitutions:

$$S_{ij}^{\wedge} \equiv C_D C_\mu f_\mu \frac{k}{\varepsilon} S_{ij}; \quad \Omega_{ij}^{\wedge} \equiv 2C_D C_\mu f_\mu \frac{k}{\varepsilon}\Omega_{ij}; \quad a_{ij}^{\wedge} \equiv C_D a_{ij}, \tag{15.49}$$

[5]In fact, Abe *et al.* prefer to use the anisotropy definition $b_{ij} \equiv \frac{1}{2}a_{ij}$, but this is of no physical consequence.

Eq. (15.48) can be written:

$$a_{ij}^{\wedge} = \left[-S_{ij}^{\wedge} - (S_{ik}^{\wedge}\Omega_{kj}^{\wedge} - \Omega_{ik}^{\wedge}S_{kj}^{\wedge}) + 2\left(S_{ik}^{\wedge}S_{kj}^{\wedge} - \frac{1}{3}S_{mn}^{\wedge}S_{mn}^{\wedge}\delta_{ij} \right) \right]. \tag{15.50}$$

This has the same form as Eq. (15.47), except for the group pre-multiplying the bracketed right-hand side sum of terms. Abe *et al.* then proposed to use Eq. (15.47), but with the substitutions of the groups in Eq. (15.49), instead of Eq. (15.46), thus self-evidently yielding a non-rigorous EARSM.

As previously observed by Gatski and Speziale (Section 15.2.3), the denominator of the pre-multiplier in Eq. (15.47) can become zero or negative at large strain values, thus requiring a 'regularisation. This is achieved, here, by first re-writing the pre-multiplier as follows[6]:

$$\frac{6}{3 + 6\{\Omega_{mn}^{\wedge 2}\} - 2\{S_{mn}^{\wedge 2}\}} = \frac{1}{\frac{1}{2} + \frac{11}{3}\left(\frac{\{\Omega_{mn}^{\wedge 2}\}}{4}\right) + \frac{1}{3}\left(\frac{\{\Omega_{mn}^{\wedge 2}\}}{4} - \{S_{mn}^{\wedge 2}\}\right)}. \tag{15.51}$$

The rationale of this split is rooted in the fact that the second bracketed term in Eq. (15.51) vanishes in simple shear for which $\Omega_{mn}^2 = S_{mn}^2$ (note the factor 2 in the definition of $\Omega_{mn}^{\wedge}$ in Eq. (15.49)). Next, the limiter, $f_B = 1 + C\left(\frac{\{\Omega_{mn}^{\wedge 2}\}}{4} - \{S_{mn}^{\wedge 2}\}\right)$ is introduced as a multiplier of the second bracketed term to give:

$$\frac{6}{3 + 6\{\Omega_{mn}^{\wedge 2}\} - 2\{S_{mn}^{\wedge 2}\}} = \frac{1}{\frac{1}{2} + \frac{11}{3}\left(\frac{\{\Omega_{mn}^{\wedge 2}\}}{4}\right) + \frac{1}{3}\left(\frac{\{\Omega_{mn}^{\wedge 2}\}}{4} - \{S_{mn}^{\wedge 2}\}\right) f_B}. \tag{15.52}$$

[6]Note that $\{\Omega_{mn}^2\} = \Omega_{mn}\Omega_{mn}$ while $\{\mathbf{\Omega}^2\} = \Omega_{mn}\Omega_{nm}$, so that $\{\Omega_{mn}^2\} = -\{\mathbf{\Omega}^2\}$.

This multiplier takes the value 1 in simple shear, $\{S_{mn}^{\wedge 2}\} = \frac{\{\Omega_{mn}^{\wedge 2}\}}{4}$, and introduces an unconditionally positive term, quadratic in $\left(\frac{\{\Omega_{mn}^{\wedge 2}\}}{4} - \{S_{mn}^{\wedge 2}\}\right)$, which counteracts the negative linear term when $\{S_{mn}^{\wedge 2}\} > \frac{\{\Omega_{mn}^{\wedge 2}\}}{4}$, the degree of opposition depending on the constant C, chosen to be 5.

As a consequence of introducing the model in Eq. (15.50), the constants C_μ and C_D need to be determined. These are chosen as $C_\mu = 0.12$ and $C_D = 0.8$, so as to give $a_{12} = -0.3$ for the equilibrium states of both near-wall shear and homogeneous shear ($S_{eq}^* \approx 3.3$ and 6, respectively).

Next, the constants in the $k - \varepsilon$ equations, used as part of this model, need to be considered. The only non-standard elements, relative to the standard $k - \varepsilon$ model, are, first, the use of the GGDH approximations for diffusion (see Section 12.5):

$$D_k = \frac{\partial}{\partial x_k}\left(C_s' f_{t1}\left(C_\mu f_\mu \frac{k}{\varepsilon}\right)\overline{u_k u_l}\frac{\partial k}{\partial x_l}\right);$$

$$D_\varepsilon = \frac{\partial}{\partial x_k}\left(C_\varepsilon' f_{t2}\left(C_\mu f_\mu \frac{k}{\varepsilon}\right)\overline{u_k u_l}\frac{\partial \varepsilon}{\partial x_l}\right), \tag{15.53}$$

in which f_{t1} and f_{t2} are viscous-damping functions (see below); and, second, the introduction of a further damping function, f_ε, in the destruction term of the ε-equation. In addition, f_μ, appearing in the eddy-viscosity relation and associated with Eq. (15.48), needs to be prescribed.

It is recalled that Launder *et al.* (1975) recommend the constants $C_s = 0.22 - 0.25$ and $C_\varepsilon = 0.15 - 0.18$ in the turbulent-transport approximations for $\overline{u_i u_j}$ and ε, respectively. Abe *et al.* chose the values $C_s' = C_\varepsilon' = 1.4$, giving $C_s = C_\varepsilon = C_s' C_\mu = 0.17$, hence yielding a lower diffusion coefficient for turbulence energy than the value implied by Launder *et al.*'s model.

The choice of viscous-damping functions is, essentially, an intuitive process that is supported by trial and error, i.e. testing and calibration for a range of flows. In particular, these functions (especially f_μ and f_ε) are designed to procure the correct wall-asymptotic variation of k, ε, ν_t and $\overline{uv}$. A notable, though not unique, feature of Abe *et al.*'s functions is their dependence on the turbulent as well as the Kolmogorov scale, the latter pertaining to the viscous region very close to the wall. Thus, Abe *et al.* proposed:

$$f_\mu = \left\{1 + \frac{35}{R_t^{\frac{3}{4}}} \exp\left[-\left(\frac{R_t}{30}\right)^{\frac{3}{4}}\right]\right\}\left\{1 - \exp\left[-\left(\frac{n_\eta}{26}\right)^2\right]\right\}, \tag{15.54}$$

where $R_t = k^2/\nu\varepsilon$ and $n_\eta = u_\eta n/\nu$, in which $u_\eta = (\nu\varepsilon)^{1/4}$ is the Kolmogorov velocity scale, and n is the wall distance. Physically, the use of the Kolmogorov scales in the immediate near-wall region is entirely sensible, for in this layer, transport processes are dictated by the fluid properties. The Kolmogorov velocity scale is also used in f_{t1} and f_{t2}, Eq. (15.53). The former function chosen by Abe *et al.* is:

$$f_{t1} = 1 + 5\exp\left\{-\left(\frac{n_\eta}{5}\right)^5\right\}, \tag{15.55}$$

its effect being to increase the turbulent diffusion in the thin near-wall portion of the viscous sublayer.

Modified variations of the above model were later proposed by Abe, Jang and Leschziner (2003) (see also Jang *et al.* (2002)). These include corrective terms that improve the representation of effects arising from strong normal straining, and also of near-wall anisotropy, one variant operating with a $k - \omega$ model. A discussion of these variations is not pursued in this text, as this would not enhance the basic understanding of the principles underpinning the model. However, we include one set of results that illustrate the performance of the model (operating in conjunction with the $k - \omega$ equations) for a challenging test case that is widely used to examine the ability of turbulence models to resolve separation from curved surfaces. This flow is contained in a channel of periodically spaced two-dimensional hill-like obstructions on one wall (see also Fig. 14.4). If the number of obstructions is high, the flow between any two consecutive obstacles is periodic, and one may restrict attention to a single segment, as shown in Fig. 15.4, with periodicity conditions applied to the inlet and outlet (i.e. the conditions are

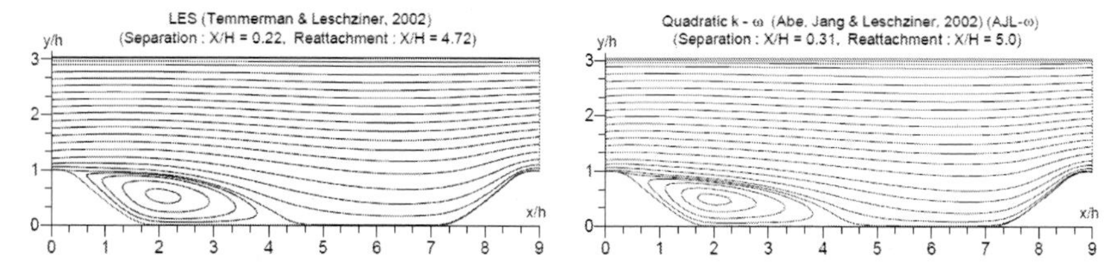

Fig. 15.4: Stream-function contours predicted for a flow in a periodic channel segment with hill-like obstacles. Comparisons with highly resolved LES data by Temmerman and Leschziner (2001) and Froehlich *et al.* (2005) (reproduced with permission of Springer from Jang *et al.* (2002)).

adjusted dynamically to be identical). Figure 15.4 shows results for the stream-function contours, whereas Fig. 15.5 gives mean-velocity and Reynolds-stress profiles at three streamwise stations: $x/H = 2, 6, 8$ (see Fig. 15.4).

Although the profiles at the first two stations agree well with the large eddy simulation (LES) solution, the stresses at the last station are clearly not well reproduced. However, around this station, the flow undergoes strong acceleration and very rapid streamwise changes in flow conditions, and these conditions pose serious challenges to any model.

15.3.2 *The Model of Apsley and Leschziner (1998)*

Starting from the implicit ARSM (Eq. (15.20)), which can be written, more compactly, as:

$$a_{ij} = -\alpha S^*_{ij} - \beta \left(a_{ik} S^*_{kj} + S^*_{ik} a_{kj} - \frac{2}{3} a_{kl} S^*_{lk} \delta_{ij} \right) - \gamma (a_{ik} \Omega^*_{kj} - \Omega^*_{ik} a_{kj}), \tag{15.56}$$

where $\alpha = b_0 b_1, \beta = b_0 b_2, \gamma = -b_0 b_3$, Apsley and Leschziner derive an explicit form by adopting an iterative approach, rather than by a formal solution of the system in Eq. (15.10). The starting point is:

$$a^{(1)}_{ij} = -\alpha \mathbf{S}, \tag{15.57}$$

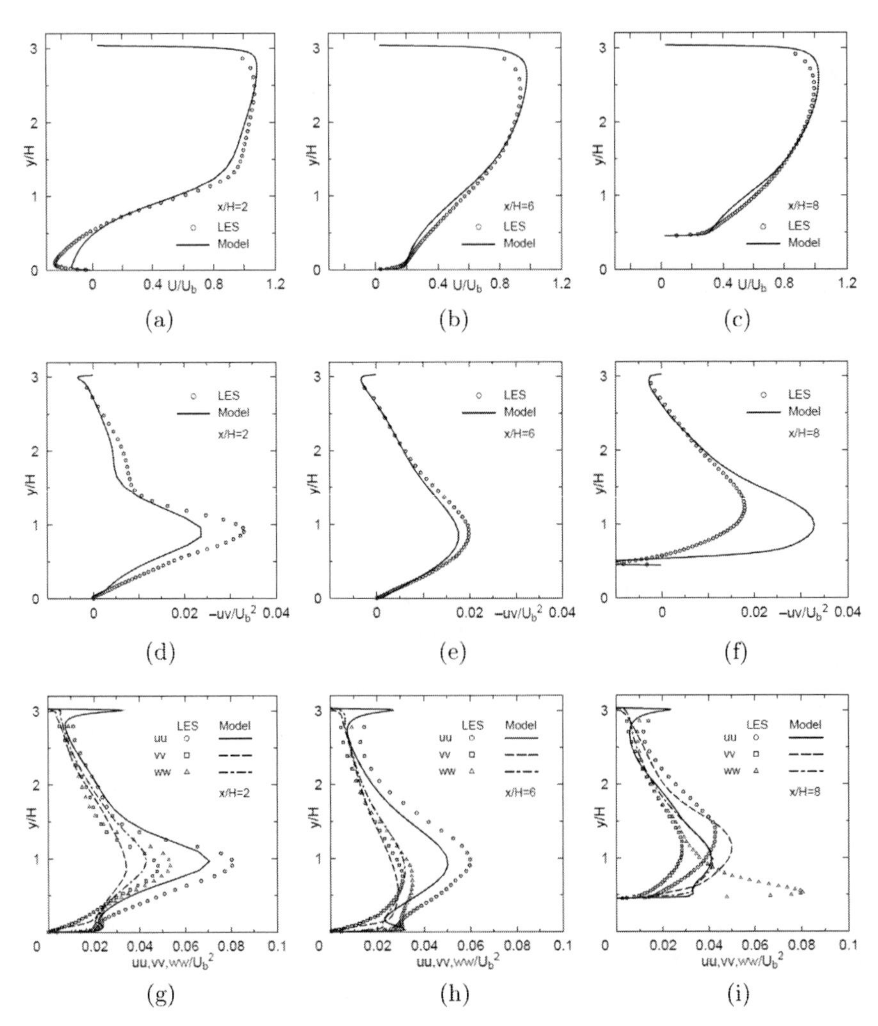

Fig. 15.5: Profiles of mean velocity and Reynolds stresses predicted for the flow in Fig. 15.4 (reproduced with permission of Elsevier from Abe, Jang and Leschziner (2003)).

i.e. the linear eddy-viscosity form. The next step is to insert Eq. (15.57) into Eq. (15.56), which gives:

$$a_{ij}^{(2)} = -\alpha S_{ij}^* - \beta\left(a_{ik}^{(1)} S_{kj}^* + S_{ik}^* a_{kj}^{(1)} - \frac{2}{3} a_{kl}^{(1)} S_{lk}^* \delta_{ij}\right)$$
$$- \gamma\left(a_{ik}^{(1)} \Omega_{kj}^* - \Omega_{ik}^* a_{kj}^{(1)}\right)$$

$$= -\alpha \mathbf{S} + 2\alpha\beta \left(\mathbf{S}^2 - \frac{1}{3} \left\{ \mathbf{S}^2 \right\} \mathbf{I} \right)$$
$$+ \alpha\gamma(\boldsymbol{\Omega}\mathbf{S} - \mathbf{S}\boldsymbol{\Omega}). \tag{15.58}$$

This may be regarded as a quadratic eddy-viscosity model, with coefficients to be determined.

As the third and last iteration, Eq. (15.58) is again inserted into Eq. (15.56), which results in:

$$a_{ij}^{(3)} = -\alpha S_{ij}^* - \beta \left(a_{ik}^{(2)} S_{kj}^* + S_{ik}^* a_{kj}^{(2)} - \frac{2}{3} a_{kl}^{(2)} S_{lk}^* \delta_{ij} \right)$$
$$- \gamma \left(a_{ik}^{(2)} \Omega_{kj}^* - \Omega_{ik}^* a_{kj}^{(2)} \right)$$
$$= -\alpha \mathbf{S} + 2\alpha\beta \left(\mathbf{S}^2 - \frac{1}{3} \left\{ \mathbf{S}^2 \right\} \mathbf{I} \right) + \alpha\gamma(\boldsymbol{\Omega}\mathbf{S} - \mathbf{S}\boldsymbol{\Omega})$$
$$- 4\alpha\beta^2 \left(\mathbf{S}^3 - \frac{1}{3} \left\{ \mathbf{S}^3 \right\} \mathbf{I} - \frac{1}{3} \left\{ \mathbf{S}^2 \right\} \mathbf{S} \right) - 3\alpha\beta\gamma \left(\boldsymbol{\Omega}\mathbf{S}^2 - \mathbf{S}^2\boldsymbol{\Omega} \right)$$
$$- \alpha\gamma^2 \left(\boldsymbol{\Omega}^2\mathbf{S} + \mathbf{S}\boldsymbol{\Omega}^2 - 2\boldsymbol{\Omega}\boldsymbol{S}\boldsymbol{\Omega} \right). \tag{15.59}$$

This may be regarded as a cubic eddy-viscosity model.

Two simplifications are now possible, due to the Cayley–Hamilton theorem. First, it is noted that Eq. (15.10) does not contain a term that involves $\mathbf{S}^3$ or $\{\mathbf{S}^3\}$. This is a consequence of the fact that these cubic terms can be expressed with lower-order products, namely:

$$\mathbf{S}^3 - \frac{1}{3} \left\{ \mathbf{S}^3 \right\} \mathbf{I} = \frac{1}{2} \left\{ \mathbf{S}^2 \right\} \mathbf{S}. \tag{15.60}$$

Second, Eq. (15.10) does not contain the product $\boldsymbol{\Omega}\boldsymbol{S}\boldsymbol{\Omega}$, for this can be expressed as:

$$\boldsymbol{\Omega}\boldsymbol{S}\boldsymbol{\Omega} - \{\boldsymbol{\Omega}\boldsymbol{S}\boldsymbol{\Omega}\} = -\left(\boldsymbol{\Omega}^2\mathbf{S} + \mathbf{S}\boldsymbol{\Omega}^2 - \frac{1}{2} \left\{ \boldsymbol{\Omega}^2 \right\} \mathbf{S} \right). \tag{15.61}$$

Substitution of Eqs. (15.60) and (15.61) into Eq. (15.59) gives:

$$a_{ij}^{(3)} = -\alpha \mathbf{S} + 2\alpha\beta \left(\mathbf{S}^2 - \frac{1}{3} \left\{ \mathbf{S}^2 \right\} \mathbf{I} \right) + \alpha\gamma(\boldsymbol{\Omega}\mathbf{S} - \mathbf{S}\boldsymbol{\Omega})$$
$$- \frac{2}{3} \alpha\beta^2 \left\{ \mathbf{S}^2 \right\} \mathbf{S} - 2\alpha\gamma^2 \left\{ \boldsymbol{\Omega}^2 \right\} \mathbf{S}$$

$$- 3\alpha\gamma^2 \left(\boldsymbol{\Omega}^2\mathbf{S} + \mathbf{S}\boldsymbol{\Omega}^2 - \{\boldsymbol{\Omega}^2\}\,\mathbf{S} - \frac{2}{3}\{\boldsymbol{\Omega}\mathbf{S}\boldsymbol{\Omega}\}\,\mathbf{I} \right)$$
$$- 3\alpha\beta\gamma\left(\boldsymbol{\Omega}\mathbf{S}^2 - \mathbf{S}^2\boldsymbol{\Omega}\right). \tag{15.62}$$

It is observed that the above equation contains four linear (or quasi-linear) terms, involving products of scalar invariants and $\mathbf{S}$. Several terms that are included in Eq. (15.62) also appear in Eq. (15.10), namely: $\mathbf{T}^{(1)}, \mathbf{T}^{(2)}, \mathbf{T}^{(4)}, \mathbf{T}^{(5)}$ and $\mathbf{T}^{(6)}$.

While Eq. (15.62) is obviously not the same model as a formal EARSM, it is an approximation of it, and the coefficients are, in principle, fully described by the RSTM. However, it is recalled that neither the Launder *et al.* model nor the Speziale *et al.* form is applicable to low-Reynolds-number near-wall regions. In fact, it is this that necessitated the introduction of the wall-related corrections discussed in Section 15.2.5. Hence, Apsley and Leschziner do not use the actual numerical values of the RSTM coefficients to determine the coefficients in Eq. (15.62), but rather employ Eq. (15.62) as a means of extracting the relationship between the coefficients.

For simple two-dimensional shear, we have (see Eq. (15.39):

$$S_{12}^* = \Omega_{12}^* = \frac{1}{2}\frac{k}{\varepsilon}\frac{\partial U}{\partial y}, \quad \{\mathbf{S}^2\} = -\{\boldsymbol{\Omega}^2\} = 2(S_{12}^*)^2, \tag{15.63}$$

and insertion into Eq. (15.62) is found to give rise to the three coupled equations:

$$\begin{aligned} a_{11} &= \frac{2}{3}(\alpha\beta + 3\alpha\gamma)S_{12}^* \\ a_{22} &= \frac{2}{3}(\alpha\beta - 3\alpha\gamma)S_{12}^* \\ a_{12} &= -\alpha S_{12}^*\left[1 - 4\left(\gamma^2 - \frac{1}{3}\beta^2\right)S_{12}^*\right]. \end{aligned} \tag{15.64}$$

This set can now be inverted to give:

$$\alpha S_{ij}^* = \frac{1}{2}\left(-a_{12} + \sqrt{a_{12}^2 + (a_{11} - a_{22})^2 - 3(a_{11} + a_{22})^2}\right)$$

$$\beta S^*_{ij} = \frac{3}{4(\alpha S^*_{ij})}\,(a_{11} + a_{22})$$

$$\gamma S^*_{ij} = \frac{3}{4(\alpha S^*_{ij})}\,(a_{11} - a_{22})\,. \tag{15.65}$$

This clarifies the meaning of 'extracting the relationship between the coefficients made just ahead of Eq. (15.63). The actual values of the coefficients are now determined by calibration: the variations $a_{11}(y^*), a_{22}(y^*), a_{12}(y^*), S^*_{12}(y^*)$, where $y^* = k^{1/2}y/\nu$, are extracted from DNS data for channel flow and boundary layers, subject to the constraints $a_{22}(0) = -\frac{2}{3}, a_{12}(0) = 0, a_{11}(0) = 0.78$. In other words, the calibration is by reference to conditions compatible with turbulence-energy equilibrium and the wall-asymptotic values of the stress components.

Within the range that includes the viscous near-wall and the log-law regions, Apsley and Leschziner show that several sets of DNS data nearly collapse, in terms of the wall-normal coordinate y^*, and can thus be represented by a curve-fitting process in the form $\frac{\overline{u^2}}{k}(y^*), \frac{\overline{v^2}}{k}(y^*), \frac{\overline{uv}}{k}(y^*), S^*_{12}(y^*)$, from which the corresponding anisotropy components can be extracted and inserted into Eq. (15.65). In particular, in the fully turbulent outer region (i.e. large values of y^*), these fits yield $(\alpha S^*_{12})_\infty = 0.48, (\beta S^*_{12})_\infty = 0.11, (\gamma S^*_{12})_\infty = 0.31$.

Clearly, the above calibration process has been performed by reference to flows at turbulence-energy equilibrium.[7] Apsley and Leschziner thus provide an extension that accounts for departures from this condition. The foundation of this extension is the fact that the coefficients α, β and γ in the ARSM Eq. (15.56) all contain the multiplier b_0 which is defined below Eq. (15.20) as $b_0 = \left(\frac{P_k}{\varepsilon} + C_1 - 1\right)^{-1}$. Hence, a clearly sensible approach, in principle, is to multiply the coefficients by a factor:

$$f_{\frac{P}{\varepsilon}} = \frac{C_1 + P_{k,eq}/\varepsilon - 1}{C_1 + P_k/\varepsilon - 1}, \tag{15.66}$$

[7] This is not to say, however, that the model itself is restricted to equilibrium flow. It is only the calibration that is at issue here.

where $P_{k,eq}$ is the turbulence-energy production applicable to the calibration flows.[8] For numerical-stability reasons, Apsley and Leschziner introduced the approximation:

$$\frac{P_k}{\varepsilon} \approx f_{\frac{P}{\varepsilon}} C_\mu (\sigma^*)^2, \tag{15.67}$$

where $\sigma^* = \sqrt{S^*_{ij}S^*_{ij} + \Omega^*_{ij}\Omega^*_{ij}}$. This is an expression similar to a linear eddy-viscosity model, except for $\sigma^* \neq \sqrt{2S^*_{ij}S^*_{ij}}$ and the inclusion of the multiplier $f_{P/\varepsilon}$, which reflects the contribution of $b_0 = \left(\frac{P_k}{\varepsilon} + C_1 - 1\right)^{-1}$ to the coefficient α in the leading term of Eq. (15.56). The inclusion of the vorticity tensor in σ^* is motivated by the need to accommodate flows with weak straining and high rotation.

Substitution of (15.66) into (15.67) and inversion for $f_{P/\varepsilon}$ gives:

$$f_{\frac{P}{\varepsilon}} = \frac{2f_0}{1 + \sqrt{4f_0(f_0 - 1)\left(\frac{\sigma^*}{\sigma^*_{eq}}\right)^2}}, \tag{15.68}$$

where $f_0 = 1 + 2C_\mu(\sigma^*)^2/(C_1 - 1)$. Clearly, $f_{P/\varepsilon} = 1$ when σ^* has the equilibrium value σ^*_{eq}, as it should be, regardless of the behaviour of f_0. Apsley and Leschziner show that the requirement for the anisotropy components to remain bounded is assured if $(f_0 - 1) > 0$, which then requires the limiter $C_\mu(\sigma^*_{eq})^2 = \max(0.09(\sigma^*_{eq})^2, 1.0)$, true only in the log-law region.

The extended model thus consists of Eq. (15.62), with coefficients determined from relations (15.65), all multiplied by the function in Eq. (15.68). In addition, the turbulence energy and the length scale have to be supplied to the model from related transport equations, and Apsley and Leschziner opted to use a slightly modified form of the $k - \varepsilon$ model of Lien and Leschziner (1994) (see Section 9.3.2).

Figures 15.6 and 15.7 give two sets of results arising from the model. The former merely verifies that the performance of the model

[8] Note that $P_{k,eq}/\varepsilon = 1$ is only valid in the outer log-law region.

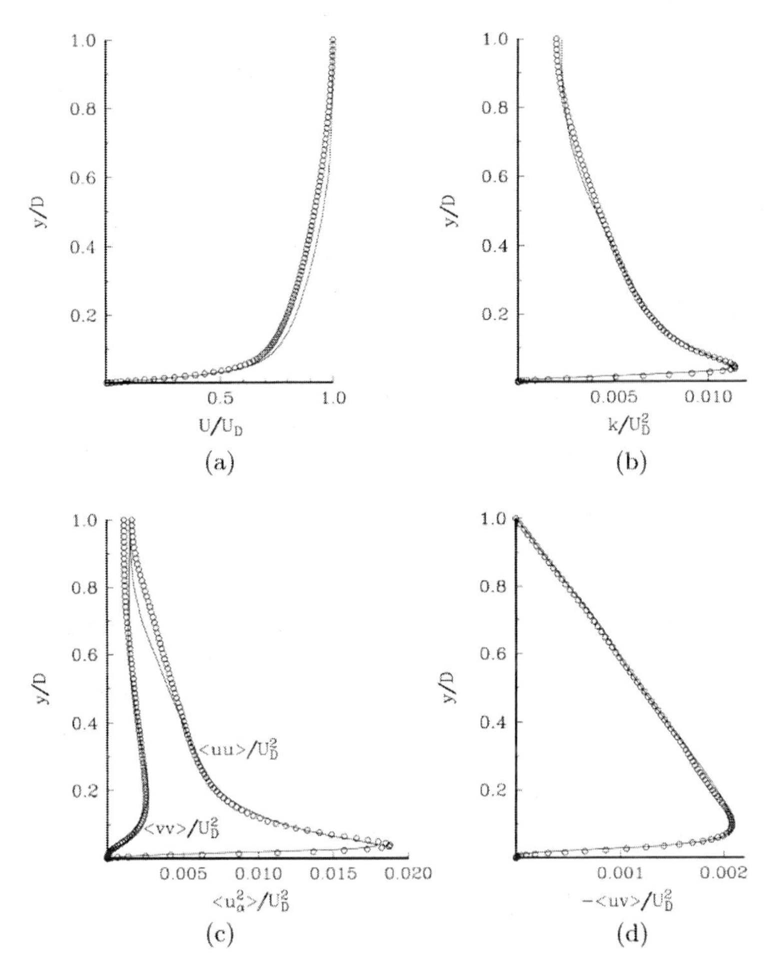

Fig. 15.6: Velocity and Reynolds-stress profiles in channel flow at mean-flow-Reynolds number 7890 (based on channel height). Lines: Apsley and Leschziner model. Symbols: DNS of Kim, Moin and Moser (1987) (reproduced with permission of Elsevier from Apsley and Leschziner (1998)).

is compatible and consistent with the calibration process. Figure 15.7 shows results for a (nominally) two-dimensional, asymmetric diffuser, with weak separation on one wall, examined experimentally by Obi, Aoki and Masuda (1993). While the model fails to resolve adequately the weak separation region on the upper wall, it returns much improved results relative to earlier models.

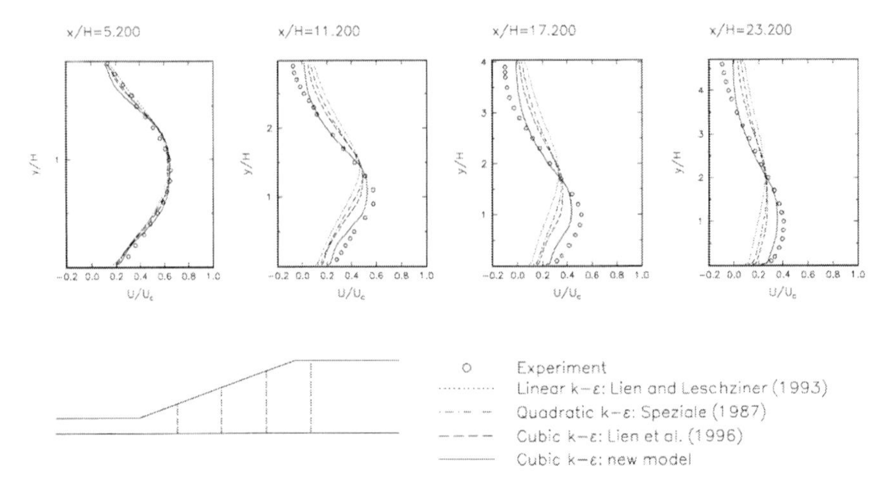

Fig. 15.7: Velocity profiles at four streamwise stations in a diffuser at inflow-Reynolds number 21200, based on channel height (reproduced with permission of Elsevier from Apsley and Leschziner (1998)).

15.4 *Non-linear eddy-viscosity models (NLEVMs)*

A distinguishing feature of NLEVMs is that they are not based on a Reynolds-stress closure. Instead they start from a non-linear relationship of the form Eq. (15.10), or truncations thereof, and then involve the imposition of flow-physical constraints and calibration by reference to DNS and experimental data to determine the unknown coefficients.

Early proposals, all quadratic in their final form, are those of Speziale (1987) (see Eq. (15.48) above), Nisizima and Yoshizawa (1987), Rubinstein and Barton (1990), Myong and Kasagi (1990) and Shih *et al.* (1995).

In two-dimensional flow, the rigorous manipulations explained in Section 15.2 show that an explicit formulation of the Reynolds-stress equations should contain only quadratic terms. However, in three-dimensional flows, the explicit model contains at least cubic and fifth-order products of strain and vorticity, as is the case in Wallin and Johansson's model (see Eq. (15.34)), while the model of Gatski and Speziale involves all groups in Eq. (15.10). As several EARSMs and NLEVMs evolved concurrently within a narrow time window — 1993 to 1997 — it is unclear whether this complexity provided the primary motivation for developing the NLEVMs; it seems to have been the driver of at least one higher-order NLEVM, namely that of Craft *et al.* (1996). Even the earliest NLEVMs — all quadratic in strain — followed, chronologically, the EARSM of Pope (1975) (which was also quadratic), and this suggests that NLEVMs were mostly pursued as interesting alternative approaches to the EARSM formalism. One credible argument in favour of this non-rigorous route is that the ARSM equations may introduce significant modelling errors by virtue of the approximation of stress transport (or omission of anisotropy transport) and the use of poorly performing pressure-strain approximations. There is, therefore, no reason to believe that EARSMs have, in general, superior predictive properties to those of well-calibrated NLEVMs.

15.4.1 *The model of Shih et al. (1995)*

This attractively simple model, already mentioned in Section 11.2, under the heading *Realisability*, is especially interesting, in so far as it illustrates how physically reasonable constraints can be invoked in efforts to simplify the task of determining unknown coefficients by reference to external data. While the model is simple, the reduction to its final minimal form is elaborate, because of the lengthy mathematical analysis associated with the imposition of the constraints. This analysis is, therefore, only outlined here in broad descriptive terms.

The starting point of this model is, essentially, the general constitutive equation (15.10), truncated after the quadratic terms, i.e. after the first line. Consistency with the eddy-viscosity concept implies that the first linear term is:

$$\beta_1 \mathbf{S} = 2C_\mu \frac{k^2}{\varepsilon} S_{ij}, \tag{15.69}$$

with C_μ not necessarily constant. Next, the remaining three quadratic terms are subjected to two constraints:

(i) Rapid distortion theory (Batchelor and Proudman (1954)) is used to relate one of the unknown coefficients to the other two. This examines the response of isotropic turbulence to rapid rotation, subject to zero straining. The theory shows that the isotropy should remain unaffected, i.e. $a_{ij} = 0$, and this gives rise to a constraint that eliminates one of the coefficients.

(ii) The requirement of realisability is imposed analytically. This has two parts:

$$\overline{u_\alpha^2} \geq 0 \quad (\alpha = 1, 2, 3)$$
$$\frac{(\overline{u_\alpha u_\beta})^2}{\overline{u_\alpha^2}\,\overline{u_\beta^2}} \leq 1 \quad (\alpha, \beta = 1, 2, 3), \tag{15.70}$$

the former demanding positive normal stresses and the latter expressing the fact that cross-correlations cannot be perfect. To impose these constraints, the realisability formalism is applied to $\overline{u_1^2}$ and $\overline{uv}$ in the cases of strong contraction, $S_{11} \gg 0$, and pure shear, $S_{12} > 0$, respectively.

The end point, prior to calibration, is an expression that contains only one non-linear group and two coefficients:

$$a_{ij} = -2C_\mu \frac{k^2}{\varepsilon} S_{ij} + 2C_1 \frac{k^3}{\varepsilon^2} (\Omega_{ik} S_{kj} - S_{ik} \Omega_{kj}). \tag{15.71}$$

The coefficients in Eq. (15.71) turn out to be themselves functions of the strain and vorticity invariants namely:

$$C_\mu = \frac{1}{A_0 + F_S \frac{k}{\varepsilon} (S_{ij} S_{ij} + \Omega_{ij} \Omega_{ij})^{\frac{1}{2}}}$$
$$C_1 = \frac{\left(1 - 9C_\mu^2 \frac{k^2}{\varepsilon^2} S_{ij} S_{ij}\right)^{\frac{1}{2}}}{C_0 + 6 \frac{k^2}{\varepsilon^2} (S_{ij} S_{ij})^{\frac{1}{2}} (\Omega_{ij} \Omega_{ij})^{\frac{1}{2}}}, \tag{15.72}$$

in which A_0 and C_0 remain to be determined by calibration, and F_S is a function of the third strain invariant $\frac{S_{ij} S_{jk} S_{ki}}{(S_{ij} S_{ij})^{3/2}}$, which arises from the imposition of the first realisability constraint $\overline{u_1^2} \to 0$ as $S_{11} \to \infty$ in Eq. (15.70) (in which case, the linear eddy-viscosity model will give the non-realisable state $\overline{u_1^2} \to -\infty$). The constants

A_0 and C_0 are determined by applying Eq. (15.71) to two simple flows: homogeneous shear and the log-law layer, both at equilibrium conditions, $S^*_{eq} = \frac{k}{\varepsilon}(2S_{ij}S_{ij})^{1/2} = \frac{k}{\varepsilon}\frac{\partial U}{\partial y} \approx 6$ and ≈ 3.3, respectively. For these conditions, $\{a_{12}, a_{11}, a_{22}\} = \{-0.33, 0.3, -0.18\}$ and $\{-0.29, 0.34, -0.26\}$, respectively, the former obtained experimentally and the latter derived from channel-flow DNS. This calibration results in $A_0 = 6.5$ and $C_0 = 1.0$.

In closing this section, we dwell briefly on the form of the relationship for C_μ. The fact that it features the strain in its denominator should not come as a surprise. Previous considerations on realisability, in Section 11.2 demonstrated the need to make C_μ indirectly proportional to the strain. Indeed, it was shown via Eq. (11.11) that Menter's SST model (1994) (see Section 9.4.3) secures its improved predictive properties by making $C_\mu \propto 0.3/S^*$ beyond the equilibrium value $S^*_{eq} = 3.3$. The difference is that the present model derives this dependence by a rigorous formalism, and it is more general than earlier forms. We shall compare $C_\mu = f(S^*)$ variations arising from different NLEVMs at the end of the next section.

15.4.2 *The model of Craft et al. (1996)*

Craft *et al.* start from the assumption that all 'essential' strain effects (including curvature) can be represented by a general cubic *Ansatz* in which the eddy viscosity is a pre-multiplier of all strain and vorticity groups:

$$
\begin{aligned}
a_{ij} = &-\frac{\nu_t}{k}S_{ij} + c_1\frac{\nu_t}{\varepsilon}\left(S_{ik}S_{jk} - \frac{1}{3}S_{kl}S_{kl}\delta_{ij}\right) + c_2\frac{\nu_t}{\varepsilon}\left(\Omega_{ik}S_{jk} + \Omega_{jk}S_{ik}\right) \\
&+ c_3\frac{\nu_t}{\varepsilon}\left(\Omega_{ik}\Omega_{jk} - \frac{1}{3}\Omega_{kl}\Omega_{kl}\delta_{ij}\right) \\
&+ c_4\frac{\nu_t k}{\varepsilon^2}\left(S_{ki}\Omega_{lj} + S_{kj}\Omega_{li}\right)S_{kl} \\
&+ c_5\frac{\nu_t k}{\varepsilon^2}\left(\Omega_{il}\Omega_{lm}S_{mj} + S_{il}\Omega_{lm}\Omega_{mj} - \frac{2}{3}S_{lm}\Omega_{mn}\Omega_{nl}\delta_{ij}\right) \\
&+ c_6\frac{\nu_t k}{\varepsilon^2}\left(S_{ij}S_{kl}S_{kl}\right) + c_7\frac{\nu_t k}{\varepsilon^2}\left(S_{ij}\Omega_{kl}\Omega_{kl}\right). \qquad (15.73)
\end{aligned}
$$

We note, first, that the groups in the upper three lines are all contained in the first two lines of Eq. (15.10), up to the term $\mathbf{T}^{(6)}$. Also, the upper row comprises stress- and vorticity-tensor combinations that are compatible with the rigorous EARSM for two-dimensional conditions. Second, the last two groups in the lowest row are not cubic, but linear is S_{ij}, because the other tensors involved (e.g. $S_{kl}S_{kl}$) contract to scalars. The reason for their inclusion is that the β-coefficients in Eq. (15.12) are functions of the scalar invariants, among them $S_{kl}S_{lk}$ and $\Omega_{kl}\Omega_{lk}$, and this applies, in particular, to the linear term $\beta_1\mathbf{S}$ is Eq. (15.10). Third, since $\nu_t \propto \frac{k^2}{\varepsilon}$, the pre-multipliers of all groups in Eq. (15.73) may simply be regarded as products of the time scale $\frac{k}{\varepsilon}$, thus rendering the strain and vorticity tensors dimensionless, i.e. $S^*_{ij} = \frac{k}{\varepsilon}S_{ij}, \Omega^*_{ij} = \frac{k}{\varepsilon}\Omega_{ij}$. The retention of ν_t in all terms in Eq. (15.73) makes the designation 'non-linear eddy-viscosity model' (NLEVM) transparent. Fourth, Eq. (15.73) is not simply a cubic tensorial 'expansion' in S_{ij} and Ω_{ij}, but it takes into account constrains associated with symmetry and contraction properties. In particular, as a_{ij} is a symmetric tensor — all right-hand-side groups must, ultimately, be symmetric tensors in (i, j). Moreover, the contraction $a_{kk} = 0$ is a constraint on all individual additive right-hand-side terms.

The distinctive feature of Eq. (15.73) is that the coefficients $c_1 \cdots c_7$ and C_μ in ν_t (which is not necessarily 0.09!) are unknown and, in contrast to EARSMs, are not determined from inverting the implicit ARSM. Rather, the coefficients are determined by calibration and 'optimisation' over a range of key flows. This is, of course, a laborious and lengthy process that cannot be described in mathematical form herein. There follow a few examples that illustrate the process.

In simple shear,

$$
\begin{aligned}
&S^*_{12} = S^*_{21} = \frac{1}{2}\frac{k}{\varepsilon}\frac{\partial U}{\partial y}, \quad \Omega^*_{12} = S^*_{12}, \quad \Omega^*_{21} = -S^*_{21} \\
&S^* = \frac{k}{\varepsilon}\frac{\partial U}{\partial y}.
\end{aligned}
\tag{15.74}
$$

Substitution into Eq. (15.73) then gives:

$$\begin{aligned}
a_{11} &= C_\mu S^{*2}\left(\frac{1}{3}c_1 + 2c_2 + \frac{1}{3}c_3\right)\\
a_{22} &= C_\mu S^{*2}\left(\frac{1}{3}c_1 - 2c_2 + \frac{1}{3}c_3\right)\\
a_{33} &= C_\mu S^{*2}\left(-\frac{2}{3}c_1 - \frac{2}{3}c_3\right)\\
a_{12} &= -C_\mu S^* + 2c_\mu S^{*3}(-c_5 + c_6 + c_7)\\
a_{13} &= a_{23} = 0.
\end{aligned} \tag{15.75}$$

Note that $a_{11} + a_{22} + a_{33} = 0$, as required.

In simple shear, a_{12} is already well represented by any well-crafted expression of the form:

$$a_{12} = -C_\mu \frac{k}{\varepsilon}\frac{\partial U}{\partial y}, \tag{15.76}$$

arising from linear eddy-viscosity models — although it is recalled that a constant C_μ gives rise to unphysical behaviour at high strain rates, requiring its sensitisation to strain and vorticity invariants (see Sections 11.2). If this is to be retained as a constraint, with C_μ appropriately changed from its constant value 0.09, it follows that:

$$-c_5 + c_6 + c_7 = 0. \tag{15.77}$$

As regards the normal stresses, we need to 'optimise' $(c_1 + c_3)$ and c_2, and the choices made for these will fully describe the anisotropy. The choices $(c_1 + c_3) = 0.16$ $c_2 = 0.1$ give:

$$\{a_{11}, a_{22}, a_{33}\} = c_\mu S^{*2}\{0.25, -0.15, -0.1\}, \tag{15.78}$$

which compare with the channel-flow values $\{0.34, -0.26, -0.08\}$ (Table 15.1) and the homogeneous-shear values $\{0.3, -0.18, -0.12\}$.

In efforts to determine the remaining coefficients, Craft *et al.* consider curved and swirling flows. In a flow with streamline curvature,

$$S_{12} = \frac{1}{2}\left(\frac{\partial U}{\partial r} - \frac{U}{r}\right), \quad \Omega_{12} = \frac{1}{2}\left(\frac{\partial U}{\partial r} + \frac{U}{r}\right), \tag{15.79}$$

where r is the radius of curvature and U is the velocity normal to that radius, i.e. along the curved streamlines. For this case, a_{12} in Eq. (15.73) becomes:

$$a_{12} = -C_\mu S^* - 2C_\mu S^* \Omega^{*2}(c_5 - c_7) + 2c_6 C_\mu S^{*3}. \tag{15.80}$$

The choice $c_5 = 0$ is now made, in which case $c_7 = -c_6$, from Eq. (15.77), and this makes Eq. (15.80) dependent on one constant only. Experimental data are then used to calibrate this constant, and the value suggested by Craft *et al.* is $c_7 = -c_6 = 5C_\mu^2$.

Next, for the case of fully developed swirling flow in a rotating pipe:

$$S_{12} = \Omega_{12} = \frac{1}{2}\frac{\partial U}{\partial r}, \quad S_{23} = \frac{1}{2}\left(\frac{\partial W}{\partial r} - \frac{W}{r}\right)$$
$$\Omega_{23} = \frac{1}{2}\left(-\frac{\partial W}{\partial r} - \frac{W}{r}\right), \tag{15.81}$$

and

$$\begin{aligned} a_{23} = &-C_\mu S_{23}^* + c_4 \frac{C_\mu}{8} S_{12}^* \Omega_{12}^* S_{23}^* + c_5 \frac{C_\mu}{8}(-\Omega_{12}^{*2} - 2\Omega_{23}^{*2}) S_{23}^* \\ &+ c_6 \frac{C_\mu}{4}(S_{12}^{*2} + S_{23}^{*2}) S_{23}^* + c_7 \frac{C_\mu}{4}(\Omega_{12}^{*2} + \Omega_{23}^{*2}) S_{23}^* \\ &+ \frac{C_\mu}{8}(c_4 S_{12}^{*2} \Omega_{23}^* + c_5 S_{12}^* \Omega_{12}^* \Omega_{23}^*). \end{aligned} \tag{15.82}$$

At fully developed conditions, the shear-stress $\overline{vw}$, and hence a_{23}, vanish, but S_{23} does not. With c_5, c_6, c_7 known, the only constant that is unknown in Eq. (15.82) is c_4. Again, measurements can be used to determine this value, and Craft *et al.* suggest $c_4 = -10C_\mu^2$.

The above represents, in principle, a rational route to determining the constants. However, in reality, the precise values of the coefficients arise from an elaborate computational-optimisation process in which the model is applied to a variety of flows (e.g. curved channel flow, impinging jet, swirling flow), with adjustments made so as to achieve a compromise in respect of the predictive performance over the range of flows examined. This optimisation includes modifications to the value of C_μ, and the inclusion of a viscous correction term that renders the model applicable to near-wall flows:

$$C_\mu = \frac{0.3}{1 + 0.35\left[\max(S^*, \Omega^*)\right]^{1.5}} \times \left\{1 - \exp\left[\frac{-0.36}{\exp\left[-0.75\max(S^*, \Omega^*)\right]}\right]\right\} f_\mu \quad (15.83)$$

$$f_\mu = 1 - \exp\left[-\left(\frac{R_t}{90}\right)^{0.5} - \left(\frac{R_t}{400}\right)^2\right].$$

The damping function aside, this form is closely related to the first equation in Eq. (15.72). It reflects the fact that, with the constraint (15.76) imposed, a constant C_μ is only applicable to flows near turbulence-energy equilibrium, and thus it gives the wrong behaviour at high shear- and normal-strain rates.

With Eq. (15.83) stated, this is an appropriate point at which we can now draw parallels between several models in respect of the variation $C_\mu(S^*)$. We recall that this dependence (or the equivalent dependence $C_\mu(P_k/\varepsilon)$ of Rodi (1993)) is a key to a realisable response of the shear stress to the strain rate, whether a model is linear or non-linear. We discussed this issue, repeatedly, in Sections 9.4.3 and 11.2, as well as in the present chapter. Figure 15.8 compares the variations returned by four models, three of them non-linear and the fourth the linear SST model of Menter (1994). Remarkably, all display virtually the same response to a strain rate — essentially of the form $C_\mu \propto 1/(\alpha + \beta S^*)$ — beyond the equilibrium value $\sigma = S^* = 3.3$. This may be taken to reflect, variously, the behaviour derived from second-moment closure, and/or the use of experimental and DNS data in the calibration process, and/or the imposition of realisability constraints, as done by Shih *et al.* (1995).

The performance of the model by Craft *et al.* is illustrated in Figs. 15.9 and 15.10. As seen from Fig. 15.9, the anisotropy in the channel flow is not represented especially well, and this reflects (in part) the basic difficulty of developing a high-fidelity model of this type (as contrasted with the more formal EARSM route), in which the determination of the constants relies entirely on calibration by reference to a small set of generic flows. The swirl-velocity profile is

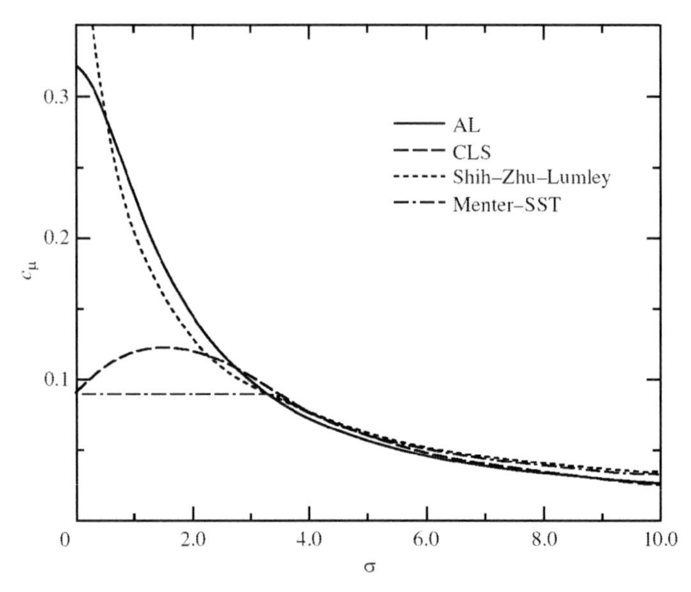

Fig. 15.8: Variation of the eddy-viscosity coefficient $C_\mu(\sigma)(\sigma = S^* = k/\varepsilon\ (\partial U/\partial y))$ for different models in simple shear AL=Apsley and Leschziner, CLS = Craft *et al.* (reproduced with permission of the Royal Society from Leschziner (2000)).

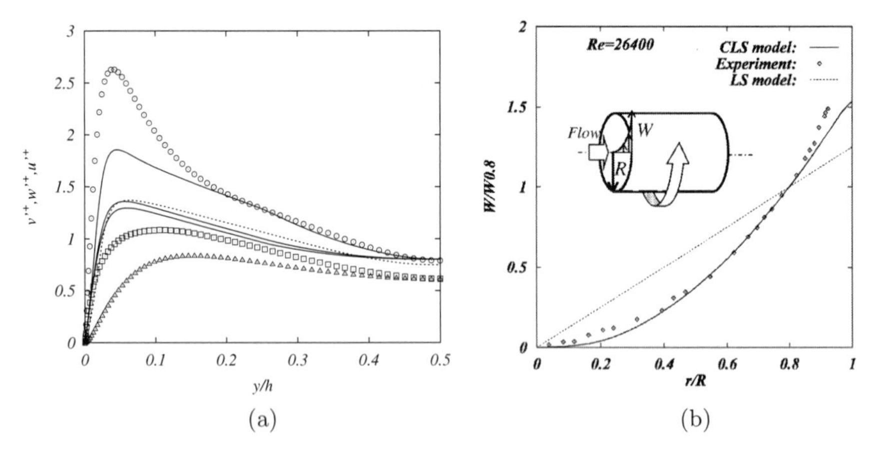

Fig. 15.9: Performance of the non-linear eddy-viscosity model of Craft *et al.* (1996): (a) Turbulence-intensity profiles in channel flow at mean-flow Reynolds number $Re = 5600$ relative to DNS data (symbols); dashed line: linear eddy-viscosity model; (b) Swirl-velocity profiles in rotating pipe relative to experiment (symbols); dashed line: standard linear eddy-viscosity model (reproduced with permission of Elsevier from Craft, Ince and Launder (1996)).

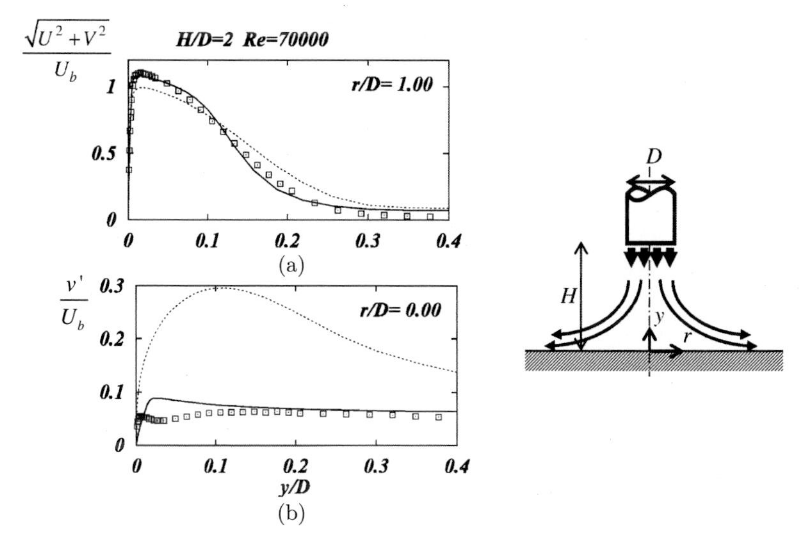

Fig. 15.10: Velocity Performance of non-linear eddy-viscosity model of Craft, Ince and Launder (1996) for a round impinging jet: (a) Velocity profile at $r/D = 1$; (b) Wall-normal turbulence intensity on centre-line. Dashed lines: linear eddy-viscosity model (reproduced with permission of Elsevier from Craft, Ince and Launder (1996)).

well returned, but this reflects the fact that the constant associated with the terms specifically sensitive to swirl was calibrated specifically for this type of flow.

Figure 15.10 relates to a jet impinging normally on a flat plate. The challenge here is to represent correctly the (weak) response of the turbulence intensity to normal straining. Again, this is one of the topics discussed, from a fundamental perspective, under the heading *Realisability* in Section 11.2. Any linear eddy-viscosity model, without any limiters of the type shown in Fig. 15.8, returns a strong amplification of the turbulence intensity close to a stagnation point, as emerges from Fig. 15.10(b), in which the profiles relate to the wall-normal turbulence intensity. As a consequence of the seriously excessive turbulence level, the turbulent diffusion is correspondingly overestimated, and this leads to the velocity profile in Fig. 15.10(a) being far too flat, relative to the measured data. The Craft *et al.* model clearly provides a far better representation of the flow, but it has to be said that this is, principally, a consequence of the model fragment (15.83), which depresses C_μ in the impingement zone.

15.5 *Summary and lessons*

A common feature of all models covered in this chapter is that they all comprise of a set of algebraic equations that link, explicitly, the stress components to the velocity gradients within the strain and vorticity tensors. In all cases, the leading term is linear in the strain, thus allowing its multiplier to be interpreted as a strain-sensitised eddy viscosity that promotes realisability. The presence of higher-order terms allows a range of properties and sensitivities to be captured that linear models are unable to represent. This includes normal-stress anisotropy and the response to curvature and swirl. The formal structure of all models permits a numerical implementation of the models that follows closely that of their linear counterparts. This is not only economically advantageous, but favours computational stability and convergence, by virtue of the fact that the leading term of the models gives rise to the usual second-order derivative of the velocity components in the RANS equations.

With this commonality highlighted, there are major differences among the models, not simply in terms of minor variations in the derivation, calibration and differences in predictive performance, but also in terms of the starting point and the modelling route that follows. At the high-fidelity end of the range are models (EARSMs) that arise from (close to) rigorous and formal 'projections' of simplified forms of the Reynolds-stress-transport equations onto a set of tensorial polynomials that depend on the strain and vorticity tensors and their invariants. The principal simplification is the omission of anisotropy transport by convection and diffusion. It follows that the resulting EARSMs are, at best, approximations of more general models. This route is driven by the wish to retain all major properties of the parent stress-transport equations at much lower cost, and to reshape the model into a form in which the turbulent-transport terms in the RANS equations are second-order derivatives of the type that arise from linear eddy-viscosity models. However, formally correct three-dimensional versions of such models are extremely complex and virtually never used in practice. In most circumstances, simplified two-dimensional forms are used, in which the number of terms is much lower. However, it is important to point out that such stripped-down versions can still be used for three-dimensional flows without a violation of system-invariance and objectivity, provided all components are included in the stress, strain and vorticity tensors.

The low-fidelity end is populated by a number of non-linear eddy-viscosity models. While also based on tensorial polynomials, truncated to different orders, such models are not formally linked to the Reynolds-stress equations, but approach the closure of the non-linear constitutive equations by a calibration of the coefficients, undertaken upon the application of the model to a range of generic or canonical flows, such as homogeneous shear, channel flow, curved shear flow and swirling flow. In general, the performance of these models does not quite match that displayed by more formal EARSMs over a broad range of flows.

There are also a number of 'middle-of-the-range' models that combine, in one way or another, elements of second-moment closure with simplified forms of the tensorial constitutive equations and calibration by reference to canonical flows.

Of the models covered in this chapter, that by Wallin and Johansson (2000) is the most recent, and probably constitutes the best compromise between fidelity and simplicity. It is closely linked to the second-moment closure of Launder, Reece and Rodi (1975), and it incorporates modifications that make the model applicable to near-wall flows. Also, its two-dimensional form is relatively simple. It is not surprising, therefore, that this particular model is the most popular, within its class, in terms of its use in codes applied to practical flows in industrial CFD.

Appendix: Basic Tensor Algebra and Rules

Coordinates

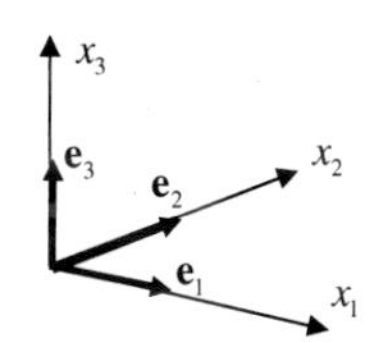

$$x_i = x_1, x_2, x_3 = x, y, z$$

Unit vectors

$$\mathbf{e}_i = \mathbf{e}_1, \mathbf{e}_2, \mathbf{e}_3 = \mathbf{i}, \mathbf{j}, \mathbf{k}$$

Boldface letter is the dyadic notation for vectors.

A vector

$$\mathbf{u} = u_i = u_1, u_2, u_3 = u, v, w$$

Summation convention

$$u_i v_i = \sum_1^3 u_i v_i$$ — a scalar quantity (equivalent to scalar product $\mathbf{u} \cdot \mathbf{v}$)

$$a_{ii} = \sum_1^3 a_{ii}$$ — a scalar quantity

A second-rank tensor (equivalent to a matrix)

$$u_i u_j = \begin{Bmatrix} u_1 u_1 & u_1 u_2 & u_1 u_3 \\ u_2 u_1 & u_2 u_2 & u_2 u_3 \\ u_3 u_1 & u_3 u_2 & u_3 u_3 \end{Bmatrix}$$

$$a_{ij} = \{\mathbf{a}\} = \begin{Bmatrix} a_{11} & a_{12} & a_{13} \\ a_{21} & a_{22} & a_{23} \\ a_{31} & a_{32} & a_{33} \end{Bmatrix}$$

$$\left(\text{e.g. the strain-rate tensor } S_{ij} = \frac{1}{2}\left(\frac{\partial U_i}{\partial x_j} + \frac{\partial U_j}{\partial x_i}\right)\right)$$

Higher-rank tensors

$$
\begin{array}{ll}
u_i u_j u_k, & u_i u_j u_k u_l \dots \\
a_{ijk}, & a_{ijkl\dots}
\end{array}
$$

Kronecker Delta (equivalent to unity matrix)

$$
\mathbf{I} = \delta_{ij} = \begin{Bmatrix} 1 & 0 & 0 \\ 0 & 1 & 0 \\ 0 & 0 & 1 \end{Bmatrix} = \mathbf{e}_i \cdot \mathbf{e}_j
$$

Contraction

$$
\delta_{ij} a_{ij} = a_{11} + a_{22} + a_{33} = a_{ii} \text{ — a scalar quantity}
$$

$$
\delta_{ij} a_{ijkl\dots} = \delta_{11} a_{11kl\dots} + \delta_{22} a_{22kl\dots} + \delta_{33} a_{33kl\dots} = c_{kl\dots}
$$

$$
a_{ik} b_{kj} = a_{i1} b_{1j} + a_{i2} b_{2j} + a_{i3} b_{3j} = c_{ij}
$$

Substitution

$$
\delta_{ij} a_{jkl\dots} = \delta_{i1} a_{1kl\dots} + \delta_{i2} a_{2kl\dots} + \delta_{i3} a_{3kl\dots} = a_{ikl\dots}
$$

Permutation tensor

$$
\varepsilon_{ijk} = \begin{cases} 0 & \text{if any two indeces are equal, e.g. } ijk = 11k, 2j2, i33 \\ 1 & \text{if all indices are different and in cyclic order,} \\ & \quad \text{e.g. 123, 231, 312} \\ -1 & \text{if all indices are different and in anti-cyclic order,} \\ & \quad \text{e.g. 321, 213, 132} \end{cases}
$$

Gradient

$$
\operatorname{grad} \phi = \nabla \phi = \frac{\partial \phi}{\partial x_i}
$$

Divergence

$$
\operatorname{div} \mathbf{u} = \nabla \cdot \mathbf{u} = \frac{\partial u_i}{\partial x_i} \text{ (= 0 for constant density flow)}
$$

Curl

$$\mathbf{v} = \operatorname{curl}\mathbf{u} = \varepsilon_{ijk}\frac{\partial u_j}{\partial x_k} = v_i \quad \left(\text{e.g. } v_1 = (\operatorname{curl}\mathbf{u})_1 = \frac{\partial u_3}{\partial x_2} - \frac{\partial u_2}{\partial x_3}\right)$$

Laplacean

$$\nabla^2\phi = \frac{\partial^2\phi}{\partial x^2} + \frac{\partial^2\phi}{\partial y^2} + \frac{\partial^2\phi}{\partial z^2} = \frac{\partial}{\partial x_i}\left(\frac{\partial\phi}{\partial x_i}\right)$$

Cross-product

$$\mathbf{w} = \mathbf{u} \times \mathbf{v} = \varepsilon_{ijk}u_j v_k = w_i$$

Triple product

$$\phi = \mathbf{u} \cdot (\mathbf{v} \times \mathbf{w}) = u_i\varepsilon_{ijk}v_j w_k = \varepsilon_{ijk}u_i v_j w_k$$

References

1. Abe, K., Kondoh, T. and Nagano, Y. (1997), On Reynolds-stress expressions and near-wall scaling parameters for predicting wall and homogeneous turbulent shear flows, *Int. J. Heat Fluid Flow*, **18**, 266–282.
2. Abe, K., Jang, Y.-J. and Leschziner, M.A. (2003), An investigation of wall-anisotropy expressions and length-scale equations for non-linear eddy-viscosity models, *Int. J. Heat Fluid Flow*, **24**, 181–198.
3. Agostini, L. and Leschziner, M.A. (2014), On the influence of outer large-scale structures on near-wall turbulence in channel flow, *Phys. of Fluids*, 26, 075107.
4. Agostini, L., Touber, E. and Leschziner, M.A. (2014), Spanwise oscillatory wall motion in channel flow: Drag-reduction mechanisms inferred from DNS-predicted phase-wise property variations at $\mathrm{Re}_\tau = 1000$, *J. Fluid Mechanics*, **743**, 606–635.
5. Amano, R.S. and Goel, P. (1986), Triple-velocity products in a channel with a backward-facing Step, *J. AIAA*, **24**, 1040–1043.
6. Apsley, D.D. and Leschziner, M.A. (1998), A new low-Reynolds-number nonlinear two-equation turbulence model for complex flows, *Int. J. Heat and Fluid Flow*, **19**, 209–222.
7. Apsley, D.D. and Leschziner, M.A. (2000), Advanced turbulence modelling of separated flow in a diffuser, *Flow, Turbulence and Combustion*, **63**, 81–112.
8. Apsley, D.D. and Leschziner, M.A. (2001), Investigation of advanced turbulence models for the flow in a generic wing-body junction, *Flow, Turbulence and Combustion*, **57**, 25–55.
9. Baldwin, B.S. and Lomax, H. (1978), Thin-layer approximation algebraic model for separated turbulent flows, *AIAA Paper 78-0257*.
10. Baldwin, B.W. and Barth, T.A. (1991), One-equation turbulence transport model for high Reynolds number wall-bounded flows, *AIAA Paper 91-0610*.

11. Batchelor, G.K. and Proudman, I. (1954), The effect of rapid distortion of a fluid in turbulent motion, *Q.J. Appl. Math*, **7**, 83–103.
12. Batten, P., Craft, T.J, Leschziner, M.A. and Loyau, H. (1999), Reynolds-stress-transport modelling for compressible aerodynamic flows, *J. AIAA*, **37**, 785–796.
13. Bentaleb, Y., Lardeau, S. and Leschziner, M.A. (2012) Large-eddy simulation of turbulent boundary-layer separation from a rounded step, *J. of Turbulence*, **13**, 1–28. (http://www.tandfonline.com/doi/full/10.1080/14685248.2011.637923).
14. Billard F. (2011), Development of robust elliptic-blending turbulence model for near-wall separated and buoyant flows, *Ph.D. thesis*, University of Manchester.
15. Billard, F., Osman, K. and Laurence, D. (2012), Development of adaptive wall functions for an elliptic-blending model, *Proc. 7th Symposium on Turbulence, Heat and Mass Transfer*, Palermo (Begell House Inc. Publ.), 465–468.
16. Blackburn, H.M., Mansour, N.N. and Cantwell, B.J. (1996), Topology of fine-scalke motions in turbulent channel flow, *J. Fluid Mech.*, **310**, 269–292.
17. Bogey, C. and Bailly, C. (2009), Turbulence and energy budget in a self-preserving round jet: Direct evaluation using large eddy simulation, *J. Fluid Mech.*, **627**, 129–160.
18. Boussinesq, J. (1877), “Essai sur la théorie des eaux courantes”, *Mémoires présentés par divers savants à l'Académie des Sciences*, **23**.
19. Bradshaw, P. (1965), The analogy between streamline curvature and buoyancy in turbulent shear flows, *J. Fluid Mech.*, **36**, 177–191.
20. Bradshaw, P. (1973), Effects of streamline curvature on turbulent flow, *AGARDograph* **169**.
21. Bradshaw, P., Ferris, D.H. and Atwell, N.P. (1967), Calculation of boundary layer development using the turbulent energy equation, *J. Fluid Mech.*, **23**, 31–64.
22. Cebeci, T. and Smith, A.M.O. (1974), Analysis of turbulent boundary layers, *Ser. in Appl. Math. & Mech, XV*, Academic Press, London.
23. Chien, K.Y. (1982), Predictions of channel and boundary-layer flows with a low-Reynolds-number turbulence model, *J. AIAA*, **20**, 33–38.
24. Chieng, C. and Launder, B.E. (1980), On the calculation of turbulent heat transport downstream from an abrupt pipe expansion, *Numerical Heat Transfer*, **3**, 189–207.
25. Craft, T.J. (1998), Developments in a low-Reynolds-number second-moment closure and its application to separating and reattaching flows, *Int. J. Heat and Fluid Flow*, **19**, 541–548.

26. Craft, T.J., Gant, S., Iacovides, H. and Launder, B.E. (2004), A new wall function strategy for complex turbulent flows, *Numerical Heat Transfer, Part B: Fundamentals*, **45**, 301–318.
27. Craft, T.J., Gerasimov, A.V., Iacovides, H. and Launder, B.E. (2002), Progress in the generalisation of wall-function treatments, *Int. J. Heat Fluid Flow*, **23**, 148–160.
28. Craft, T.J., Graham, L.J.W. and Launder, B.E. (1993), Impinging jet studies for turbulence model assessment, part ii: An examination of performance of four turbulence models, *Int. J. Heat Mass Transfer*, **36**, 2685.
29. Craft, T.J., Ince, N.Z. and Launder, B.E. (1996), Recent developments in second-moment closure for buoyancy-affected flows, *Dynamics of Atmospheres and Oceans*, **23**, 99–114.
30. Craft, T.J. and Launder, B.E. (1989), A new model for the pressure scalar gradient correlation and its application to homogeneous and inhomogeneous free shear flows, *Proc. 7th Symp. on Turbulent Shear Flows*, Stanford University, 17.1.1–17.1.6.
31. Craft, T.J. and Launder, B.E. (1996), Reynolds stress closure designed for complex geometries, *Int. J. Heat Fluid Flow*, **17**, 245–254.
32. Craft, T.J., Launder, B.E. and Suga, K. (1996), Development and application of a cubic eddy viscosity model of turbulence, *Int. J. of Heat and Fluid Flow*, **17**, 108–115.
33. Cresswell, R., Haroutunian, V., Ince, N.Z., Launder, B.E. and Szczepura, R.T. (1989), Measurement and modelling of buoyancy modified elliptic turbulent shear flows, *Proc. 7th Symp. on Turbulent Shear Flows*, Standford, USA.
34. Crow, S.C. (1968), Viscoelastic properties of fine-grained incompressible turbulence, *J. Fluid Mech.*, **33**, 1–20.
35. Daly, B.J. and Harlow, F.H. (1970), Transport equations in turbulence, *Phys. of Fluids*, **13**, 2634–2649.
36. Davidson (2004), Turbulence, An Introduction for Scientists and Engineers, Oxford University Press.
37. del Àlamo, J.C. and Jimenez, J. (2003), Spectra of the very large anisotropic scales in turbulent channels. *Phys. Fluids*, **15**, 41–44.
38. del Àlamo, J.C., Jimenez, J., Zandonade, P. and Moser, R.D. (2004), Scaling of the energy spectra of turbulent channels. *J. Fluid Mech.*, **500**, 135–144.
39. Dejoan, A and Leschziner, M.A. (2005), Large eddy simulation of a plane turbulent wall jet, *Phys. of Fluids*, **17**, 025102.
40. Demuren, A.O. and Sarkar, S. (1993), Perspective: Systematic study of Reynolds-stress closure models in the computation of plane channel flows, *ASME J. Fluids Eng.*, **115**, 5–12.

41. Dong, S. (2007), Direct numerical simulation of turbulent Taylor-Couette flow, *J. Fluid Mech.*, **587**, 373–393.
42. Durbin, P.A. (1991), Near-wall turbulence closure modelling without 'damping functions', *Theoret. Comput. Fluid Dynamics*, **3**, 1–13.
43. Durbin, P.A. (1993), A Reynolds stress model for near-wall turbulence, *J. Fluid Mech.* **249**, 465–498.
44. Durbin, P.A. (2011), Review: Adapting scalar turbulence closure models for rotation and curvature, *ASME J. Fluids Engng.*, 133–140.
45. Durbin, P.A. and Pettersson-Reif, B.A. (2001), Statistical Theory and Modelling for Turbulent Flows, John Wiley & Sons.
46. Frisch, U. (1996), Turbulence, Cambridge University Press.
47. Fu, S. (1988), Computational modelling of turbulent swirling flows with second-moment closures, *PhD Thesis*, University of Manchester.
48. Froehlich, J., Mellen, C., Rodi, W., Temmerman, L. and Leschziner, M.A. (2005), Highly resolved large eddy simulation of separated flow in a channel with streamwise periodic constrictions, *J. Fluid Mech.*, **526**, 19–56.
49. Fu, S. (1988), Computational modelling of turbulent swirling flows with second-moment closures, *PhD Thesis*, University of Manchester.
50. Fu, S., Launder, B.E. and Tselepidakis, D.P. (1987), Accommodating the effects of high strain rates in modelling the pressure-strain correlation. *Report TFD/87/5* Mechanical Engineering Dept., UMIST, Manchester.
51. Fu, S., Leschziner, M.A. and Launder, B.E. (1987), Modelling strongly swirling recirculating jet flow with Reynolds-stress transport closure. *Proc. 6th Symp. on Turbulent Shear Flow*, Toulouse, 17.6.1–17.6.6.
52. Gatski, T.B. and Speziale, C.G. (1993), On explicit algebraic stress models for complex turbulent flows, *J. Fluid Mech.* **254**, 59–78.
53. Gibson, M.M. and Launder, B.E (1978)., Ground effects on pressure fluctuations in the atmospheric boundary layer. *J. Fluid Mech.*, **86**, 491–511.
54. Gilbert, N. and Kleiser, L. (1991), Turbulence model testing with the aid of direct numerical simulation results, *Proc. 8th Symp. on Turbulent Shear Flows*, Munich, 26.1.1–26.1.6.
55. Goldberg, U.C. and Ramakrishnan, S.V. (1994), A pointwise version of the Baldwin-Barth turbulene model, *Int. J. Comp. Fluid Dynamics*, **1**, 321–338.
56. Goldberg, U.C. (1994), Towards a pointwise turbulence model for wall-bounded and free shear flows. *ASME J. Fluids Engrg.*, **116**, 72–79.
57. Granville, P.S. (1987), Baldwin-Lomax factors for turbulent boundary layers in pressure gradients, *J. AIAA*, **25**, 1624–1627.

58. Hanjalić, K. (1970), Two-dimensional asymmetric turbulent flow in ducts, Ph.D. Thesis, University of London.
59. Hanjalić, K. (1984), Modelling of turbulent transport processes — recent advances and further research trends, *Proc. Academy of Sciences and Arts of Bosnia and Herzegovina*, Dept. of Technical Sciences, **13**.
60. Hanjalić, K. (1994), Advanced turbulence closure models: A view of current status and future prospects, *Int. J. Heat Fluid Flow*, **15**, 178–203.
61. Hanjalić, K. and Jakirlić, S. (1993), A model of stress dissipation in second moment closures. *Appl. Scientific Research*, **51**, 513–518.
62. Hanjalić, K. and Kenjeres, S. (2001), T-RANS simulation of deterministic eddy structure in flows driven by thermal buoyancy and Lorentz Force, *Flow, Turbulence and Combustion*, **66**, 427–451.
63. Hanjalić, K., Kenjeres, S. and Durst, F. (1996), Natural convection in partitioned two-dimensional enclosures at higher Rayleigh numbers, *Int. J. Heat Mass Transfer*, **39**, 1407–1427.
64. Hanjalić, K. and Launder, B.E. (1972), A Reynolds stress model and its application to thin shear flows, *J. Fluid Mech.*, **52**, 609–638.
65. Hanjalić, K. and Launder, B.E. (1976), Contribution towards a Reynolds-stress closure for low-Reynolds-number turbulence. *J. Fluid Mech.*, **74**, 593–610.
66. Hanjalić, K., Popovac, M. and Hadziabdić, M. (2004), A robust near-wall elliptic-relaxation eddy-viscosity turbulence model for CFD. *Int. J. of Heat Fluid Flow*, **25**, 1047–1051.
67. Harlow, F.H. and Nakayama, P.I. (1968), Transport of turbulence energy decay rate, *University of California Report LA-3854*, Los Alamos Science Laboratory.
68. Hassid, S. and Poreh, M. (1975), A turbulence energy model for flows with drag reduction, *ASME J. of Fluids Engineering*, **97**, 234–241.
69. Hellsten, A. (1998), Some improvements in Menter's k-ω SST turbulence model, *AIAA Paper* 98-2554.
70. Hinze, J.O. (1959), Turbulence. McGraw Hill, 2nd edn., New York.
71. Hoffman, G.H. (1975), Improved form of low Reynolds-number k-ε turbulence model, *Phys. of Fluids*, **18**, 309–312.
72. Hogg S., Leschziner M.A., (1989), Computation of highly swirling confined flow with a Reynolds-stress turbulence model, *J. AIAA*, **27**, 57–67.
73. Hoyas, S. and Jiménez, J. (2006), Scaling of the velocity fluctuations in turbulent channels up to $Re_\tau = 2003$, *Phys. of Fluids*, **18**, paper 011702.

74. Howard, R. (2012), A wall heat transfer function for equilibrium flows — a combination of Reichardt and Kader profiles with Wang type scaling, *Proc. 7th Int. Symp. on Turbulence, Heat and Mass Transfer*, Palermo (Begell House Inc. Publ.), 345–348.
75. Huang, P.G (1986), The computation of elliptic turbiulent flows with second-model-closure models, *Ph.D. Thesis*, University of Manchester.
76. Huang, P.G. and Leschziner, M.A. (1985), Stabilisation of recirculating flow computations performed with second-moment closure and third-order discretisation, *Proc. 5th Symp. on Turbulent Shear Flow*, Cornell University, 20.7.
77. Hussain, A. (1983), Coherent structures: Reality and myth, *Phys. of Fluids*, **26**, 2816.
78. Jakirlić, S. (1997), Reynolds-spannungs-modellierung komplexer turbulenter Stroemungen, *Ph.D. Thesis*, University of Erlangen-Nuernberg.
79. Jakirlić, S. and Hanjalić, K. (1995), A second-moment closure for non-equilibrium and separating high- and low-Re-number flows. *Proc. 10th Symp. on Turbulent Shear Flows, Pennsylvania State University*, 23.25–23.31.
80. Jakirlić, S. and Hanjalić, K. (2002), A new approach to modelling near-wall turbulence energy and stress dissipation, *J. Fluid Mech.*, **459**, 139–166.
81. Jang, Y.J., Leschziner, M.A., Abe, K. and Temmerman, L. (2002), Investigation of anisotropy-resolving turbulence models by reference to highly-resolved LES data for separated flow, *Flow, Turbulence and Combustion*, **69**, 161–203.
82. Jayatilleke, C. (1969), The influence of Prandtl number and surface roughness on the resistance of the laminar sublayer to momentum and heat transfer, *Prog. Heat Mass Transfer*, **1**, 193–329.
83. Jeffreys, H. (1969), Cartesian Tensors, Cambridge University Press.
84. Johansson, A.V. and Hallbaeck, M. (1995), Modelling of rapid pressure-strain in Reynolds-stress closure. *J. Fluid Mech.*, **269**, 143–168.
85. Johansson, S.H., Davidson. L. and Olsson, E. (1993), Numerical simulation of vortex shedding past triangular cylinders at high Reynolds number using a k-ε turbulence model, *Int. J. for Num. Meths. in Fluids*, **16**, 859–878.
86. Jongen, T. and Gatski, T.B. (1998), General explicit algebraic stress relations and best approximations for three-dimensional flows, *Int. J. Eng. Sci.*, **36**, 739–763.
87. Jones, W.P. and Launder, B.E. (1972), The prediction of laminarization with a two-equation model of turbulence. *Int. J. Heat and Mass*

Transfer, **15**, 301–314.

88. Jovanovic, J., Ye, Q.-Y. and Durst, F. (1995), Statistical interpretation of the turbulent dissipation rate in wall-bounded flows, *J. Fluid Mech.*, **293**, 321–347.
89. Kader, B. (1981), Temperature and concentration profiles in fully turbulent boundary layers, *Int. J. Heat and Mass Transfer*, **24**, 1541–1544.
90. Kalitzin, G., Medic, G., Iaccarino, G. and Durbin, P. (2005), Near-wall behaviour of RANS turbulence models and implications for wall functions, *J. Comput. Phys.*, **204**, 265–291.
91. Kasagi, N., Tomita, Y. and Kuroda, A. (1992), Direct numerical simulation of passive scalar field in a turbulent channel flow, *Trans. ASME*, **114**, 598–606.
92. Kawamura, H., Abe, H. and Matsuo, Y. (1999) DNS of turbulent heat transfer in channel flow with respect to Reynolds and Prandtl number effects, *Int. J. Heat and Fluid Flow*, **20**, 196–207.
93. Kawamura, H. and Kawashima, N. (1994), A proposal for a k-ε model with relevance to the near-wall turbulence, Paper IP.1, *Proc. Int. Symp. on Turbulence, Heat and Mass Transfer*, Lisbon.
94. Kebede, W., Launder, B.E. and Younis, B.A. (1985), Large-amplitude periodic pipe flow: A second-moment closure study, *Proc. of 5th Symp. on Turbulent Shear Flows*, 16.23, Cornell University, Ithaca, New York USA.
95. Kim, J., Moin, P. and Moser, D. (1987), Turbulence statistics in fully developed channel flow at the low Reynolds number, *J. Fluid Mech.*, **177**, 133–166,
96. Kolmogorov, A.N. (1941), The local structure of turbulenceb in incompressible viscous fluids for very large Reynolds number, *Dokl. Akad. Nauk.*, **30**, 301–305.
97. Kolmogorov, A.N. (1942), The equations of turbulent motion in an incompressible fluid, *Izvestia Acad. Sci., USSR, Phys.*, **6**, 56–58.
98. Kok, J.C. (2000), Resolving the dependence on freestream values for the k-ω turbulence model, *J. AIAA*, **38**, 1292–1295.
99. Lam, C.K.G. and Bremhorst, K. (1981), A modified form of the k-ε model for predicting wall turbulence. *J. Fluids Engrg.*, **103**, 456–460.
100. Langtry, R.B and Menter, F.R. (2009), Correlation-based transition modeling for unstructured parallelized computational fluid dynamics codes, *J. AIAA Journal*, **47**, 2894–2906.
101. Lardeau, S. and Leschziner, M.A. (2006), Unsteady RANS modeling of wake-induced transition in linear LP-turbine cascades, *J. AIAA*, **44**, 1854–1865.

102. Lardeau, S. and Leschziner, M.A. (2013), The streamwise drag-reduction response of a boundary layer subjected to a sudden imposition of transverse oscillatory wall motion, *Phys. of Fluids*, **25**, paper 075109.
103. Lardeau, S., Li, N. and Leschziner, M.A. (2007), Large eddy simulation of transitional boundary layer at high free-stream turbulence intensity and implications for RANS modelling, *ASME J. of Turbomachinery*, **129**, 311–317.
104. Laufer, J. (1954), The structure of turbulence in fully developed pipe flow, *NACA Report 1174*.
105. Launder, B.E. (1989), Second-moment closure: Present and future? *Int. J. Heat Fluid Flow*, **10**, 282–300.
106. Launder, B.E. (1975), A Reynolds-stress closure model of turbulence applied to the calculation of a highly curved mixing layer, *J. Fluid Mech.*, **67**, 569–581.
107. Launder, B.E. and Kato, M. (1993), Modelling of flow-induced oscillations in turbulent flow around a square cylinder, *ASME FED*, **157**, 189–199.
108. Launder, B.E. and Sharma, B.I. (1974), Application of the energy-dissipation model of turbulence to the calculation of flow near a spinning disc. Letters in *Heat Mass Transfer*, **1**, 131–138.
109. Launder, B.E. and Shima, N. (1989), Second-moment closure for the near-wall sublayer, *J. AIAA*, **27**, 1319–1325.
110. Launder, B.E. and Tselepidakis, D.P. (1993), Contribution to the modelling of near-wall turbulence. *Turbulent Shear Flows 8*, (F. Durst *et al.*, Eds.) Springer-Verlag, 81–96.
111. Launder, B.E., Priddin, C.H. and Sharma, B. (1977), The calculation of turbulent boundary layers on curved and spinning surfaces, *ASME J. Fluids Eng.*, **99**, 237–239.
112. Launder, B.E., Reece, G.J. and Rodi, W. (1975), Progress in the development of Reynolds-stress turbulence closure, *J. Fluid Mech.*, **68**, 537–566.
113. Laurence, D., Uribe, J. and Utyuzhnikov, S., (2004), A robust formulation of the v2 − f model, *Flow, Turbulence and Combustion*, **73**, 169–185.
114. Lee, M.J. and Reynolds, W.C. (1985), Numerical experiments on the structure of homogeneous turbulence, *Technical Report TF-24*, Stanford University.
115. Leschziner, M.A. (2000), Turbulence modelling for separated flows with anisotropy-resolving closures, *Phil. Trans. Royal Soc. London A*, **358**, 3247–3277.

116. Leschziner, M.A., Batten, P. and Craft, T.J. (2001), Reynolds-stress modelling of afterbody flows, *The Aeronautical Journal*, **105**, 297–306.
117. Leschziner, M.A., Fishpool, G.M. and Lardeau, S. (2009), Turbulent shear flow — a paradigmatic multi-scale phenomenon, *J. Multiscale Modelling*, **1**, 197–222.
118. Leschziner, M.A. and Rodi, W. (1981), Calculation of annular and twin parallel jets using various discretization schemes and turbulence models, *ASME J. Fluids Eng.* **103**, 352–360.
119. Leschziner M.A. and Rodi W. (1983), Calculation of a heated water discharge, *ASCE J. of Hydraulic Engineering*, **109**, 1380–1384.
120. Lien, F.S. and Durbin, P.A. (1996), Non-linear k-v2 modelling with application to high-lift, *Proc. Summer Prog., Centre For Turbulence Research*, Stanford University, 5–26.
121. Lien, F.S. and Leschziner, M.A. (1994), Modelling the flow in a transition duct with a non-orthogonal FV procedure and low-Re turbulence-transport models, *Proc. ASME FED Summer Meeting, Symposium on Advances in Computational Methods in Fluid Dynamics*, 93–106.
122. Lien, F.-S. and Leschziner, M.A. (1994), A general non-orthogonal finite volume algorithm for turbulent flow at all speeds incorporating second-moment closure. Part I: Numerical implementation. *Comput. Meth. Appl. Mech. Engnrg.*, **114**, 123–148.
123. Lien F.S. and Leschziner M.A. (1996), Second-moment closure for three-dimensional turbulent flow around and within complex geometries", *Computers and Fluids*, **25**, 237–262.
124. Lien, F.-S. and Leschziner, M.A. (2002), Numerical aspects of applying second-moment closure to complex flows, *Closure Strategies for Turbulent and Transitional Flows* (B.E. Launder and N. Sandham, eds.), Cambridge University Press, 153–187.
125. Liou, W.W., Huang, G. and Shih T.-H. (2000), Turbulence model assessment for shock wave/turbulent boundary-layer interaction in transonic and supersonic flow, *Computers and Fluids*, **29**, 275–299.
126. Lumley, J.L. (1978), Computational modelling of turbulent flows, *Adv. Appl. Mech.*, **18**, 123–176.
127. Magnaudet, J. (1992), The modelling of inhomogeneous turbulence in the absence of mean velocity gradients, *Proc. 4th European Conference* (1992), Delft, The Netherland.
128. Manceau, R. and K. Hanjalić, K. (2002), Elliptic blending model: A new near-wall Reynolds stress turbulence closure, *Phys. of Fluids*, **14**, 744–754.
129. McGuirk, J.J. and Rodi, W. (1979), The calculation of three-dimensional turbulent free jets, *Turbulent Shear Flows I*, Springer.

130. Mellor, G.L. and Herring, H.J. (1973), A survey of the mean turbulent field closure models, *J. AIAA*, **11** (1973), 590–599.
131. Menter, F.R. (1994), Two equation eddy viscosity turbulence models for engineering applications, *J. AIAA*, **32**, 1598–1605.
132. Menter, F.R. (1997), Eddy viscosity transport equations and their relation to the k-ε model, *ASME J. Fluids Engineering*, **119**, 876–884.
133. Menter, F.R. (1992), Influence of freestream values on k-ω turbulence model predictions, *J. AIAA*, **30**, 1657–1659.
134. Menter, F.R., Garbaruk, A.V. and Egorov, Y. (2009), Explicit algebraic Reynolds stress models for anisotropic wall-bounded flows, *Proc. EUCASS — 3rd European Conference for Aero-Space Sciences*, Versailles, France.
135. Moser, R., Kim, J. and Mansour, N.N. (1999), Direct numerical simulation of turbulent channel flow up to $\mathrm{Re}_\tau = 590$, *Phys. of Fluids*, **11**, 943–945.
136. Myong, H.K. and Kasagi, N. (1990), A new approach to the improvement of k-ε turbulence model for wall bounded shear flows. *JSME Int. J.*, Series II, **33**, 63–72.
137. Nagano, Y. (2002), Modelling heat transfer in near-wall flows, *in Closure Strategies for Turbulent and Transitional Flows* (B.E. Launder and N. Sandham, eds.), Cambridge University Press.
138. Nagano, Y. and Hishida, M. (1987), Improved form of the k-ε model for turbulent shear flows. *Trans. ASME*, **109**, 156–162.
139. Nagano, Y. and Kim, C. (1988), A two-equation model for heat transport in wall turbulent shear flows, *Trans. ASME, J. Heat Transfer*, **110**, 583–589.
140. Nagano, Y., Tagawa, M. and Tsuji, T. (1991), An improved two-equation heat transfer model for wall turbulent shear flows, *Proc. ASME/JSME Thermal Engineering Joint Conference*, Reno, USA, **3**, 233–240.
141. Nagano, Y. and Tagawa, M. (1990), "An Improved k-epsilon Model for boundary layer flows", *ASME Journal Fluids Engineering*, **112**, 33–39.
142. Naji, H., Mompean, G. and Yahyaoi, O.E. (2004), Evaluation of explicit algebraic stress models using direct numerical simulations, *J. of Turbulence*, **5**, 1–25.
143. Naot, D., Shavit, A. and Wolfshtein, M. (1973), Two-point correlation model and the redistribution of Reynolds stresses, *Phys. of Fluids*, **16**, 738–740.
144. Nisizima, S. and Yoshizawa, A. (1987), Turbulent channel and Couette flow using an anisotropic k-ε model, *J. AIAA*, **25**, 414–420.

145. Norris, L.H. and Reynolds, W.C. (1975), Turbulence channel flow with a moving wavy boundary, *Rep. FM-10*, Dept. of Mech. Engrg., Stanford University, USA.
146. Obi, S. Aoki, K. and Masuda, S. (1993), Experimental and computational study of turbulent separated flow in an asymmetric diffuser, *Proc. 9th Symp. on Turbulent Shear Flows*, Kyoto, 305.
147. Omranian, A, Craft, T.J. and Iacovides, H. (2014), The computation of buoyant flows in differentially heated inclined cavities, *Int. J. Heat Mass Transfer*, **77**, 1–16.
148. Orszag, S.A., Yakhot, V., Flannery, W.S., Boysan, F., Choudhury, D., Maruzewski, J. and Patel, B. (1993), Renormalisation group modelling and turbulence simulations. *Near-Wall Turbulent Flows* (R.M.C. So, C.G. Speziale and B.E. Launder, Eds.), Elsevier, 1031–1046.
149. Patel, C.V., Rodi, W. and Scheuerer, G. (1985), Turbulence models for near-wall and low Reynolds number flows: A review, *J. AIAA*, **23**, 1308–1319.
150. Pettersson-Reif, B.A., Durbin, P. and Ooi, A. (1999), Modelling rotational effects in eddy-viscosity closures, *Int. J. Heat and Fluid Flow*, **20**, 563–573.
151. Pope, S.B. (1975), A more general effective-viscosity hypothesis, *J. Fluid Mech.*, **72**, 331–340.
152. Pope, S.B. (2000), Turbulent Flows, Cambridge University Press.
153. Prandtl, L. (1925), Bericht ueber die Enstehung der Turbulenz. *Z. Angew. Math. Mech.*, **5**, 136–139.
154. Prandtl, L. (1945), Ueber ein neues Formalsystem fuer die ausgebildete Turbulence, *Nach. Akad. Wiss. Goettingen, Math.-Phys.*, **K1**, 6–19.
155. Proud'homme, M. and Elghobashi, S. (1983), Prediction of wall-bounded turbulent flows with an improved version of a Reynolds-stress model, *Proc. 4th Symp. on Turbulent Shear Flows*, Karlruhe, Germany, 1.11.
156. Reichardt, R. (1951), Vollstaendige Darstellung der turbulenten Geschwindigkeitsverteilung in glatten Leitungen, *ZAMM Journal of Applied Mathematics and Mechanics*, **31**, 208–219.
157. Reynolds, O. (1895), On the dynamical theory of incompressible viscous fluids and the determination of the critertion, *Phil. Trans. Roy. Soc.* **186**, 123–164.
158. Reynolds, W.C. (1976), Computation of turbulent flows, *Annu. Rev. of Fluid Mechanics*, **8**, 283–208.
159. Reynolds, W.C. (1987), Fundamentals of Turbulence for Turbulence Modeling and Simulation, *Rep. AGARD-CP-93*, NATO.

160. Rodi, W. (1972), The prediction of free turbulent boundary layers by use of a two-equation model of turbulence, *Ph.D. Thesis*, University of London.
161. Rodi, W. (1976), A new algebraic relation for calculating the Reynolds stresses, *Z. angew. Math. Mech.*, **56**, 219–221.
162. Rodi, W. (1979), Influence of buoyancy and rotation on equations for the turbulent length scale, *Proc. 2nd Symp. on Turbulent Shear Flows*, London, 10.37–10.42.
163. Rodi, W. (1993), Turbulence models and their application in hydraulics, a state-of-the-art review, IAHR Monograph, Taylor & Francis.
164. Rodi, W. and Mansour, N.N. (1993), Low Reynolds number k-ε modelling with the aid of direct numerical simulation data, *J. Fluid Mech.*, **250**, 509–529.
165. Rodi, W. and Scheuerer, G. (1983), Calculation of curved shear layers with two-equation turbulence models, *Phys. of Fluids*, **26**, paper 1422.
166. Rotta, J.C. (1951), Statistische Theory nichthomogener Turbulenz, *Zeitschrift der Physik*, **129**, 547–572 (and **131** (1951) 51–77).
167. Rubinstein, R. and Barton, J.M. (1990), Non-linear Reynolds stress models and the renormalisation group, *Phys. of Fluids A*, **2**, 1472–1476.
168. Saffman, P.G. (1970), A model for inhomogeneous turbulent flow, *Proc. Royal Soc. London, A*, **317**, 427–433.
169. Saffman, P.G. and Wilcox, D.C. (1974), Turbulence model predictions for turbulent boundary layers, *J. AIAA*, **12**, 541–546.
170. Shih, T.-H. and Lumley, J.L. (1985), Modelling of pressure correlation terms in Reynolds-stress and scalar-flux equations. *Report FDA-85-3*, Sibley School of Mech. and Aerospace Eng, Cornell University.
171. Shih, T.-H., Zhu, J. and Lumley, J.L. (1995), A realisable Reynolds stress algebraic equation model, *Comp. Methods Appl. Mech. Engrg.* **125**, 287–302.
172. Shih, T.-H., Liou, W.W., Shabbir, A., Yang, Z. and Zhu, J. (1995), A new k-epsilon eddy-viscosity model for high Reynolds number turbulent flows, *Computers and Fluids*, **3**, 227–238.
173. Shih, T.-H. and Mansour, N.N. (1990), Modelling of near-wall turbulence, *NASA TM103222.*
174. Shima, N. (1988), A Reynolds-stress model for near-wall and low-Reynolds-number regions, *ASME J. of Fluids Engineering*, **110**, 38–44.
175. Shir, C.C. (1973), A preliminary numerical study of atmospheric turbulent flows in the idealised planetary boundary layer. *J. Atmos. Sci.* **30**, 1327–1339.

176. Smith, G.F. (1971), On isotropic functions of symmetric tensors, skew-symmetric tensors and vectors, *Int. J. of Engineering Sciences*, **9**, 899–916.
177. So, R.M.C., Lai, Y.G., Zhang, H.S. and Hwang, B.C. (1991), Second-order near-wall turbulence closures: A review, *J. AIAA*, **29**, 1819–1835.
178. So, R.M.C., Lai, Y.G., Hwang, B.C. and Yoo, G.J. (1988), Low Reynolds number modelling of flows over a backward facing step, *ZAMP*, **39**, 13–27.
179. So, R.M.C., Zhang, H.S. and Speziale, C.G. (1991), Near-wall modelling of the dissipation rate equation, *J. AIAA*, **29**, 2069–2076.
180. Sommer, T.P., So, R.M.C. and Lai, Y.G. (1992), A near-wall two-equation model for turbulent heat fluxes, *Int. J. Heat Mass Transfer*, **35**, 3375–3387.
181. Spalart P.R. and Allmaras, S.R. (1992), A one-equation turbulence model for aerodynamic flows. *AIAA Paper 92-0439.*
182. Spalart, P.R. and Shur, M.L. (1997), On the sensitization of turbulence models to rotation and curvature, *Aerosp. Sci. Technol.*, **1**, 297–302.
183. Spalding, D.B. (1967), Monograph on turbulent boundary layers, *Technical Report TWF/TN/33*, Mech. Eng. Dept., Imperial College London.
184. Spencer, A.J.M. and Rivlin, R.S. (1959), The theory of matrix polynomials and its application to the mechanics of isotropic continua, *Arch. Rat. Mech. Anal.*, **2**, 309–336.
185. Speziale, C.G. (1987), On non-linear k-l and k-ε models of turbulence, *J. Fluid Mechanics*, **178**, 459–475.
186. Speziale, C.G., Sarkar, S. and Gatski, T.B. (1991), Modelling the pressure-strain correlation of turbulence: An invariant dynamical systems approach. *J. Fluid Mech.* **227**, 245–272.
187. Taulbee, D.B. (1992), An improved algebraic Reynolds stress model and corresponding nonlinear stress model, *Phys. of Fluids A*, **4**, 2555–2561.
188. Temmerman, L. and Leschziner, M. (2001), "Large Eddy Simulation of separated Flow in a channel with corrugated wall", *Proc. 2nd Int. Conf. on Turbulence and Shear Flow Phenomena*, KTH Stockholm, **3**, 399–404.
189. Tennekes, H. and Lumley, J.L. (1972), A First Course in Turbulence, MIT Press.
190. Timoshenko, S., Theory of Elasticity, McGraw-Hill Ltd.
191. Tselepidakis, D.P. (1991), Development and Application of a New Second-Moment Closure for Turbulent Flows Near Walls, *Ph.D. Thesis*, The University of Manchester, UK.

192. Uribe, J.C. (2006), An industrial approach to near-wall turbulence modelling for unstructured finite volume methods, *Ph.D. thesis*, The University of Manchester, UK.
193. von Karman, T. (1930), Mechanische Aenlichkeit und Turbulenz, *Proc. Third Int. Congr. Applied Mechanics*, Stockholm, 85–105.
194. Wallin, S. (2000), Engineering turbulence modelling for CFD with a focus on explicit algebraic Reynolds stress models, *Ph.D. Thesis*, KTH Stockholm.
195. Wallin, S. and Johansson, A.V. (2000), An explicit algebraic Reynolds stress model for incompressible and compressible turbulent flows, *J. Fluid Mech.*, **403**, 89–132.
196. Wallin, S. and Johansson, A.V. (2001), Modelling of streamline curvature effects on turbulence in explicit algebraic Reynolds stress turbulence models, *Proc. 2nd Int. Symp. on Turbulence and Shear Flow Phenomena*, Stockholm, 223–228.
197. Weinmann, M. and Sandberg, R.D. (2009), Suitability of explicit albegraic stress models for predicting complex three-dimensional flows, *paper AIAA-2009–3663*, 19th AIAA Computational Fluid Dynamics Conference, San Antonio, Texas.
198. Wilcox, D.C. and Rubesin, M.W. (1980), Progress in turbulence modelling for complex flow field including effects of compressibility, *Rep. NASA TP1517*.
199. Wilcox, D.C. (1988), Reassessment of the scale-determining equation for advanced turbulence models. *J. AIAA*, **26**, 1299–1310.
200. Wilcox, D.C. (1993), *Turbulence Modelling for CFD*, DCW Industries, La Canada, California (1993).
201. Wilcox, D.C. (1994), Simulation of transition with a two-equation turbulence model. *J. AIAA*, **32**, 247–255.
202. Wolfshtein, M. (1969), The velocity and temperature distribution in one-dimensional flow with turbulence augmentation and pressure gradient. *Int. J. Heat Mass Transfer*, **12**, 301–318.
203. Yakhot, V., Orszag, S.A., Thangham, S., Gatski, T.B. and Speziale, C.G. (1992), Development of turbulence models for shear flows by a double expansion technique, *Phys. of Fluids A*, **4** (1992), paper 1510.
204. Yap, C.R. (1987), Turbulent heat and momentum transfer in recirculating and impinging flows, *Ph.D. Thesis*, University of Manchester.
205. Yeung, P.K., Donzis, D.A. and Sreenivasan, K.R. (2005), High-Reynolds-number simulation of turbulent mixing, *Phys. of Fluids*, **17**, paper 081703.
206. Zhu, Y. and Antonia, R.A. (1997), On the correlation between enstrophy and energy dissipation rate in a turbulent wake, *Appl. Sci. Research*, **57**, 337–347.

Index

ω-transport equation, 165
$\overline{v^2} - f$ model, 315
$k - \omega$ model, 164
  ω at free-stream boundaries, 167
  mixed-derivative term, 168
  pathological behaviour, 168
  prediction of transition, 170
$k - \omega$-model — low-Reynolds-number extensions, 169
$k - \omega$ model
  boundary condition, 170
$k - \phi$ models, 161
$k - \varepsilon$ models, 131, 137
  Chien, 155
  Jones-Launder, 138
  Lam–Bremhorst, 155
  Launder–Sharma, 155
  Lien–Leschziner, 159
$k - \varepsilon$–model — low-Reynolds-number extensions, 149
$v^2 - f$ model
  $\overline{v^2}$ — transport equation, 317, 321
  $\varphi - \alpha$ model, 321, 322
  f-relaxation equation, 317, 321
  code-friendly variable $\varphi = \overline{v^2}/k$, 321
  code-friendly version, 320, 321
  BL-v^2/k model, 322, 324
  blending concept, 321
  blending formula for f, 322
  channel flow, 324
  eddy viscosity, 318
  separated flow in a one-sided plane diffuser, 325
  separated flow in a periodically constricted channel, 325
  wall-asymptotic variation of f, 318, 322
  wall-boundary conditions, 322

algebraic length-scale prescription, 97
algebraic Reynolds-stress models, 327
  algebraic (ARSM) equivalent form of the differential parent (RSTM), 330
  algebraically scaled transport of the turbulence energy, 328
  anisotropic state in simple shear, 339
  anisotropy components in channel flow — model predictions, 344
  boundary between EARSMs and NLEVMs , 332
  convective contributions, 329
  cylindrical-polar coordinate system, 329
  dichotomy of the convection/production terms, 329
  explicit algebraic Reynolds-stress models (EARSMs), 332
  iterative strategy, 333
  Launder *et al.*'s RSTM, 330, 338
  normal Reynolds stresses in channel flow, 353
  normal-stress components of the anisotropy tensor, 339

numerically stabilizing eddy-viscosity terms, 331
Rodi's algebraic stress model (ASM) proposal, 328
Speziale *et al.*'s (SSG) RSTM, 330, 340
stability-promoting algorithmic aids, 331
strain- and vorticity-sensitised eddy-viscosity coefficient, 338
streamline-aligned co-ordinates, 329
three-dimensional conditions, 341, 345, 348
transport of the anisotropy, 328
two-dimensional conditions, 343
anisotropic turbulence, 15
anisotropy, 25, 57, 67
anisotropy components, 69
anisotropy invariants for channel flow, 239
anisotropy of dissipation, 265
Apsley & Leschziner's model, 359
asymptotic analysis, 67
asymptotic near-wall behaviour, 137
asymptotic near-wall variation of turbulence-energy production, 71
asymptotic variation of $\tilde{\varepsilon}$, 151

Baldwin & Barth's one-equation model, 176
Biot–Savart law, 9
body forces — buoyancy, electromagnetic effects, rotation, 201, 219
body-force terms, 308
body-force-induced stress-generation term, 219
boundary condition for ω, 170
box turbulence, 46
Bradshaw *et al.*'s model, 130
Bradshaw's relation, 131, 173, 177
Brownian motion, 9, 88
budget of exact dissipation-rate equation, 142
buffer layer, 71, 75, 158, 181
buoyancy, 27, 29
buoyancy parameter, 222
buoyancy-driven flows in cavities, 306
buoyancy-induced production, 220–222, 233, 307
buoyancy-related correction in ε equation, 222
buoyancy-related flux-generation terms, 308
buoyancy-related scalar-variance-generation terms, 310
buoyant flows, 79, 220
butterfly effect, 8

cascade, 14
Cayley–Hamilton theorem, 263, 265, 334, 335, 337
channel flow, 57, 67, 69, 76
channel flow subjected to orthogonal rotation, 218
channel flow — heated, 309
chemical species, 79
Chien's model, 155
closure of the Reynolds-stress-transport equations, 230
coherent unsteadiness, 39
concave curvature, 209
contraction, 380
convex curvature, 209
correlations of turbulent velocity fluctuations, 4
cross-correlation, 23
cross-flow flux of momentum, 93
cross-product, 381
curl, 381
curvature, 208
curvature correction, 211
curvature strain, 208
curvature-dependent C_μ, 212
curvature-related body forcing, 29

curvature-related correction terms in ε equation, 214
curvature-related production, 210, 214
curvature-related strain, 213
cycle of turbulence, 62

damping functions, 178
decay of isotropic turbulence, 146, 166
decaying wind-tunnel turbulence, 89
deviatoric parts of the stresses, 90
density fluctuations, 220
detached-eddy simulation, 123
diffusive flux transport, 310
diffusive shear-stress flux, 131
diffusive turbulence-energy flux, 131
dissipation, 14, 62, 103, 104
dissipation of Reynolds stresses, 244
  anisotropic-dissipation-rate model, 247
  anisotropy-promoting blocking effect of the wall, 246
  dissipation anisotropy in channel flow, 250, 254
  dissipation-anisotroy invariants, 247
  dissipation-rate anisotropy parameter, E, 247
  dissipation-rate components, 247
  dissipation-rate profiles in channel flow, 256
  GGDH approximation, 255
  isotropic-dissipation model, 244
  normal-stress-dissipation components, 246
  transport equation for turbulence-energy dissipation, 254
  viscous-damping functions in dissipation-transport equation, 255
  viscous-damping functions for dissipation tensor, 250
  wall-asymptotic values for the normalised dissipation-tensor components, 249
  wall-asymptotic variations of the components ε_{ij}, 248
  wall-corrected anisotropic-dissipation-rate models, 248, 252
  wall-limiting dissipation anisotropy, 253
  wall-orientation indicators, 252
dissipation of Reynolds stresses, 244
dissipation of scalar fluxes, 309
dissipation-rate equation — exact form, 138
dissipation-rate equation — modelled form, 145
dissipation-rate equilibrium, 142
dissipation-rate tensor, 232
divergence, 380
double-diffusion, 308
duct — orthogonal-mode rotation, 237
dynamical system, 8

eddies, 7, 8, 16, 20
eddy diffusivity, 304
eddy size, 18, 19
eddy viscosity, 87, 96
  asymptotic variation, 107
eddy-size 'pipeline', 104
eddy-size range, 17
eddy-turnover-time scale, 162
eddy viscosity, 92
eddy-viscosity coefficient C_μ — sensitivity to strain, 373, 374
eddy-viscosity relation, 138, 177
eddy-viscosity-transport equation, 132
eddy-viscosity-transport models, 115
effects of curvature, 208
eigenvectors of the stress and strain tensors, 202, 228
Einstein's summation convention, 98
elliptic relaxation, 315

elliptic relaxation of pressure-strain correlation, 291
  asymptotic variation of viscous-stress diffusion, 295
  blending-parameter equation, 299
  boundary conditions for blending-parameter equation, 299
  elliptic-relaxation equation, 293, 295, 297
  homogeneous pressure-strain term, 293
  length scale L, 297
  scalar-blending model, 298
  wall-asymptotic conditions, 296
  wall-boundary conditions, 295, 297
energy density, 12
energy-transfer rate, 15
ensemble-averaging, 34
enstrophy, 139
enstrophy equation, 143
enstrophy of turbulence, 162
enstrophy-transport equation, 139
explicit algebraic Reynolds-stress models (EARSMs), 332
  $k-\omega$ model of Wilcox, 352
  $k-\varepsilon$ model of Chien, 352
  $k-\varepsilon$ model of Lien and Leschziner, 364
  Abe *et al.*'s model, 354
  anisotropy components in channel flow, 351
  approximations to rigorous EARSMs, 353
  channel flow — model predictions, 365
  Gatski & Speziale's model, 339
  general formalism, 334
  general single-tensor function, 334
  general stress-strain depencence, 334
  general twin-tensor function , 335
  GGDH approximations for diffusion, 357
  iterative strategy, 359
  Kolmogorov scale, 358
  near-wall behaviour, 348
  Pope's model, 337
  regularisation, 341, 356
  scale-governing equations, 352
  separated asymmetric-diffuser flow — model predictions, 366
  separated channel flow — model predictions, 359
  strain/vorticity invariants, 335
  van Driest-type damping functions, 349
  viscous damping functions, 357
  wall-asymptotic variation of anisotropy components, 350
  Wallin & Johansson's model, 342
exponential damping function, 108
exponential decay function, 157

filtered velocity, 4, 36
finest-scale structures, 18
flat-plate boundary layer, 57, 70, 72
flatness parameter, 238
flux equations, 79
flux invariant, 312
flux Richardson number, 214, 217
flux vector, 39
flux-transport equations, 80, 306, 308
forcing turbulence, 25
Fourier–Fick law, 86, 220
free-stream turbulence, 116
friction Reynolds number, 73
friction velocity, 73

Galilean-invariant, 234
general realisability constraint, 204
generalised gradient-diffusion hypothesis (GGDH), 305
gradient Richardson number, 213, 217

gradient-diffusion hypothesis, 145
gravity, 220
grid-turbulence decay, 148

heat-flux components, 83, 84
heat transfer, 79
heat-flux budgets, 83, 309
heat-flux vector, 233
heat-flux-transport modelling, 303
heat-transfer coefficient, 306
heated mixing layer, 83
heated wall, 306
heated/cooled vertical surfaces, 308
higher-rank tensors, 380
homogeneous isotropic turbulence, 8
homogeneous plane strain — temporal evolution of anisotropy, 280
homogeneous shear flow, 148
homogeneous streamwise straining, 203
horizontal mixing/thermal layer, 82
hybrid $k-\omega/k-\varepsilon$ model, 172
hybrid LES-RANS modelling, 123

inertial sub-range, 12, 104
instabilities, 26
invariants map, 238
invariants of the anisotropy tensor, 334
inviscid wall blocking, 69, 109, 150
isotropic dissipation, 136, 149, 255

jet — impinging, 307
joint probability-density function — velocity fluctuations, 24
Jones & Launder's model, 131

kinematic constraints, 234
Kolmogorov, 13, 14, 18
Kolmogorov length scale, 19, 151
Kolmogorov time scale, 19, 154
Kronecker Delta, 380

Lam and Bremhorst's model, 155
laplacean, 381
large eddy simulations (LES), 15, 123
Launder and Sharma's model, 155
Launder and Shima's wall-proximity correction, 288
Launder *et al.*'s RSTM — simplified, 342
length scale, 15, 92, 102, 104
    asymptotic wall behaviour, 109
length-scale corrections, 223
length-scale relations, 118
length-scale surrogate, 112, 135
length-scale surrogate combining k and ε, 161
length-scale transport, 110
length-scale-related correction in ε equation, 224
length-scale-surrogate equation, 128
length-scale-surrogate variables, 136
length-scale-transport equation, 112
Lien and Leschziner's model, 159
life time of scalar fluctuations, 304
life time of the energy-containing eddies, 304
linear eddy-viscosity models — defects, 199
log-law, 121
log-law region, 146, 152
logarithmic layer, 73
logarithmic profile, 76
low-Reynolds-number models, 149, 153
low-Reynolds-number modifications, 152
Lumley's realisability map, 238

macro-scales, 22, 103
mass-specific enthalpy, 38
massively separated flow, 128
mean-scalar-transport Equation, 83
Menter's one-equation model, 176
mesh-resolution requirements, 3
mixed derivatives, 163
mixing across a shear layer, 93
mixing length, 106
mixing-length model, 94–96, 103, 105, 107, 121

Mohr's stress circle, 235
multi-scale phenomenon, 10

near-wall behaviour, 136
near-wall damping, 152
near-wall dissipation, 70
near-wall eddy structure, 66
near-wall interaction, 65
near-wall layer, 73, 181, 182
near-wall turbulence-energy maximum, 71
near-wall variation of the dissipation rate, 69
near-wall variations of the Reynolds-stresses, 68
negative normal stresses, 203
non-linear eddy-viscosity models (NLEVMs), 327, 366
- channel flow — model predictions, 374
- coefficients calibration/optimisation process, 370
- Craft *et al.*'s model, 369
- cubic *Ansatz*, 369
- curved and swirling flows, 371
- impingement, 376
- impinging jet — model predictions, 375
- realisability, 368, 375
- rotating pipe — model predictions, 374
- Shih *et al.*'s model, 367
- viscous correction, 373

non-locality, 21, 90
normal straining, 90
normal straining due to deceleration or acceleration, 205
normal-stress anisotropy, 57, 205
normal-stress anisotropy in shear, 269, 271
normal-stress components, 56
normal-stress isotropy, 90
Norris and Reynolds' model, 120, 159
Nusselt number, 306

one-dimensional momentum equation, 119
one-equation eddy-viscosity models, 105, 115
- eddy-viscosity-transport models, 122
- turbulence-energy-based models, 117

orthogonal-mode rotation, 219
Osborne Reynolds, 33

periodic motion, 42
permutation tensor, 380
phase average, 43
Prandtl–Schmidt number, 82, 116, 145, 221, 304
pressure diffusion, 55
pressure-diffusion terms, 232
pressure-scalar interaction, 82
pressure-strain interaction, 55, 232, 310
pressure-velocity interaction, 232
pressure-velocity interaction in Reynolds-stress equations, 257
- anisotropy of turbulence-intensity components in channel flow, 290
- asymptotic variation of pressure-strain interaction, 286
- asymptotic variation of the pressure-velocity interaction, 286
- asymptotic variations of the pressure diffusion, 287
- body forces, 290
- Craft and Launder's model, 264, 272
- Craft and Launder's wall-proximity correction, 285
- Crow's rapid-straining relationship, 268
- Durbin's model, 292
- general expansion of fourth-rank tensor, 265

general fourth-rank tensor, 265
general second-rank tensor function, 275
Gibson and Launder's model, 268
Gibson and Launder's wall-proximity correction, 285
Hanjalić and Launder's model, 272
inviscid wall effects, 281
isotropic tensor, 276
isotropization-of/by-production (IP) model, 267
Jakirlić and Hanjalić's model, 263, 273
Launder *et al.*'s model, 270
modelling $\Phi_{ij,1} + \Phi_{ij,2}$ collectively, 275
modelling of the rapid term $\Phi_{ij,2}$, 265
modelling of the slow term $\Phi_{ij,1}$, 262
Naot *et al.*'s model, 270
near-wall budgets of the wall-normal and shear-stress components in a separated channel flow, 288
near-wall effects, 280
normal-stress anisotropy in shear, 283
pressure-Poisson equation for the pressure fluctuation, 258
Prud'homme and Elghobashi's wall-proximity correction, 288
rapid and slow terms, 260, 310
Rotta's model, 262, 267
Shir's wall-proximity model, 283
Speziale *et al.*'s (SSG) model, 263, 275
viscous wall effects, 285
wall-proximity corrections, 282
pressure/scalar-gradient interaction, 303, 309, 310
principal axes of the stress tensor, 236
principal stresses, 236
probability-density function, 11
production of Reynolds stresses, 232
production of shear stress, 62

rapid term, 310
RANS equation, 88
Rayleigh–Bénard convection, 27, 42, 45, 307
realisability, 91, 202, 205, 235
realisability constraint, 203
realisability — duct in orthogonal-mode rotation, 237
realisable state, 202, 235
redistribution by pressure-velocity correlations, 62
renormalisation-group-(RNG-) theory-based $k - \varepsilon$ model, 224
Reynolds decomposition, 51
Reynolds number, 16
Reynolds shear-stress equation, 50
Reynolds-averaged mass-conservation equation, 38
Reynolds-averaged Navier–Stokes (RANS) equations, 33, 37
Reynolds-number-transport equation, 129
Reynolds-shear-stress equation, 52
dissipation-rate, 54
Reynolds-stress budgets for channel flow, 240, 246
Reynolds-stress budgets for separated flow, 241, 245
Reynolds-stress closure — basic rules, 234
Reynolds-stress equilibrium, 329
Reynolds-stress invariants, 235, 238
Reynolds-stress-transport equation, 294
Reynolds-stress-transport equations — exact, 231
Reynolds-stress-transport equations — production of the stresses, 201
Reynolds-stress-transport modelling, 227

Reynolds stresses, 37
Richardson number, 222
Richtmyer–Meshkov instability, 28
Rossby number, 219
rotating channels, 219
rotation, 217
rotation-related production terms, 217
rotation-related correction in ε equation, 219
rotationality, 8
round jet, 60

scalar property, Φ, 79
scalar time scale, 304
scalar variance, 303, 305
scalar-dissipation-transport equation, 311
scalar-flux equations, 80, 303
scalar-flux vector, 79, 303
scalar-flux-transport modelling, 303
scalar-flux/scalar-gradient interactions, 79
scalar-variance-transport equation, 311
scale separation, 17, 34, 41
Schwartz inequalities, 202
second-rank tensor, 379
self-convection, 9
semi-viscous sublayer, 107
shear- and normal-stress equations — simple shear, 54
shear-strained turbulence, 22
shear-stress-transport equation, 131
shock-induced separation, 127
simple-shear flow, 54, 62
    stress components, 54
slow term, 310
solid-body rotation, 9
Spalart–Allmaras model, 123
spatial averaging, 45
species concentration, 38
specific dissipation, 112, 163
specific dissipation — Wilcox's model, 165
specific dissipation — modelled transport equation, 164
spectral gap, 41
SST model, 172
    blending factor F, 173, 175
    blending function, 172
    mixed-derivative term, 173
    shear-stress limiter, 174
    shear-stress transport, 175
statistical homogeneity, 38, 45
statistical two-dimensionality, 38
statistical viewpoint, 1
steady flow, 37
strain invariants, 235
strain tensor, 130
strain-dependent C_μ, 203
streaks, 23
streamwise fluctuations, 25
streamwise straining, 223
stress anisotropy, 57
stress tensor, 37
stress-anisotropy parameter, A, 247
stress-flux analogy, 309
stress/strain interactions, 49
summation convention, 379
swirl, 214
swirl-related-production fragments, 215
system rotation, 217
system-objective, 234

Taylor microscale, 19
Taylor vortices, 30, 42
Taylor–Couette flow, 30
temperature fluctuation, 28, 84
temperature or mass-concentration fluctuations, 220
temporal evolution of the anisotropy, 206
temporal mixing layer, 167
tensor algebra and rules, 379
tensorial consistency, 234
tensorial symmetry, 234
term-by-term approximation process, 143
thermal stratification, 220

third invariant of the strain tensor, 204
time scale, 16, 91, 148, 150
time-averaged value, 10
time-averaged velocities, 38
time-averaging, 34
time-filtered signals, 35
time-integration, 33
time-scale ratio, 304, 312
transition to turbulence, 27
transport equation for the eddy viscosity, 122
transport equation for the shear stress, 91
transport equations for the turbulent Reynolds-number, 123
transport of intermittency, 171
transport of scalar quantities, 38
triple decomposition, 43
triple moments, 241
triple product, 381
turbulence energy, 5, 11, 34
turbulence energy and its rate of dissipation in channel flow, 185
turbulence-energy equation, 56
turbulence-energy-transport equation — simple shear
    convection, 56
    dissipation, 56
    generation, 56
    turbulent diffusion, 56
    viscous diffusion, 56
turbulence organisation, 9, 20
turbulence Reynolds-number, 128
turbulence-dissipation spectrum, 13
turbulence-energy amplification, 207
turbulence-energy budget, 70
turbulence-energy dissipation, 13
turbulence-energy equation, 119, 128
    pressure-strain term, 57
turbulence-energy equilibrium, 106, 107, 145, 146, 184, 269, 271
turbulence-energy generation — simple shear, 26
turbulence-energy production, 71
turbulence-energy spectrum, 13
turbulence-energy-based models, 115
turbulence-energy-transport equation, 98, 102, 117
    convection, 98
    decaying turbulence, 99
    dissipation, 98
    duct or pipe flow, 99
    generation, 98
    isotropic exchange coefficient, 101
    jet/wake axes of symmetry, 99
    Prandtl–Schmidt number, 101
    pressure-velocity interaction, 101
    rapid distortion, 99
    turbulence-energy equilibrium, 99
    turbulent diffusivity, 101
    turbulent transport, 98
    turbulent-diffusion vector, 101
    viscous diffusion, 98
turbulence-equilibrium concept, 121
turbulence-viscosity equilibrium, 125
turbulent diffusivity of heat, or species concentration, 116
turbulent enstrophy, 141
turbulent fluxes, 80
turbulent frequency, 162
turbulent mixing, 17, 88
turbulent Reynolds number, 156
turbulent time scale, 305
turbulent transport of Reynolds stresses, 239, 243
    Daly and Harlow's approximation, 242
    eddy-diffusivity approximation, 243
    Hanjalić and Launder's approximation, 242
    Mellor and Herring's approximation, 242
    simple gradient diffusion, 243
turbulent transport of Reynolds stresses, 232
turbulent velocity scale, 11
turbulent viscosity, 88, 92, 108

turbulent-length scale, 102
turbulent-velocity scale, 15, 94, 97
two-component state, 67
two-component turbulence, 51, 53, 238
two-equation $k - \varepsilon$ model, 132
two-equation models, 128, 135
two-equation models — reductions to one-equations forms, 176
two-layer strategies, 118, 120
two-point correlation tensor, 21

universal law of the wall, 73
universal logarithmic profile of the mean velocity, 77
universal near-wall description, 74
universal velocity profiles, 76
universality, 8
unsteady flow, 39
unsteady RANS, 10, 42

van Driest damping, 126
van Driest damping function, 120, 121, 152
variance of the temperature fluctuations, 82
velocity scale, 16, 92, 97, 102
viscosity-affected region, 96
viscosity-affected sublayer, 120
viscosity-dominated layer, 73
viscosity-related length scale, 121
viscous-damping function, 109, 121, 126, 152
viscous destruction, 141
viscous near-wall flow, 74
viscous sublayer, 18, 66, 67, 96, 109, 118
volumetric expansion coefficients, 220
von Kármán constant, 75, 95
von Kármán vortices, 10
vortex stretching, 12, 140
vortices, 7, 8
vorticity, 8, 9, 16
vorticity fluctuations, 139
vorticity tensor, 130, 204

wall functions
- analytical wall functions, 194
- Billard *et al.*'s method, 194
- Craft *et al.*'s method, 190, 191
- Kalitzin *et al.*'s method, 194
- near-wall finite-volume cell, 183
- Reichardt general law of the wall, 193
- cell-averaged production and dissipation rates, 186
- Chieng and Launder's method, 187
- eddy-viscosity-based wall functions, 189
- finite-volume approach, 182
- general wall functions, 193
- Howard's thermal law of the wall, 196
- Jayatilleke's function, 196
- log-law-based wall functions, 182
- numerical wall functions, 191
- semi-analytical near-wall bridges, 182
- thermal viscous sublayers, 196
- turbulence-energy equation in the near-wall cell, 186
- wall functions for heat transfer, 195

wall functions for linear eddy-viscosity models, 181
wall jet, 57, 60
wall variation of dissipation rate, 158
wall variation of production-to-dissipation ratio, 158
wall-asymptotic behaviour, 136, 150
wall-asymptotic stress behaviour, 67
wall-asymptotic variation, 151, 155
wall-boundary condition for $\tilde{\varepsilon}$, 137
wall-normal fluctuations, 25, 69
wall-normal heat flux, 84
wall-normal mixing, 150
wall-normal stress, 67
wall normal variation of the turbulent fluctuations, 66

wall-normal velocity variation, 65
wall-parallel normal stresses, 67
wall-scaled dissipation rate, 70
wave number, 11, 12
wind-tunnel turbulence, 89
wing/body junction — model predictions, 274
Wolfshtein's model, 118, 159

zero-pressure-gradient boundary layer, 73